GENDER AND FORESTS

D0139025

This enlightening book brings together the work of gender and forestry specialists from various backgrounds and fields of research and action to analyze global gender conditions as related to forests. Using a variety of methods and approaches, the authors build on a spectrum of theoretical perspectives to bring depth and breadth to the relevant issues and address timely and under-studied themes.

Focusing particularly on tropical forests, the book presents both local case studies and global comparative studies from Africa, Asia and Latin America, as well as the US and Europe. The studies range from personal histories of elderly American women's attitudes toward conservation, to a combined qualitative/quantitative international comparative study on REDD+, to a longitudinal examination of oil palm and gender roles over time in Kalimantan. Issues are examined across scales, from the household to the nation state and the global arena, and reach back to the past to inform present and future considerations.

The collection will be of relevance to academics, researchers, policy makers and advocates with different levels of familiarity with gender issues in the field of forestry.

Carol J. Pierce Colfer is a Senior Associate, Center for International Forestry Research (CIFOR) and Visiting Scholar, Southeast Asia Program, Cornell University, USA.

Bimbika Sijapati Basnett is Gender Coordinator at CIFOR, based at its headquarters in Indonesia.

Marlène Elias is a Gender Specialist at Bioversity International, based in Malaysia.

The Earthscan Forest Library

This series brings together a wide collection of volumes addressing diverse aspects of forests and forestry and draws on a range of disciplinary perspectives. Titles cover the full range of forest science and include the biology, ecology, biodiversity, restoration, management (including silviculture and timber production), geography and environment (including climate change), socio-economics, anthropology, policy, law and governance. The series aims to demonstrate the important role of forests in nature, peoples' livelihoods and in contributing to broader sustainable development goals. It is aimed at undergraduate and postgraduate students, researchers, professionals, policy-makers and concerned members of civil society.

Series Editorial Advisers:

John L. Innes, Professor and Dean, Faculty of Forestry, University of British Columbia, Canada.

Markku Kanninen, Professor of Tropical Silviculture and Director, Viikki Tropical Resources Institute (VITRI), University of Helsinki, Finland.

John Parrotta, Research Program Leader for International Science Issues, US Forest Service - Research & Development, Arlington, Virginia, USA.

Jeffrey Sayer: Professor and Director, Development Practice Programme, School of Earth and Environmental Sciences, James Cook University, Australia, and Member, Independent Science and Partnership Council, CGIAR (Consultative Group on International Agricultural Research).

Gender and Forests: Climate Change, Tenure, Value Chains and Emerging Issues
Edited by Carol J. Pierce Colfer, Bimbika Sijapati Basnett and Marlène Elias

Forests, Business and Sustainability
Edited by Rajat Panwar, Robert Kozak and Eric Hansen

Climate Change Impacts on Tropical Forests in Central America: An Ecosystem Service Perspective
Edited by Aline Chiabai

Rainforest Tourism, Conservation and Management: Challenges for Sustainable Development
Edited by Bruce Prideaux

Large-scale Forest Restoration
David Lamb

Additional information on these and further titles can be found at http://www.routledge.com/books/series/ECTEFL.

GENDER AND FORESTS

Climate change, tenure, value chains and emerging issues

Edited by Carol J. Pierce Colfer, Bimbika Sijapati Basnett and Marlène Elias

Routledge
Taylor & Francis Group

LONDON AND NEW YORK

from Routledge

First published 2016
by Routledge
2 Park Square, Milton Park, Abingdon, Oxon OX14 4RN

and by Routledge
711 Third Avenue, New York, NY 10017

Routledge is an imprint of the Taylor & Francis Group, an informa business

British Library Cataloguing in Publication Data
A catalogue record for this book is available from the British Library

Library of Congress Cataloging in Publication Data
A catalogue for this book has been requested

ISBN: 978-1-138-95503-5 (hbk)
ISBN: 978-1-138-95504-2 (pbk)
ISBN: 978-1-315-66662-4 (ebk)

Typeset in Bembo
By Swales & Willis Ltd, Exeter, Devon, UK

MIX
Paper from
responsible sources
FSC
www.fsc.org   FSC® C013056

Printed and bound in Great Britain by
TJ International Ltd, Padstow, Cornwall

This book is dedicated to the many women and men who live in forests and whose capabilities, interests, and long term goals have been inadequately acknowledged.

"The bold experimentation and collaborative learning conveyed in this volume trace an odyssey through which conversations around gender and forests moved from initial dismissive laughter to powerful research and practice ongoing across scales and contexts, scientific disciplines, and diverse ways of knowing and being."
– *Susan Paulson, University of Florida, USA.*

"Amid growing recognition of the importance of gender in many aspects of development, many researchers and practitioners are unsure what this means in practice. This volume provides clear guidance, with a clear conceptual framework and case study applications. Although focusing on forestry, it is also relevant to others working on climate change, tenure, and value chains."
– *Ruth Meinzen-Dick, International Food Policy Research Institute, USA.*

"The multiple roles of women in the forest sector are frequently underestimated. This excellent book provides hard and convincing evidence of the need for significant changes in the way that gender issues are dealt with by the forest sector and should be mandatory reading for all current and future forest professionals."
– *John Innes, University of British Columbia, Canada.*

CONTENTS

FIGURES

TABLES

ACKNOWLEDGEMENTS

We express our gratitude for the collaborative spirit of the authors who worked with us to develop this book. We also thank Bioversity International and the Center for International Forestry Research for the funding and support through the CGIAR Research Program on Forests, Trees and Agroforestry that made this book possible; and Cornell University for ongoing access to their marvelous libraries. We thank Tim Hardwick, Ashley Wright and Megan Smith at Earthscan, Laura Christopher at Swales & Willis and Jeanne Brady, for their technical expertise, patience and cooperative spirit, in editing and finalizing this book. We are most fundamentally grateful to the many women and men in villages and forests around the world who voluntarily contributed their time and thoughts to our analyses.

CONTRIBUTORS

Lorena Aguilar *Global Senior Gender Advisor-IUCN, 1630 Connecticut Ave NW – Suite 300, Washington, DC 20009, USA*
Email: lorena.aguilar@iucn.org

A Global Senior Gender Advisor, with more than thirty years of experience in initiatives involving public policy development, building local institutions and the incorporation of social and gender issues into development.

Seema Arora-Jonsson *Department of Urban and Rural Development, Swedish University of Agricultural Sciences, Uppsala, Sweden*
Email: Seema.Arora.Jonsson@slu.se

Seema Arora-Jonsson is Associate Professor at the Swedish University of Agricultural Sciences. Her research lies at the intersection of gender, development and environmental governance, including analysis of questions from a North-South perspective in the globalizing context of environmental governance and feminist participatory research. Recent publications include articles in *Antipode, Women's Studies International Forum* and the 2013 book, *Gender, Development and Environmental Governance: Theorizing Connections.*

Kiran Asher *Senior Scientist, Forests and Livelihoods, CIFOR, Bogor, Indonesia and Associate Professor, Amherst University, Amherst, MA*
Email: kasher@umass.edu

Asher's diverse research interests focus on the gendered and raced dimensions of social and environmental change in the Global South, particularly Latin America and South Asia (e.g. the monograph, *Black and Green: Afro-Colombians, Development, and Nature in the Pacific Lowlands,* Duke University Press, 2009). She is currently working on a theoretical and political critique of development theories and proposals, drawing on feminism and marxism. At CIFOR 2013–15, she is now Associate Professor, Department of Women, Gender, and Sexuality Studies at the University of Massachusetts, Amherst.

Stibniati Atmadja *Center for International Forestry Research, Jl. CIFOR, Situ Gede, Sindang Barang, Bogor 16680, Indonesia*
Email: s.atmadja@cgiar.org

Stibniati Atmadja is a natural resource economist at CIFOR. Since 2008, she has done extensive research on REDD+ implementation, with a focus at the local level in Indonesia. She started a new geographical focus in East Africa, with her assignment in CIFOR Ethiopia, beginning in mid-August 2015.

Abdon Awono *Center for International Forestry Research, P.O. Box 2008, Yaoundé, Cameroon, University of Montpellier 3, Paul-Valérya*
Email: abdon@cgiar.org

Abdon Awono is a senior researcher at the Center for International Forestry Research. He has been working for the past 18 years on forest policy and non-timber forest products (NTFP) valuation in the Congo Basin. He is currently conducting his PhD study with University of Montpellier 3, Paul-Valery in France on socio-economic impacts of NTFP exploitation in central Africa.

Solange Bandiaky-Badji *Rights and Resources Initiative, Regional Program Director, Africa, 1238 Wisconsin Avenue NW, Suite 300, Washington, DC 20007*
Email: sbandiaky@rightsandresources.com

Bandiaky-Badji leads the development of RRI's strategy for engagement in Africa with a focus on tenure rights and gender in Africa. She holds a PhD in Women's and Gender Studies from Clark University in Massachusetts and Master's degrees in Environmental Sciences and in Philosophy from Cheikh Anta Diop University in Senegal. She previously worked as the Regional Expert on gender and climate change for the Africa Adaptation Programme (AAP) and the UNDP/BDP Gender Team. She has published work on gender in relation to natural resource management, decentralization/local governance, and forest and land reforms.

Bimbika Sijapati Basnett *CIFOR, Jl. CIFOR, Situ Gede, Sindang Barang, Bogor 16680, Indonesia*
Email: b.basnett@cgiar.org

Bimbika coordinates CIFOR's gender research and is the CIFOR gender focal point for the CGIAR Research Program on Forests, Trees and Agroforestry. She also contributes to CIFOR's research programs on migration and multi-local livelihoods and commercial crop expansion in forested landscapes. She holds a PhD in Development Studies from the London School of Economics and Political Science.

Her thesis examined the gender dimensions of decentralization of forest govern-ance with case studies from community forestry in Nepal.

Mekou Youssoufa Bele *Center for International Forestry Research, P.O. Box 2008, Messa, Yaoundé, Cameroon*
Email: *yoube_bele@yahoo.com; b.youssoufa@cgiar.org*

Bele is a forest researcher with extensive experience working with people and their forests to develop their lives and their communities. He particularly relishes the climate change adaptation ethos, which combines adaptation for forests and forests for adaptation.

Priyanka Bhalla *National University of Singapore, Lee Kuan Yew School of Public Policy, 469C Bukit Timah Road, Singapore 259772*
Email: *Piya.Bhalla@gmail.com*

Priyanka Bhalla is currently a PhD candidate at the Lee Kuan Yew School of Public Policy in Singapore. Her research interests include gender and property rights in the South Asia region. She has formerly worked with the UN and several NGOs in Nepal, India and the US.

Maria Brockhaus *CIFOR, Jl. CIFOR, Situ Gede, Sindang Barang, Bogor 16680, Indonesia*
Email: *m.brockhaus@cgiar.org*

Maria Brockhaus is a senior scientist at the Center for International Forestry Research (CIFOR). Her background is in agricultural economics and forest pol-icy. Her research focuses mainly on policy and institutional change and policy and social network analysis. Since 2009, she has been leading the research on REDD+ policies in CIFOR's global comparative study (GCS-REDD+).

Carol J. Pierce Colfer *CIFOR and Cornell University, 21 Etna Lane, Etna, New York 13062–0280, USA*
Email: *c.colfer@cgiar.org*

Colfer is an anthropologist with over four decades of research on rural peoples. Her foci have included variously health, agriculture, devolution, education and gender, mostly in the tropics. Her gender and forestry work began in the early 1970s in a rural American logging community; her most recent research is on gen-der and forested landscapes in southern Sulawesi, Indonesia.

Eugene Loh Chia *CIFOR, P.O. Box 2008, Messa, Yaoundé, Cameroon*
Email: *lohchia@gmail.com*

Eugene Loh Chia is an environmentalist with an interdisciplinary background that cuts across environmental and development issues. He holds a MSc Degree in International Environmental Studies, from the Department of International

Development and Environmental Studies, Norwegian University of Life Sciences. He has years of experience on climate change adaptation and mitigation in the Congo Basin region.

Leigh Cobban *African Climate and Development Initiative (ACDI), University of Cape Town, Geological Sciences Building, Rondebosch, Cape Town, 7701, South Africa*
Email: leigh.cobban@uct.ac.za

Leigh Cobban works on project development and research coordination for the African Climate & Development Initiative (ACDI) at the University of Cape Town. She obtained an MSc from Rhodes University, South Africa, in which she analyzed rural livelihoods and household assets to assess vulnerability to interacting stressors, focusing on climate change and HIV/AIDS. She has worked in academic, private and non-profit organizations in South Africa and the United Kingdom. Her research interests include climate-resilient development, capacity building and alternative economics.

Marina Cromberg *Center for International Forestry Research, Rua do Russel 450/601 CEP: 22210–010, Glória, Rio de Janeiro, Brazil*
Email: mcromberg@gmail.com

Marina Cromberg is a field research supervisor at the Center for International Forestry Research (CIFOR). She has extensive fieldwork experience in Brazil and her work focuses on analysis of local livelihoods and local perceptions on conservation initiatives. She currently analyzes sub-national REDD+ initiatives in Brazil as part of CIFOR's Global Comparative Study on REDD+.

Peter Cronkleton *Center for International Forestry Research, c/o CIP, Av. La Molina 1895, Lima 12, Peru*
Email: p.cronkleton@cgiar.org

Peter Cronkleton is an anthropologist with CIFOR's Forest and Livelihoods Program and leads CIFOR's global research on smallholder forestry and markets. Dr Cronkleton is a specialist in community forestry development, forest tenure, social movements and participatory approaches to research. Currently based in Peru, his work has concentrated on the western Amazon but has also involved comparative research with forest users in Central America, Africa and Asia.

Houria Djoudi *Center for International Forestry Research, Jl. CIFOR, Situ Gede, Sindang Barang, Bogor 16680, Indonesia*
Email: H.djoudi@cgiar.org

Houria Djoudi is a scientist at the Center for International Forestry Research (CIFOR). Her background is in pastoralism and natural resources management science with strong focus on socio-ecological systems analysis and environmental changes. The focus of her current work is vulnerability, resilience and climate change adaptation at the interface of research and development. Houria has a particular interest on gender analysis as a part of linkages and feedbacks of institutional processes in socio-ecological systems.

Therese Dokken *School of Economics and Business, Norwegian University of Life Sciences, P.O. Box 5003, NO-1432, Aas, Norway*
Email: theresedokken@gmail.com

Therese Dokken is an economist and associated researcher at the Norwegian University of Life Sciences. Her research focuses on East Africa, and she has fieldwork experience in Tanzania and Ethiopia. Her recent work includes analyses of sub-national REDD+ initiatives in Tanzania, household poverty, forest resource use, land tenure rights and gender.

Amy E. Duchelle *Center for International Forestry Research, Rua do Russel 450/601 CEP: 22210–010, Glória, Rio de Janeiro, Brazil*
Email: a.duchelle@cgiar.org

Amy Duchelle is a scientist at the Center for International Forestry Research (CIFOR). Her work focuses on Latin America, and she has extensive fieldwork experience in Brazil, Bolivia, Peru and Ecuador. She currently analyzes the effectiveness, efficiency, equity and co-benefits of sub-national REDD+ initiatives in six countries as part of CIFOR's Global Comparative Study on REDD+.

Marlène Elias *Bioversity International, P.O. Box 236, UPM Post Office, Serdang, 43400 Selangor Darul Ehsan, Malaysia*
Email: marlene.elias@cgiar.org

Marlène Elias is a Gender Specialist at Bioversity International. A geographer and feminist political ecologist, her research interests include gender and tree-based livelihoods, local ecological knowledge and alternative trade. She has conducted research in West and Central Africa, and Central, South and South East Asia.

Rebecca Elmhirst *School of Environment and Technology, University of Brighton, Lewes Road, Brighton, BN2 4GJ, UK*
Email: R.J.Elmhirst@brighton.ac.uk

Elmhirst is Principal Lecturer in Human Geography at the University of Brighton, UK. Her research interests lie in feminist political ecology, and the gender dimensions of natural resource management, resettlement and migration in Indonesia.

Carlos Grijalva-Eternod *UCL Institute for Global Health, UK, 30 Guilford Street, London, WC1N 1EH, UK*
Email: c.eternod@ucl.ac.uk

Carlos Grijalva-Eternod is a researcher at University College London. His work focuses on understanding the developmental origins of undernutrition and chronic diseases, the assessing of nutritional and well-being status among infants and children living in vulnerable settings in Central Asia, Africa and Latin America, and the impact evaluation of nutrition-sensitive interventions aimed at improving such status.

Helen Harris-Fry *UCL Institute for Global Health, UK, 30 Guilford Street, London, WC1N 1EH, UK*
Email: h.fry.11@ucl.ac.uk

Helen Harris-Fry is based at University College London and is currently conducting her PhD research on intra-household food allocation in rural Nepal. Previous research has focused on nutrition and reproductive health, evaluation of participatory projects in rural Bangladesh, and analysis of empowerment and gender dynamics within forestry programs in Central and West Africa.

Merel Haverhals *Wageningen University & Research Centre, Institute for Agricultural Economics (LEI), and Sociology of Development and Change Group, P.O. Box 29703, 2502 LS Den Haag, The Netherlands*
Email: merel87@gmail.com, merel.haverhals@wur.nl

Haverhals holds a BA degree in Cultural Anthropology and is finishing a master's in International Development Studies with a focus on gender issues and ethnographic research analysis.

Sara Holmgren *PO Box 7008, SE-75007 Uppsala, +46 18 673804. Department of Forest Products, Swedish University of Agricultural Sciences, Uppsala, Sweden.*
Email: sara.holmgren@slu.se

Sara Holmgren is a researcher at the Swedish University of Agricultural Sciences. Her research is focused on forest-related policy issues and national-global interlinks from a critical theoretical perspective. Her most recent publication is her dissertation, "Governing Forests in a Changing Climate – Exploring Patterns of Thought in the Climate Change-Forest Policy Intersection" (2015).

Marilyn W. Hoskins *1735 New Hampshire Ave NW #402, Washington, DC 20009*
Email: marilynndc@aol.com

Hoskins, a retired anthropologist, worked with USAID, the World Bank and for over a decade with the UN Food and Agriculture Organization, raising awareness within forestry of gender issues and improving community management of natural resources. She was awarded the Distinguished Service to Rural Life award by the Rural Sociology Society, and two honorary doctorates for outstanding contributions to wise use of natural resources and the field of international rural development.

Verina Ingram *CIFOR, Jl. CIFOR, Situ Gede, Sindang Barang, Bogor 16680, Indonesia; and Wageningen University & Research Centre, Institute for Agricultural Economics (LEI), and Forest and Nature Conservation Policy Group, P.O. Box 29703, 2502 LS Den Haag, The Netherlands*
Email: *verina.ingram@wur.nl, v.ingram@cgiar.org*

Ingram is assistant professor at Wageningen University & Research Centre and associate with CIFOR, where she focuses particularly on the interactions between people and natural resources management, particaurly focusing on governance, livelihoods and sustainable trade.

Virginia Kennedy *Otsego Land Trust, PO Box 173, Cooperstown, NY*
Email: *Virginia@otsegolandtrust.org/Vmk6@cornell.org*

Kennedy holds a PhD in English and American Indian Studies from Cornell University, where her research and writing focused on environmental ethics. She is currently the Executive Director of Otsego Land Trust, a conservation land trust in central New York protecting land and water through conservation easements.

Richard Sufo Kankeu *CIFOR, P.O. Box 2008, Messa, Yaoundé, Cameroon*
Email: *r.sufo@cgiar.org*

Kankeu is a forester, an expert in MRV and GIS, currently pursuing his doctoral research at the University of Maine, in the US. He has been working for CIFOR since 2013 and previously worked for a logging group in Central Africa as a technical assistant in charge of forest management and GIS aspects.

Gina LaCerva *Yale University, School of Forestry and Environmental Studies, 195 Prospect St, New Haven, CT 06511*
Email: *gina.lacerva@yale.edu*

LaCerva has an MPhil in Geography from the University of Cambridge and a MSc from Yale's School of Forestry and Environmental Studies. Previous research has included studying hazards ecology in Indonesia and the emergent behavior of vulnerability within complex social-ecological systems. Her current work examines the philosophy of domestication and the changing role of wild foods in society.

Anne M. Larson *CIFOR, c/o CIP, Av. La Molina 1895, Lima 12, Peru*
Email: *a.larson@cgiar.org*

Anne Larson is a principal scientist in the Forests and Governance program at the Center for International Forestry Research. She currently leads three comparative research projects in Latin America, Africa and Asia, on multilevel governance, indigenous and community land rights and gender.

Jonah Meyers *Department of International Development, University of Maryland, College Park, Maryland*
Email: jonahmmeyers@gmail.com

Meyers is a Masters candidate at the Maryland School of Public Policy, specializing in international development, nonprofit management and program evaluation. Before and during his studies at the University of Maryland, Jonah worked with InterAction's Humanitarian Policy & Practice department, with Tostan International in northern Senegal, and most recently with the Rights and Resources Initiative's Country and Regional Programs department. He is most interested in strengthening agricultural and forest product value chains and evaluation of food security projects in sub-Saharan Africa.

Cécile Ngo Ntamag-Ndjebet *African Women's Network for Community Management of Forests (REFACOF), Founder and President, P.O. Box 791 Edea, Cameroon*
Email: cecilendjebet28@gmail.com

Ndjebet is the president of the African Women's Network for Community Management of Forests (REFACOF), the executive director of Cameroon Ecology, a national NGO based in Edéa, Cameroon, and the National Coordinator of Cameroon's civil society platform on REDD+ and Climate Change. Devoted to women's rights issues, Njebet holds a Master's in social forestry and is a gender specialist, providing training and advice in women's leadership. She is also a member of the Steering Committee of Forests Dialogue, a focal point of the Women's Major Group of the United Nations Forest Forum (UNFF), and a representative of the Women of the Global South Group in the World Bank Forest Carbon Partnership Facility (FCPF).

Sjoerd Petersen *Wageningen University & Research Centre, Institute for Agricultural Economics (LEI), and Forest and Nature Conservation Policy Group, P.O. Box 29703, 2502 LS Den Haag, The Netherlands*
Email: sjoerd.pietersen@wur.nl

Petersen is a Master of Science student at Wageningen University where he studies international land and water management. His major interests are how diverse compositions of trees, shrubs and other NTFP's contribute to farmers' livelihoods, and investigating the ecological and socio-economic functions of agroforestry systems in developing countries.

Ida Ayu Pradnja Resosudarmo *Center for International Forestry Research, Jl. CIFOR, Situ Gede, Sindang Barang, Bogor 16680, Indonesia*
Email: d.resosudarmo@cgiar.org

Daju Resosudarmo is a senior scientist at the Center for International Forestry Research (CIFOR). Her work focuses on Indonesia, including on causes of deforestation, decentralization, local governance and tenure. She is currently working on REDD+.

Galia Selaya *Center for International Forestry Research, P.O. Box 2909, Santa Cruz, Bolivia*
Email: gselaya@yahoo.com

Galia Selaya is an environmental scientist interested in synergies between forest conservation for non-timber forest products and additional income streams via PES programs. Her research bridges forest monitoring and public policies. She is actively involved in research projects in Bolivia, Brazil and Peru.

Sheona Shackleton *Department of Environmental Science, Rhodes University, Grahamstown, 6140, South Africa*
Email: s.shackleton@ru.ac.za

Sheona Shackleton, PhD, is Professor and Head of the Department of Environmental Science at Rhodes University. She has worked extensively at the human-environmental interface for the last 35 years, undertaking research in such spheres as community conservation, rural livelihoods and vulnerability, ecosystem services and human well-being, forest product use and commercialization, and climate change adaptation. She has participated in several large international, interdisciplinary and inter-institutional research programs, and teaches aspects of inter- and transdisciplinarity and complex social-ecological systems in her undergraduate and postgraduate courses.

Mia Siscawati *Gender Studies Graduate Program, Graduate Program of Multidisciplinary Studies, University of Indonesia, Gedung Rektorat Lt. 4 Jl. Salemba Raya No. 4, Jakarta 10430, Indonesia*
Email address: miasisca@gmail.com

Siscawati holds a PhD in Anthropology from the University of Washington, Seattle. She is Lecturer and Head of the Gender Studies Graduate Program, Graduate Program of Multidisciplinary Studies, and Associate Lecturer, Department of Anthropology, Faculty of Social and Political Sciences, University of Indonesia. Her work focuses on feminist political ecology, the linkages between gender, forest and land tenure, and the politics of natural resources governance, as well as gender dimensions of the environmental and community rights movements in Indonesia.

Sola Phosiso *World Agroforestry Centre (ICRAF), United Nations Avenue, Gigiri PO Box 30677, Nairobi, 00100, Kenya.*
Email: p.sola@cgiar.org

Sola obtained her PhD from the University of Wales, Bangor focusing on impacts of commercialization of non-timber forest products. With twenty years' work experience in Africa and Asia, her current research, as a programme coordinator at ICRAF, focuses on governance of agro and forest products value chains in eastern and southern Africa.

William Sunderlin *Center for International Forestry Research, Jl. CIFOR, Situ Gede, Sindang Barang, Bogor 16680, Indonesia*
Email: w.sunderlin@cgiar.org

William Sunderlin is a principal scientist in the Forest and Livelihoods portfolio at CIFOR. He is currently leading CIFOR's research on REDD+ subnational initiatives through its Global Comparative Study on REDD+.

Alba Saray Perez Teran *CIFOR, P.O. Box 2008, Messa, Yaoundé, Cameroon*
Email: a.perezteran@cgiar.org

Pérez-Terán holds a MSc in Environmental Sciences. Currently a Research Officer with CIFOR in Central Africa, she has previously conducted research with the World Agroforestry Center in South East Asia. She is currently focusing on gender and climate change adaptation.

Anne-Marie Tiani *CIFOR, P.O. Box 2008, Messa, Yaoundé, Cameroon*
Email: a.tiani@cgiar.org

Tiani is a socio-ecologist and senior scientist, currently working at CIFOR, coordinating a project related to synergy between mitigation and adaptation to climate change in the forests areas of Central Africa.

Julie T. B. Weah *Executive Director, Foundation for Community Initiatives (FCI), Duazon Village, Robertsfield Highway, Monrovia, Liberia*
Email: fcommunityinitiatives@yahoo.com

Weah currently serves as executive director of the Foundation for Community Initiatives (FCI). For the past seven years, her work has focused on securing the rights of women to land and forest tenure. Julie is a civil society representative on various platforms, including the working group on the Voluntary Partnership Agreement (VPA), REDD+, and Climate Change. She also serves as the vice president of the African Women's Network on Community Management of Forest (REFACOF). She is actively engaged in the activities of the NGO Coalition of Liberia, a network of civil society institutions working on human rights and natural resource issues.

FOREWORD

Lorena Aguilar

Approximately 1.2 billion people around the world—mainly in tropical regions—depend on agro-forestry farming and forest resources for their livelihoods (FAO n.d.). At least 50 percent of these are women and in some communities where men have migrated, their presence has increased substantially. Thus, it is reasonable to think that the simple fact that "600 million women live [in] and depend on forests" should be reflected in every aspect of forest governance. Nothing, however, could be further from the truth.

One of the recently developed datasets under the International Union for Conservation of Nature (IUCN) Environment and Gender Index (EGI)[1] proves that women who rely on forest resources are often underrepresented in forest governance at the local, national and international levels (see Tiani et al. on Cameroon and Bhalla on India). For example, less than a quarter of the 173 Focal Points to the UN Forum on Forests (UNFF)—an intergovernmental forest policy forum—are women.

Likewise, a recently conducted survey (IUCN 2015) assessing whether or not gender considerations, as well as gender focal points, are being included in national-level policies and programs across various environmental sectors and ministries shows that of 26 survey responses, only *seven* nations indicated that their ministry or agency of forestry has a formal gender policy, and *ten* nations indicated that their forestry ministry or agency incorporates gender considerations into its policies and

---

1 The first accountability and monitoring mechanism of its kind, the EGI is an index developed by IUCN that brings together gender and environment variables. In its pilot phase, the Index scored and ranked 73 countries along 27 dimensions in six categories: Ecosystems, Gender Based Education, Governance, Country Reported Activities, Livelihoods, and Gender Based Rights and Participation. (See the full dataset report, *Women's Participation in Global Environmental Decision Making: An EGI Supplemental Report,* or visit genderandenvironment.org/EGI).

programs. Of 65 survey responses, *seventeen* nations confirmed that their ministry or agency of forestry has a gender focal point (see Bandiaky-Badji,[2] on these issues in Liberia and Cameroon). Sijapati Basnett's chapter notes that even where there are gender-relevant policies, these may ignore the changing context of women's lives and livelihoods.

These numbers illustrate the lack of women in environmental decision-making positions—ultimately resulting in a lost opportunity for capitalizing on their unique knowledge and experiences of natural resources, a topic addressed by Elias. True conservation and management of forests will require a further paradigm shift, particularly in regards to leadership roles, where all forest work incorporates gender considerations from its initial stages.

However, full participation in forest governance is not the only aspect that is affected due to the lack of recognition of the value of gender considerations in the forestry sector. Due to socio-cultural norms, women and men (of different ages and ethnic groups) in forest-dependent communities, have differentiated needs, uses and knowledge in relation to their ecosystems. Failure to take full account of this diverse societal context has hindered accountable initiatives in the forest sector.

Gender blindness is a real, and detrimental, thing and the result is lost opportunities to achieve multiple benefits. In many places and in most cases, women's role in forestry is not acknowledged, let alone understood. A gender-equitable approach to forest governance starts with an increased understanding of the unique role that women play in the management of forest resources.

Recently, a study identifying unconventional edible plants was undertaken in the Amazon using gender-sensitive methodologies. The resulting data showed 45 new species that had never previously been documented (Marin 2014)—species that could be an alternative source of food and improve nutritional intake for local communities as they are forced to adapt their livelihoods in the face of climate change. These groundbreaking findings highlight women's role as knowledge sources—sharing information that is crucial for forest management and biodiversity/conservation policies, as well as strengthening food sovereignty (see Shackleton and Cobban). However, data also show that implementation of native species as crops in this region has declined significantly for elderly women and youth, with only seven native species being grown by elderly women and just one on the farms of young women. This example underscores the need for identifying and addressing women's and men's roles, responsibilities and specific knowledge separately, as well as promoting women as decision makers, ensuring the success of environmental policy and planning.

Likewise, economic, social, cultural, political and legal environments affect the rights of women and men to control forest resources and own land. Even where women have ownership rights to land, their access to forest products and opportunities for forest-generated income may not be ensured. Different members of the community may have established informal rights to use of different parts of

2 Authors listed without date refer to chapters in this volume.

the forest or even of a tree—women may have access to the leaves but not to the wood. This differentiation by gender has major implications for the ownership and usufruct rights to the forest and its by-products. As pointed out in several articles in this book (e.g. Bandiaky-Badji et al., Bhalla, Asher), there is a need to promote equal access of women to land ownership and to other resources necessary for effective socio-economic participation (e.g. land, capital, technical assistance, technology, tools, equipment, markets and time).

An example of the impacts of addressing such topics was discussed in May 2014 at a technical workshop on gender and REDD+ (Reducing Emissions from Deforestation and forest Degradation), which brought together 52 participants from more than twenty countries. The workshop discussions addressed the impact that REDD+ processes are triggering at the national level in relation to land and forest tenure reforms, among other crucial topics (also see Larson et al.).

By comparing countries engaged in REDD+ with the list of countries where women have no/few legal rights to access or own land or access is severely restricted by discriminatory practices, of the seven lowest ranked countries (Sri Lanka, Ghana, Benin, Gambia, Uganda, Cameroon and Burundi), four are 'REDD+' countries. While this might seem like a dismal statistic at first, as one of the workshop participants pointed out, we can be optimistic about the opportunities as we now see in REDD+ a means to introduce reforms in much-needed areas (see Harris-Fry and Grijalva-Eternod, for a framework for doing this).

The African Women's Network for Community Management of Forests (Réseau des Femmes Africaines pour la Gestion Communautaire des Forêts, REFACOF) aims to change that by creating a network of women involved in sustainable forest resource management in Africa. REFACOF strives to make concrete, meaningful and effective contributions to forest governance in order to influence national policies and international frameworks regarding women's rights and tenure in member countries. Under a variety of African customary laws, women seldom own or inherit land, and the only way they can access it is by marriage or through their male children. REFACOF has realized some impressive results, specifically in regards to its remarkable progress in reforming national land tenure laws through the lens of gender and REDD+ by presenting women's legislation for land tenure reform and using REDD+ as a window for opportunity. Now in Cameroon, 30–40 percent women are included in decision-making positions at the village, district, regional and national levels—contributing toward integrating gender into REDD+ policies and planning, as well as other processes. Incredible indeed: if environmental initiatives fully embrace the principles of gender equality and women's empowerment, they can have an unprecedented impact.

From Costa Rica to Cameroon, women are the primary users of forest resources—playing essential roles in forest management, sustainable conservation and climate change adaptation and mitigation. Globally, women's heavy dependence on forests and their associated products also means that they often have more at stake than men when forests are degraded or when forest access is denied (cf. Shackleton et al. on a South African example).

When it comes to solving complex problems or making innovations, such as solutions for adapting to and mitigating climate change, a diverse group of competent performers almost always outperforms a homogenous group by a significant margin.[3] The more diverse the stakeholders are, the more likely they will succeed in the face of uncertainty and ambiguity. Each person categorizes people, things and events based on their background and experience. For example, in the context of climate change and forest-related initiatives, this can mean the difference between seeing a woman as vulnerable and immobilized or as an empowered agent of change.

The World Bank and the United Nations Food and Agriculture Organization have published evidence documenting the transformative impact of women's empowerment on the economy (World Bank 2012) and agricultural production (FAO 2011). These revelations have signaled to the world that women's limited access to land, forest, energy, water and other natural resources is a fundamental obstacle to securing their social and economic rights. In 2014, the UN Women's World Survey affirmed this again—moreover emphasizing that women's access to, use of and control over natural resources has a direct link to environmental sustainability and meeting global goals on sustainable development more broadly.

Building on the insights gained into the role of women in forest-based livelihoods and forest governance structures (Hoskins, Elmhirst et al.), supporting the inclusion of women in multi-stakeholder processes (Larson et al., Asher) and giving attention to their interests in national strategies (Holmgren and Arora-Jonsson) are all imperative. There is a growing need to assess the strengths and the weaknesses of the global forest management models. It is important to take note of past failures to involve women in natural resource governance and to find effective ways to integrate women into decision-making and planning processes related to land-use and natural resource management—as amply discussed in Hoskins' historical contribution. Ignoring the gender dimension in forest conservation and management is not an option. The loss of these ecosystems will have a gender-differentiated impact that could jeopardize the livelihoods of women and men around the world.

Therefore, we need to move towards being not only gender-sensitive, but gender-*responsive*. Being gender-responsive means that rather than only identifying gender issues or working under the "do no harm" principle, we need a process that will substantially help to overcome historical gender biases—to "do better," so to speak—in order for women to truly engage and benefit from these actions. In addition, gender-responsive planned actions in forestry should integrate measures for promoting gender equality and women's empowerment, foster women's inclusion and provide equal opportunities for women and men to derive social and economic benefits. With this approach, women's and men's concerns and experiences equally become a fundamental element in the design, implementation, monitoring and evaluation of forest-related projects and policies.

---

3 See e.g. http://www2.deloitte.com/global/en/pages/public-sector/articles/the-gender-dividend.html.

The publishing of *Gender and Forestry: Climate Change, Tenure, Value Chains and Emerging Issues* could not be timelier. This book is not only a compilation of research that meets at the intersection of gender within forest management and conservation, it is a window into a bright future where women are leaders and decision makers and their voices are not only heard, but respected and celebrated.

We have progressed a long way, as demonstrated by Hoskins, yet much remains to be done. As we move towards more gender-responsive policies and implementation, it is not unreasonable to hope for a time when gender and forest governance are much more fully integrated. Until then, all relevant actors should pursue this goal with urgency. We are living in a time when forest ecosystems are diminishing at an unprecedented rate and forest loss is rapidly becoming one of the main development issues of this century. Preserving, managing and improving the health of forest ecosystems around the world requires equitable leadership and participation from both women and men. In these critical times, international, national and local forest and conservation initiatives must address and integrate gender in order to fully attain equality in the environment arena, achieve sustainable conservation, and succeed in adapting to and mitigating the negative effects of climate change.

References

FAO (Food and Agriculture Organization). n.d. *Gender: Forests*. Food and Agriculture Organization (forests), Rome, Italy. Accessed 2 April 2015. http://www.fao.org/gender/gender-home/gender-programme/gender-forests/en/.

——. 2011. *The state of food and agriculture 2010–2011: Women in agriculture: closing the gender gap in development.* Rome: FAO.

IUCN (International Union for Conservation of Nature). 2015. *Gender in national environmental policies & programs: Dataset for the Environment and Gender Index (EGI).* Gland: IUCN (August).

Marin, N. 2014. *Plantas alimenticias no convencionales en tres departmanetos de la Amazonia Colombiana como soporte a la soberania alimentaria, aportes desde el enfoque de género.* Instituo Amazónico de Investigaciones cientificas. Bogota: SINCHI.

World Bank. 2012. *World Development Report 2012: gender equality and development.* Washington, DC: The World Bank.

PART I

Introduction

1

A GENDER BOX ANALYSIS OF FOREST MANAGEMENT AND CONSERVATION

Carol J. Pierce Colfer, Marlène Elias and Bimbika Sijapati Basnett

Why this book?

When gender is mentioned in forestry circles, questioning looks of confusion are sometimes the response. Most foresters have been trained—for a very long time—to focus on trees. But there has been a growing recognition in some circles that women and men make different use of trees and have different knowledge and preferences about them. The facts that forests are made up of plants and animals other than trees and can be used for purposes other than timber are increasingly recognized in the field, one that has also encouraged greater attention to gender, as, in many areas, women are more likely to use non-timber forest products and are the primary collectors of fuelwood and charcoal. The physical burden of such collection affects women's health and reduces the time available to devote to other activities. Ignoring the crucial role that gender relations play in forestry not only undermines local resource conditions, it also prevents women and girls from realizing a full range of their capabilities.

We have produced this book in response to an expressed need for case materials and cross-site analyses on gender issues in forested contexts. Our own institutions are pressured to improve their record on gender in natural resource contexts; our scientists show willingness but repeatedly ask for help. Colfer conducted a survey of 150 experts in forestry-related fields, from a variety of countries and institutional contexts in August 2014, asking what sorts of materials would respond to their need for additional materials on gender and forestry. The most popular format (with 39 percent of the responses) was "an edited collection of academic analyses focused on the tropics." This book represents our response to this expressed need.

We have assembled a diverse group of authors, from varied backgrounds. The forty authors represent fifteen nationalities, nicely scattered geographically: Africa (three countries), Asia (two), Latin America (three), North America (two) and Europe (five). We also have a broad disciplinary spread, including experts in the social sciences (anthropology, geography, economics, government/policy analysis), natural resources (forestry, ecology, animal sciences), animal and human nutrition, English, and American Indian and Women's Studies. This diversity shows that a gender lens is not confined to specific disciplines, but can be applied within and across disciplinary boundaries.

We strove to include gender analyses by (and of) men in this collection. We believe we have managed to address gender relatively equitably in our analyses; but we have failed to obtain a good gender balance among our authors: 32 are women, 8, men—with men in no case serving as first authors. This imbalance sadly reflects the persistent, uneven interest in gender as a subject of study.

Our analyses range in scale from intra-household and community through landscape and country-specific studies to regional/global comparisons and reviews. Detailed analyses are provided on gender and forestry issues in 18 countries, 16 of which are in the Global South. Our aim is to converse and connect across countries and contexts; demonstrate that processes taking place in the developed world may not be as dissimilar to those in the developing, and vice-versa, and encourage readers to identify overlaps between seemingly distinct countries and contexts.

We begin here with a brief analysis of the materials themselves, using the Gender Box (Colfer and Minarchek 2013) as an organizing framework. One purpose of the Gender Box was to contribute to the perceived shortage of expertise on gender issues; another was to reflect the topics that gender and forestry specialists have examined and found to be important in maintaining and/or enhancing human and forest well being. Here we examine the topics covered in this book, as they relate to those identified as important in the framework.

Gender Box analysis

The Gender Box (Figure 1.1) is a three-dimensional figure composed of eleven issues, three scales (micro, meso, macro), and a time dimension. It was designed to serve as a mnemonic device to remind us of the influences of these key issues— scale and time—in our attempts to address gender in forestry. The key issues emerged from extensive reviews of the literature, identifying how gender was being addressed. We recognize that the boundaries between scales (micro, meso and macro) and between times (past, present and future) are fluid and fuzzy. Both represent continua more than discrete categories and they mutually interact. Still, we have found this framework helpful in comprehending the elements that influence gender in specific places. Our ex-post analysis presents the attention devoted to each of these dimensions in the corpus of materials in this book. One purpose has been to highlight the areas that are being well-covered by researchers; another has been to identify other issues that might warrant inclusion in an evolving Gender Box.

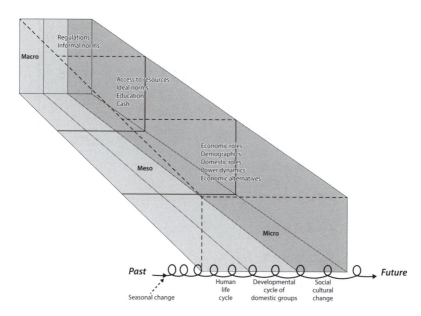

Regulations
Informal norms
Macro

Access to resources
Ideal norms
Education
Cash

Economic roles
Demographics
Domestic roles
Power dynamics
Economic alternatives

Meso

Micro

Past — Future

Human
life
cycle

Developmental
cycle of
domestic groups

Social
cultural
change

Seasonal change

FIGURE 1.1 The Gender Box

Table 1.1 presents the simple frequencies, indicating which chapters included significant discussion of which issues.[1] The table includes a Part A, which focuses on the original issues mentioned in the Gender Box, and Part B, which identifies possible additional issues, based on the discussions in these chapters. The cells are marked with 1 when the issue, scale, or time is addressed, and left blank, if not.

Looking first at Part A in this corpus of material, the dominance of issues like access to natural resources (addressed in 100 percent of the chapters), day-to-day economic activities (81 percent), access to cash (75 percent) and available economic alternatives (69 percent) is unsurprising. These are the kinds of topics on which much gender research across various disciplines has focused. The fairly high proportion of chapters that address the following issues is both surprising and encouraging: Norms of behavior (88 percent), formal laws/policies, cultural/religious trends and access to education (all 63 percent). This suggests a broadening of interest in the research community, a recognition of the holistic nature of people's lives (including those living in and around forests), and a growing understanding of interactions between scales; for instance, as norms are nested in, influence and are influenced by formal policies.

The lesser attention to intra-household dynamics (50 percent) may reflect the challenges involved in studying these and the difficulty of making policy-relevant recommendations based on such studies. The lack of focus on domestic roles and

---

1 These judgments are qualitative in nature and therefore subject to potential disagreement. See the 2015 special issue of the *International Forestry Review* (*IFR*) on gender and agroforestry, in which we have done a parallel analysis for that set of ten articles (Colfer et al. 2015).

TABLE 1.1 Gender Box analysis: issues by chapter

First authors

Gender Box issues	ASHER	BANDY	BASNET	BHALLA	DJOUDI	ELIAS	ELMHIRST	HARRIS-FRY	HOLMGREN	HOSKINS	INGRAM	KENEDY	LACERDA	LARSON	SHACKLETON	TOTAL	PERCENTAGE
Part A																	
Formal laws/policies	1	1	1	1	1	1	1	1	1	1				1	1	10	63%
Cultural/religious trends	1	1	1	1	1	1	1	1	1	1				1	1	10	63%
Access to natural resources	1	1	1	1	1	1	1	1	1	1	1	1	1	1	1	16	100%
Norms of behavior	1	1		1	1	1	1	1	1	1	1	1	1	1	1	14	88%
Access to education	1	1			1	1	1	1	1	1	1	1		1	1	10	63%
Access to cash	1	1		1	1	1	1	1	1	1	1	1		1	1	12	75%
Day-to-day economic roles	1	1	1	1	1	1	1	1	1	1	1	1	1	1	1	13	81%
Demographic issues	1	1			1						1	1			1	6	38%
[Migration]*	1	1										1			1	4	25%
Domestic roles	1				1					1		1		1	1	6	38%
Intra-household power dynamics	1	1		1	1	1		1				1		1	1	8	50%
Available economic alternatives	1	1		1	1	1	1	1	1	1	1	1	1		1	11	69%
Part B																	
Knowledge**	1	1		1	1				1	1		1		1	1	6	38%
Mgmt process involvement	1	1	1	1	1	1	1	1	1	1	1	1	1	1	1	15	94%
Leadership	1	1	1													5	31%
Women's networks/groups	1	1	1	1							1		1	1	1	10	63%
Violence against women	1														1	4	25%

*Migration has been extracted here from the total "Demographic issues" to show its preponderance in the data.

** Potentially relevant for "Access to education," but focused more on indigenous/experiential knowledge.

demographic issues (both 38 percent) reflects in part the fact that many in natural resource fields have not until recently recognized their relevance (as with household dynamics). However, these latter issues are more fully addressed in fields like demography and health, reflecting in this case also a lack of dialogue across disciplines. Where much gender research in the past has focused on exactly what men and women do, men's and women's *relationships* have been found to be central to making desirable and equitable change. The significance of domestic roles and demographic/reproductive matters pertain most fundamentally to the high demands on women's time, which compete with their time available to pursue own-account, forest-related activities. We need to address such time pressures if women are to play equal roles in such spheres as education, employment, community action and conservation.

We highlighted migration (25 percent) within Part A's demographic issues (total, 38 percent), because of its increasing incidence worldwide and its influence on gender relations as well as agriculture and natural resource management, which are becoming progressively more feminized. We remain a bit perplexed at the reluctance of many gender and forestry researchers (among others)[2] to address population issues—indeed, only one chapter addressed population growth. This seems odd, given the implications of childbearing and rearing for women's (and men's) lives and women's roles on the one hand, and the relevance of population pressure and growth for forest maintenance on the other.

On the whole, there is a good match between the key themes of the compilation and those covered in the Gender Box and listed in Table 1.1, Part A. Part B— representing topics not initially covered in the Gender Box—suggests how the Gender Box can be expanded to better reflect issues of current relevance and interest in the fields of gender and forestry. As shown, involvement in management processes (94 percent) and women's groups and social capital (63 percent) are common topics, with the related issue, leadership, addressed in 31 percent of our chapters. Knowledge(s), which differ by gender, were addressed in 38 percent of the chapters. The recognition that women and men typically hold distinct and overlapping sets of knowledge and perceptions—much experiential and/or indigenous—is the basis for many other analyses and for engaging with, and hearing from, both women and men during research and management processes.

Table 1.2 indicates the scales addressed in these papers (identified by first author; some articles addressed multiple scales). These authors have examined issues across scales fairly evenly (see Colfer et al. 2015: 80 percent of the IFR papers on gender and agroforestry focused on the meso level, with 30 percent addressing each of the other two scales).

Table 1.3 shows the periods of time addressed in each article (again, some articles addressed multiple time periods). Although the present is the most consistently addressed period, there is a fair amount of attention addressed to the past

---

2 This issue was even more pronounced in the IFR set on gender and agroforestry, with *only* migration discussed under demographic issues.

TABLE 1.2 Gender Box analysis: by scale

Scale	ASHER	BANDIAKY	BASNETT	BHALLA	DJOUDI	ELIAS	ELMHIRST	HARRIS‑FRY	HOLMGREN	HOSKINS	INGRAM	KENNEDY	LACERVA	LARSON	SHACKLETON	TIANI	TOTAL	PERCENTAGE
Micro		1	1	1	1	1	1		1		1	1	1	1			11	69%
Meso	1		1	1	1				1			1	1			1	8	50%
Macro		1	1		1	1			1	1	1		1	1			9	56%

TABLE 1.3 Gender Box analysis: by time

	ASHER	BANDIAKY	BASNETT	BHALLA	DJOUDI	ELIAS	ELMHIRST	HARRIS‑FRY	HOLMGREN	HOSKINS	INGRAM	KENNEDY	LACERVA	LARSON	SHACKLETON	TIANI	TOTAL	PERCENTAGE
Past	1	1	1				1	1	1	1		1	1				9	56%
Present	1	1	1	1	1	1	1	1	1		1	1	1	1	1	1	15	94%
Future				1		1					1	1				1	5	31%

(56 percent) and the future (31 percent). Again, this differs from the focus of the IFR collection on gender and agroforestry, where all authors looked at the present, 30 percent to the past and 10 percent to the future.

We are encouraged by the wide span of topics covered—which continue to expand as the field of "gender and forestry" gains maturity—as well as the attention to the various scales and the widespread adoption of a historical perspective to inform research on the present and visions for the future. This represents a growing recognition among researchers of the interconnectedness of people's ways of life (both in terms of issues and scales of action), the influence of history, and the relevance of ideas about the future in shaping current tree use and management strategies.

Introduction to chapters

This book is organized as follows: in Part I, this introduction is followed by a retrospective on gender and forestry by Marilyn Hoskins, one of the *grandes dames* of gender and forestry. Hoskins led the work of FAO (Food and Agriculture Organization) on community forestry—one of the first of its kind—for twelve years, beginning in 1984. She tells a fascinating tale, documenting the early and then-new understandings of women's roles in forestry and in forests. The chapter highlights a number of troubling issues that sadly remain today, despite significant progress (in which Hoskins has played a crucial role).

Part II turns to climate change, and is followed by Parts III and IV on tenure and value chains, respectively. As discussed below, we focus on these three central themes because of their timeliness and the significance of these issues—and of gender within them—to forestry research for development. In Part V, three chapters cover a potpourri of longstanding and emerging issues in gender and forestry research; a brief conclusion follows.

It is worth noting that there is an element of arbitrariness about the thematic division of chapters. People's lives are interconnected, and climate change issues can be affected by tenure and vice versa; value chains, similarly. Any given chapter may well include information about other (or even all) sections of the book.

Part II: Gender and climate change

Nations all over the world are trying to prepare for, mitigate and cope with climate change. Forest-based policies and schemes to mitigate climate change, such as REDD+, have received considerable attention, yet there has been little concern for gender in related deliberations and actions. Terry's collection (2009) is a valued exception. The seven chapters here represent a needed update. Their approaches vary enormously, ranging from Kennedy's focus on the perspectives of a small number of elderly women in upstate New York, in the US, to an international comparison of local-level responses to REDD+ implementation in six tropical countries. We begin with two chapters on the Global North for two reasons: first, placing them adjacent to each other made sense, as the issues in these parts of the world differ somewhat from those in the Global South; and second, because we wanted to introduce early on two kinds of issues: the significance of people's non-monetary values in their decision making and the power of implicit values in policy narratives. These issues are central to shaping forest management processes, but they have rarely been dealt with in the gender and forestry literature.

In recognition of the importance (and frequent dismissal) of values in forest management, we begin with Kennedy's moving chapter, which focuses on the views of 14 older women (another group often dismissed) from rural and forested areas in New York State. Early on in her chapter, she interrogates two views of gender itself (that of Butler and of Colfer) in ways that should interest and inform our readers. The chapter, rich in its explication of conservation-related values and their bearing on climate change, builds on her two years of continuing interactions with these owners of forested lands. As the director of an NGO, she has worked to codify and formalize "conservation easements" for lands the women are donating to the State of NY for conservation-oriented management "in perpetuity." For this analysis, Kennedy worked closely with the women to understand what motivates their approaches to conservation and how these relate to climate change. She beautifully weaves in the women's narratives about their feelings and their commitments to conservation of their lands. In this way, she is able to address the difficult topic of values in a manner that, we imagine, will "speak" to many from diverse cultural backgrounds—despite the acknowledged homogeneity of her sample. Her ultimate argument is that land ethics are an essential backdrop to legal protections, which, as she demonstrates for a nearby Apache (American Indian) reservation, can always be circumvented.

Remaining in the developed world but switching to the national level, Holmgren and Arora-Jonsson examine Swedish approaches to forest management—focusing on both gender and climate change policies—from a cultural perspective. They analyze the gendered values reflected in higher-level policy determination and implementation. They differentiate kinds of knowledge, pointing to that which is "credible" and "consequential," that which is implicit and explicit, and how these different types of knowledge influence and shape gender and climate change policy. The "taken for granted" aspects of Swedish forestry are illustrative of widespread assumptions in the forestry world that need to be questioned, if we are to address gender issues effectively.

The next chapter switches to a more statistical and macro approach likely to be more familiar for forest managers. Larson et al. report research results from forest sites in Brazil, Cameroon, Indonesia, Peru, Tanzania and Vietnam, where REDD+ projects have begun. Besides providing a valuable update on what's happening in the world of REDD+, this chapter discusses the variations in kinds and levels of participation through a gender lens. Specifically, Larson et al. discuss the variations, ranging from simple presence of women to a real voice in decision making. The authors examine four conditions that could render women's participation equal to men's in REDD+. Their results are variable across countries, with some interesting parallels: In general women have stronger voices within households than in village-wide decision making; and stronger voices in village decision making on topics other than matters of forest use. Not unexpectedly, men are shown to venture further into forests than women, though there are exceptions. Women are found to have less knowledge of REDD+, across the board, than men. The study's evidence that those women who use forests equally or more than men still do not participate adequately in making the rules and are less informed than men about REDD+ is worrying.

Harris-Fry and Grijalva-Eternod hone in on West and Central Africa, where they have conducted a review of the literature pertaining to conservation and intra-household decision making. These authors were struck by how little attention was being paid to such factors, as reported in REDD+ programming, despite significant amounts of REDD+ funding. In their literature review, the authors show the comparatively disempowered situation of forest dwellers, and particularly women forest dwellers, in this region, emphasizing land tenure and gender roles. They then provide a conceptual framework that can be used to improve our understanding of equity in REDD+ efforts. After defining the elements in the framework (sources of power, degree of empowerment, bargaining processes, and outcomes), the authors illustrate its elements with a case study from the Gambia. In the conclusion, they outline REDD+-related risks and opportunities for local women and men.

In Chapter 7, Tiani et al. drill down to tackle gender issues in climate change in one region of one country. They analyze gender equity with a focus on women in policy and practices of forest decentralization there, and the consequences that these can have on women's capacity to adapt to climate change. Besides using content analysis of key documents, the team examines gender and women's involvement in forest management committees in 24 community forests and four council forests in two Divisions in the Eastern Region of Cameroon. They conclude that facilitating and supporting women's leadership and equal participation in decision making can strengthen women's adaptive capacity, and they call for gender mainstreaming at all levels, especially in forest management and decision making, to ensure that adaptation policy is responsive to women's needs and aspirations. These findings are interesting and somewhat discouraging, in comparison with those of Larson et al. who found Cameroonian women expressing the greatest involvement in REDD+ discussions/decisions in the sites they studied.

Shackleton and Cobban build on a suite of methods, including household surveys in two South African communities in the Coastal Belt Forests and Savanna of the Eastern Cape. They add nuance to the common practice of looking at male- and female-headed households, dividing these further into male-only households, male-headed households with adult females present, female-only households, and female-headed households with adult males present. They identify different patterns of forest use and dependency for men and women, and examine the mental models of vulnerability of both genders in these various household types. They show that the dependence of women-only households particularly on forest products is evident in times of crisis. Interestingly, food security is reported to be highest among the women-only households (precisely those households with the lowest incomes) and lowest among men-only households. Women-only households are the only group in which doing nothing in response to a shock is never reported as the course of action (or non-action).

Part III: Tenure

Efforts to implement REDD+ have reinforced the recognition that tenure issues play a central role in shaping the management of forests, which are typically common pool resources (e.g. Larson et al. 2010; Kelly 2009). Many authors have since expressed the view that sustainable forest management will be impossible without more clarity about land and tree tenure (e.g. Richardson and Jhaveri 2013; Siscawati and Mahaningtyas 2012; Giri 2012; Buchy 2012; Kelly 2009, and more). And yet, the central role gender relations play in determining rights to and control over forest resources (e.g. Fortmann and Bruce 1988) remained, until recently, of interest only to a narrow group of researchers, with little policy attention.

The chapter by Bandiaky-Badji et al. is a practical assessment—by authors involved directly in both research and advocacy—of the recent historical and current situation with respect to women and forest management in Liberia and Cameroon, two well-forested countries. These authors first introduce some of the traditional/cultural factors that inhibit women's access to land and forests, even when policies specify equity. They then explain the relevant policies in the two countries that support (and in a few cases inhibit) women's access and rights to forests and their products. Next, they document a variety of steps taken by national and international actors to address gender equity through specific policy formulation and the activation of influence networks in both countries. Their final section is a series of recommendations for influencing gender-relevant policies on land and forest tenure and access.

Moving to South Asia, Bhalla looks at India's Forest Rights Act, summarizing useful studies on women's involvement in groups in India and elsewhere. The centerpiece of her study is a series of qualitative interviews and focus group discussions with both policy makers and local women and men in the Indian state of Odisha. Within a frame of "critical actors, junctures and acts," she examines the divergence between policy and its implementation, including specific barriers to women's (and to a lesser degree, men's) involvement in substantive policy making about forests. Her study shows that despite provisions for including women in committees with decision-making authority over local claims to forest land, their active participation in this process remains low. Yet, there are spaces and junctures within which critical actors are moving women's agenda forward in forest tenure reform; these entry points can be capitalized upon to increase women's voice in this process.

Asher's chapter reminds us that the academic and policy scholarship on tenure are too narrowly concerned with the devolution of rights to the local level. The ways in which women and men are organizing themselves to demand greater rights, and how these are a product of and linked to broader social movements for equality and change are usually inadequately considered. Drawing on women's activism in coastal Colombia, Asher stresses the importance of recognizing women as proactive agents in organizing to secure their livelihoods, and how these efforts are a part of broader movements for ethnic and territorial rights. Her case study focuses on grass-roots development among predominantly black women of forested, coastal Colombia in recent history, as it has interacted with formal state policies. She documents the women's evolving interests, which moved back and forth among concerns that were strictly economic in nature to issues of empowerment, health, education and intra-familial relations. One issue with broader applicability is the oft-perceived conflict between gender and ethnic concerns; another is found in the juxtaposition of what some call strategic and practical needs (Molyneux 1985), and their coalescing with identity factors in this context. This chapter also shows clearly the varying interests among and within groups and how these intersect with variations in power.

Part IV: Value chains

As is the case with tenure, the relevance of gender to the study of value chains has received scant attention despite the important role of gender relations in shaping constraints and opportunities for income generation. Although value chain studies have gained traction across various disciplines, few researchers have examined these "network[s] of labour and production processes whose end result is a finished commodity" (Hopkins and Wallerstein 1986, 159) from a gendered perspective. What gender-focused value chain work does exist has largely converged upon products that are specifically of importance for women (e.g. Elias and Carney 2007), or examined the division of labor between men and women along the value chains (e.g. Ruiz Pérez et al. 2002; Shackleton et al. 2011). More recently, researchers have examined how value chains are embedded in gendered norms, ideologies and power relations, operating across scales, that shape women's and men's ability to participate in and benefit from value chain development (Rubin and Manfre 2014). Here, we add to this analysis by featuring two analyses on gender and forest- or tree-product value chains; a large-scale literature review by Ingram et al. and a case study of the wild-meat value chain in the Democratic Republic of Congo (DRC).

Ingram et al. report the results of a thorough and systematic literature survey of around two hundred studies conducted across the Global South. The authors aim to demonstrate (1) the nature of gender differences in value chains pertaining to forest, tree and agroforestry products; (2) where these differences are concentrated within these value chains; (3) how to explain these differences, and (4) the types of interventions carried out to make value chains more gender-equitable and sustainable. Using examples from cocoa, shea, gums, resins and wood fuel chains, they assess the gendered outcomes resulting from these interventions. Lessons are drawn to help improve the equity and impacts of such interventions for policy makers, the private sector, development practitioners and the research community.

LaCerva takes a historical and ethnographic view of the wild meat value chain in the DRC. She traces the historical roots of the current gender configuration of this value chain, beginning with the harvest of animals in the forest, tracing the meat's trip to villages, thence to the city (Kinshasa), and on in many cases to Paris. Wild meat has become a luxury item in the DRC, valued as healthier, tastier, more prestigious, and reminiscent of an earlier perhaps more wholesome era. She finds that whereas men are the hunters, particularly of large game, women are involved subsequently in processing and marketing at all scales. This gender organization of the trade dates back to pre-colonial and colonial times, to conservative Christian patriarchy and to years of civil war that marginalized Congolese women. Yet, within a heavily patriarchal context, women carved out spaces of participation in this value chain, which persist today. She concludes that addressing the crisis in wildlife harvesting will require the involvement of women, some of whom hold positions of power and responsibility within the value chain.

Part V: A potpourri of longstanding and emerging issues

In our last section, we provide analyses from Africa and Asia that give a taste of other critical perspectives on gender and forestry. The three selected case studies touch on issues of intra-household knowledge sharing, migration and the impacts of land grabs. All three represent valuable and holistic analyses of gender-related issues in their respective parts of the world.

First, Elias examines the intra-household knowledge-sharing that underlies the management of shea (*Vitellaria paradoxa*) trees in Burkina Faso. She demonstrates the significant overlap between local knowledge of the shea tree and that of external scientists who have studied the tree. She shows how much local knowledge is shared between men and women, while at the same time reflecting the differences in their conventional roles. Using Sen's model of intra-household bargaining, she shows the social connectedness of spouses and how this manifests in some shared and some differentiated knowledge. She concludes that intra-household sharing or withholding of knowledge, in this case about agroforestry, can constitute a deliberate part of bargaining among spouses and contribute to better-informed agroforestry management strategies.

The next chapter by Sijapati Basnett examines the links between migration and forest-related policy in Nepal. She examines the ways the growing trend of male out-migration for employment affects gender dynamics within household and communal levels and in the process, shapes women's voice and influence in forest govern-ance processes. She emphasizes the importance of looking beyond gender alone to address the importance of other social factors like caste, occupation and ethnicity in shaping forest management. Her investigation of community forest-user groups in two distinct regions of Nepal elaborates on several issues. First, she emphasizes the importance of differentiating the contexts in and among different villages—using the gendered variations that characterized the forest management groups she studied. Her discussion of policies nicely complements that of Asher on policy history in Colombia. The attention to civil conflict, literacy, childcare, life-cycle differences, quotas are all recurring and important themes that play out somewhat differently in different contexts (as exemplified throughout this collection).

While Sijapati Basnett's study is situated in areas that are undergoing out-migration from forested landscapes, Elmhirst et al. consider the gender-differen-tiated impacts of agribusiness expansion in previously forested areas in our final contribution. The growing global demand for food and fiber has increased the demand for land for crop and tree plantations across the tropical world, particu-larly in Indonesia where growth in commercial oil palm production has been a major driver of deforestation and land alienation for local communities over the past thirty years. Elmhirst et al. build on Colfer's more than thirty years' experi-ence with the community of Long Segar in East Kalimantan (Colfer 2009) and a return visit in 2014, looking at the changes in landscapes and human relations over time. Elmhirst and Siscawati interviewed people and observed the current situation, in which oil palm companies have taken over much of what had been a forest-swidden mosaic dominated by forests. Although a few community members have prospered under this "land grab," a serious differentiation between rich and poor has emerged; gender stereotypes, which were barely evident in earlier times, appear to have hardened; and agriculture, central to women's responsibilities, has become more difficult due to inaccessibility of land.

Together, the studies compiled in this volume[3] yet again demonstrate the relevance of forests for the lives of the women living in or near them; and women's actual and under-acknowledged involvement in activities related to forest management and health. It is our hope that these findings may highlight women's potential contribution to, and more equitable sharing of, benefits from forests in the future. We believe that attention to gender, as shown in this volume, can benefit men, children and forests, as well as women.

References

Buchy M. 2012. *Securing women's tenure and leadership for forest management: A summary of the Asian experience.* Washington, DC: Rights and Resources Initiative. 1–10.

Colfer CJP. 2009. *Longhouse of the Tarsier: Changing landscapes, gender and well being in Borneo.* Philipps, ME: Borneo Research Council/CIFOR/UNESCO.

——— and Minarchek RD. 2013. Introducing "the gender box": A framework for analysing gender roles in forest management. *International Forestry Review* 15(4): 1–16.

———, Catacutan D, Naz F and Pottinger A. 2015. Introduction: Contributions and gaps in gender and agroforestry. *International Forestry Review* 17(S4): 1–10.

Elias M and Carney J. 2007. African shea butter: A feminized subsidy from nature. *Africa* 77(1): 37–62.

Fortmann L and Bruce J, eds. 1988. *Whose trees: Proprietary dimensions of forestry.* Boulder, CO: Westview Press.

Giri K. 2012. Gender in forest tenure: Prerequisite for sustainable forest management in Nepal. *Rights and Resources Initiative Brief* 1: 1–22.

Hopkins TK and Wallerstein I. 1986. Commodity chains in the world-economy prior to 1800. *Review* 10(1): 157–70.

Kelly JJ. 2009. Reassessing forest transition theory: Gender, land tenure insecurity and forest cover changes in rural El Salvador [PhD dissertation, Geography]. New Brunswick, NJ: Rutgers University.

Larson A, Barry D, Dahal GR and Colfer CJP, eds. 2010. *Forest for people: community rights and forest tenure reform.* London: Earthscan/CIFOR.

Molyneux M. 1985. Mobilization without emancipation? Women's interests, the state, and revolution in Nicaragua. *Feminist Studies* 2(2): 227–54.

Richardson A and Jhaveri N. 2013. *Tenure and global climate change: Gender, forest tenure, and REDD+.* Washington, DC: USAID. 9.

Rubin D and Manfre C. 2014. Promoting gender-equitable agricultural value chains: Issues, opportunities, and next steps. In A. Quisumbing, R. Meinzen-Dick, T. Raney, A. Croppenstedt, J.A. Behrman and A. Peterman, eds. *Gender in agriculture and food security: Closing the knowledge gap.* New York: Springer and FAO.

---

3 We will be editing a second volume, *The Earthscan Reader on Gender and Forests,* which will provide readers with access to the classics in this field, hopefully in 2016.

Ruiz Pérez M, Ndoye O, Eyebe A and Ngono DL. 2002. A gender analysis of forest product markets in Cameroon. *Africa Today* 49(3): 97–126.

Shackleton S, Paumgarten F, Kassa H, Husselman M and Zida M. 2011. Opportunities for enhancing poor women's socio-economic empowerment in the value chains of three African non-timber forest products (NTFPs). *International Forestry Review* 13(2): 136–51.

Siscawati M and Mahaningtyas A. 2012. Gender justice: Forest tenure and forest governance in Indonesia. *Rights and Resources Initiative Brief* 3: 1–18.

Terry G, ed. 2009. *Climate change and gender justice: Working in gender and development.* Warwickshire, UK: Practical Action Publishing.

2

GENDER AND THE ROOTS OF COMMUNITY FORESTRY

Marilyn W. Hoskins

Once in West Africa I took a photo of a woman who looked very, very tired and at least 8½ months pregnant. Holding a heavy watering can, she was tending tree seedlings. I put the photo on my desk to remind me of the many times I had heard, "Women must become more involved in projects." But involved how and with what goal? This chapter relates to my work from the late 1960s to the late 1990s on improving the conditions of people in non-industrial countries. Projects were not always helpful to either disadvantaged men or women. They did not always reflect local cultural or physical realities.

In the 1960s, in South East Asia, I heard women complain that many projects were considered successful to outsiders because of increased local production or income. In fact, they had disadvantaged women by adding more labor, changing the timing of women's inputs, changing traditional products from women's to men's control, or re-allocating income from women to men. The women wanted to be involved in activity selection and planning, and they wanted increased benefits for their labor.

In West Africa in the early 1970s, I ran a series of workshops on current and future issues for women in fields such as health, education and law. Although trees and forests were not listed as topics, they came up in many discussions. The women in this arid region knew a great deal about trees, forest products and the environmental impact of trees. They taught me that trees and forests were a life-and-death issue to them and their families, for fuel, food, fodder, shade, micro-climate and income. Forestry projects disadvantaged them by classifying and planting land close to the communities, making it off-limits to non-foresters. Local people had suddenly become poachers on land they had used for generations.

I am a cultural anthropologist, not a forester. However, since the 1970s, I have focused largely on a movement called Forestry for Local Community Development and a follow-up program, Forests, Trees and People. These United Nations programs, managed by the Food and Agriculture Organization (FAO), collaborated with many organizations and individuals in an effort to find ways for forestry to improve the lives of forest-dependent families. Layers of change in forestry goals of ministries and projects and in forestry training were needed to make women and local uses of forest products visible. This chapter focuses on these issues and the struggle of initial programs to address them. It considers not only women but gender, that is, how men and women fulfill their roles within their families through use of forest products.

Forestry goals

In many countries, forestry ministries were funded largely by profit from the sale of high-value timber and fast-growing pulp wood trees. Project goals were usually based on the number of trees planted. Foresters made career advancements partly through adding value to the treasury from trees they managed and sold, or by reaching or surpassing the production goals of projects. Foresters were taught to plant trees mostly in straight lines (Scott 1998), and to manage and protect the plantations and forests from poachers, including women and men collecting fuelwood or other forest products. In many countries, foresters were not paid a living wage. They told us they were expected to survive partly by fining people, often taking away people's axes, saws, or bicycles. Forestry students learned the Latin names of valued tree species and were given rigorous training, which fostered a sort of male bonding. Many wore uniforms complete with guns.

The approaches and goals of foresters contrasted with those of agricultural agents whose job was to convince men and women farmers to plant and manage agricultural products considered desirable by their ministry. Unlike forestry, it was not the agents, but the farmers who planted and managed and who benefited most from increased production. Although agricultural extension officers usually worked more with men than women, they understood differences in tasks and in crops planted by men and women.

Invisible women

In 1980, FAO helped organize a seminar, "The Role of Women in Community Forestry," in a famous forestry school in Dehra Dun, India. I saw a group of Indian women students rush up to a woman from Fiji, treating her as a heroine. I later learned that, when this woman from Fiji applied to the forestry school, the admission board did not know her Fijian name was a woman's name, and they admitted her. She broke the barrier and this year the school had accepted a small group of women students. It was generally neither common nor accepted that women could become foresters.

That same year, the Committee on Forestry (COFO), a high-level meeting of heads of forestry from FAO member nations, included a topic on women in the agenda. Several papers were presented all about women having trouble becoming foresters or not having good assignments. It was only in the following meeting five years later in which the second issue, that of forest-dependent women, was represented in the papers.

In the early 1980s, I traveled a week in Senegal on an FAO mission with a very nice and competent regional forester. He assured me over and over that women had no role in forestry and *never* planted trees. Other foresters backed him up. One evening, invited to his home, I asked his wife why women never planted trees. She burst out laughing. "Look around you. How do you think all these fruit, fodder and shade trees got in the compounds around the village?" She assured me she had planted many trees, as had her friends, her mother and grandmother. Perhaps one reason that her husband had not seen her as planting trees is that fruit trees are often considered horticulture. Seedlings for them were found in the agriculture, not forestry nurseries. But the women had also raised non-fruit trees from seed and had transplanted trees to be closer to their homes.

Invisible trees and plants

It was not only the women who were invisible, but also many local species that they and their families wanted and used. Most people outside rural communities do not recognize that it takes a different species of tree to make a functioning mortar than to make a pestle for pounding grain. Some communities in West Africa build granaries using a tree with a forked trunk to hold up the floor. When people tried to substitute crossing straight tree stems into an X shape, the granaries collapsed. Local beliefs are also critical. One project had planted cashew trees without learning they were taboo. The plantation was completely burned down. An elderly man told me the community had a taboo against that species and were fearful of bad luck if it started to fruit. In a project in Yemen, a forester worked with communities selecting trees that gave products they wanted in relation to the work it took to raise them. In making the final selection, local people asked, "What color flowers do the different trees have?" The foresters never expected a value on beauty in such a poverty-stricken area.

In Nepal, forestry inventories did not consider the daphnia growing in the understory. This product provided a major source of income for women who made it into paper. Without knowing how much daphnia there was, and the yearly off-take, there could be no useful rules for sustainable harvest. There could be no way to protect traditional users or the resource from outside entrepreneurs. Understory and other non-timber forest products frequently used by women are not usually considered in forest inventories.

A workshop in Thailand examined how forestry and health ministries could plan together. Participants were amazed at all the local knowledge when rural people made an exhibit of plants and small animals used for food and medicine. In Peru, a forestry project planted trees in women's potato land. This was done on the basis of planners seeing few trees and assuming a scarcity of fuelwood. Women told me they had no scarcity of fuel as they used bushes that easily spread through vegetative propagation. What was in short supply was good potato-growing land close to their village.

When gardens became forest-land, many women walked farther to get fodder for small animals, plant vegetables, gather fuelwood, or harvest forest products traditionally obtained from the woodland or fallow. During times of drought, women in West Africa relied on food made from roots and leaves of lesser-used plants growing in the natural succession following harvest. Usually these plants are more difficult to process so they are not harvested in times of plenty. Foresters often saw these plants as weeds and cleared them from the forest floor.

Initial programs

In the 1970s, international concern was raised about the sustainability of natural resources worldwide. This was especially true in the energy sector. People in the United States and Europe started buying smaller cars and even wood-burning stoves. Eric Eckholm, a journalist and political scientist with concern about the environment, wrote dramatically about the global environment and the need in less prosperous countries for fuelwood to cook meals without causing deforestation (Eckholm 1975). When international attention was drawn to problems of access to fuelwood at community level, the number of projects related to community woodlots and wood-conserving stoves multiplied and expanded. But positive results from increased planting and introduction of new stoves did not automatically follow.

A number of non-governmental organizations (NGOs) and bilateral programs, as well as international agencies, searched for ways to increase availability and efficiency of fuelwood. The following are examples from a few of the major efforts with which I worked.

The US Agency for International Development (AID) was involved in a number of community-level forestry projects, especially in arid zones of West Africa. In 1978, Arvonne Fraser, Director of AID's Women in Development Office at that time, contracted me to write a programming guide for women in forestry (Hoskins n.d.). As far as I know, it was the first publication on the topic. It stressed the need to consider gender roles in relation to trees and forests, land use and tenure, migration, timing of inputs, and so on.[1] It included formats for management-plan agreements requiring detailed planning and communication between project staff and local women as well as men.

The World Bank planted a number of woodlots to increase the availability of wood fuel near communities. The goal behind most of these projects was protection of forests and plantations from local use. Stimulated by the failure of many initial woodlots, John Spears, lead forester at the World Bank at that time, studied these projects with the help of social science information. He described destruction of woodlots, including local herders cutting through fences to let their animals graze on saplings. He noted traditional serial tenure usage in that area. Animals annually cleaned and fertilized the fields after local men and women had

---

1 Editors' note: See Part III, Gender and tenure; Sijapati Basnett; and Djoudi and Brockhaus, all in this volume, for the continuing relevance of these topics.

harvested their crops. He stressed the need to communicate more with the farmers and herders during project design, but did not specifically mention the need to consult women.

In the early 1970s, the FAO focused on how their projects could bring more benefits to rural communities. J.E.M. Arnold, then-Director of the Policy and Planning Division of the FAO's Forestry Department, organized an FAO interdepartmental committee to study how forestry could work with other sectors. He also organized researchers from diverse countries to learn how forestry was addressing this issue globally. Researchers contributed information to a book that marked a new focus for forestry in the FAO, but also showed how difficult it was to find successful examples of projects improving the lives of forest dependent people (FAO 1978).

There was an effort to increase awareness of women's dependence on forest resources but it was slow in being adopted, especially by field foresters. In 1985, an international meeting held in Kenya included discussions on women and forestry. Women said they were not being involved enough in forestry planning or benefits (see Part II, this volume, for examples of the continuing relevance of this issue). A Tanzanian forester got up and proudly stated he hired women in forestry nurseries because they had slim, delicate fingers, good for working with seedlings. Then he added that women did not need so much money as men—women only needed to buy a little salt and coffee—so he could pay them less and make the money go further. It was his mistake to report his experiences at this meeting, as many women stood up and shouted at him. They were beginning to find their voice.

The same was true at a workshop on stoves. Women were reporting on the health issues of constantly breathing smoke and the importance of flues in some types of homes. An Indian forester was doubtful that flues were necessary. He said he did not see the problem because when the cooking stove was lit, everyone just went outside. The women together said, "Who is it that goes outside?" He suddenly looked as though he saw the situation in a new light.

FAO and forestry for local community development

I did a number of consultancies for the FAO and finally joined in 1984 as the first social scientist in the Forestry Department. I had great support from my two directors of the Policy and Planning Service while there, J.E.M. Arnold and Marc de Montalembert. The Swedish aid agency, SIDA, which largely funded the early community forestry efforts, was very supportive of social science and a gender focus. A number of other foresters, however, were not sure what I was doing there. I was usually the only woman and non-forester in forestry meetings.

In the beginning it was necessary to have a good sense of humor. My office was always buzzing with a hard-working, committed group of interns and consultants, many of whom, though certainly not all, were women. One day a forester entered my office, looked around and said, "Should I call this 'the women's room'?" To which I laughed and replied, "If I can call all the other offices in the department 'men's rooms'." Another day a male colleague looked into my office and said, "How do you get all these cute girls to work in your office?" I looked at the professional young women there and replied, "I didn't know getting a master's in forestry and working in the field were supposed to make you ugly." Little by little, as we worked with increasing numbers of FAO staff and as we learned forestry terminology, the teasing diminished and collaboration increased.

The very first day in the office, I was asked to comment on a socio-economic study as the first step in initiating a fuelwood project. The questions were well designed, but the document said they were to be asked of "heads of households." I noted that in the project area it was women who collected and used the wood fuel, but only men were considered heads of household. Both men and women should be queried. The following day I saw a copy of the memo to the project staff saying the questionnaire was to be given to both men and women. I knew from that moment that through "backstopping" FAO projects, I was in a position to make some real changes.

I stayed at the FAO for a dozen years, managing first the Forestry for Local Community Development (FLCD) program and later, a much more participatory program, Forests, Trees and People (FTP). Goals of both were to improve the lives of forest-dependent people. From the beginning it was obvious that these would have to be learning programs. There had not been a lot of success in forestry projects improving the lives of women and their families and it would be necessary to learn why, and how to improve activity design and results.[2]

New information on topics found to be relevant was needed if forestry was to change. The relationship of many topics to forestry was unclear. The process we initiated was first, to identify problems or areas that appeared to need improvement from both local livelihood and forestry perspectives. Second, we studied what information was known or not known on the topic. We then researched, developed workshops and conferences, and created material in formats to make new information useful for policy makers, project managers, schools of forestry, local communities and the young. Altogether we published over fifty books and audio-visuals, most of which were released in English, Spanish and French, among many other languages. Although this chapter talks about a few of them, a more complete set can be seen on the FAO's website. It is important to realize gender, in this context, requires understanding issues from the perspectives of both men and women as well as topics often of special interest to women, such as fuel for cooking.

Learning from the field

In the 1980s, most forestry projects were focused on wood fuel—both increasing the supply and using it more conservatively. Many proved to be based on false or inaccurate assumptions or incomplete information. For example, there were

2 SIDA funded a community forestry unit in the University of Uppsala which collaborated directly with the FAO program.

disagreements on the comparative economy of burning wood and charcoal. Later research showed it depended largely on the cost of transportation. Because charcoal is more concentrated and lighter weight, it was more economical to carry it to towns distant from the source of heavy fuelwood. In some cases, saving forests was the project goal but fuelwood was not coming from the forests, as the project had assumed. It was waste from clearing farm fields. Some women reported they selected species of wood by burning qualities, such as temperature of the flame. They continued using their traditional source when projects introduced trees unsuitable for cooking.

One Asian project caused an uprising by the people it was designed to help. The project goal was for the forest service to take over scrub land to produce and sell fuelwood, thereby "lowering the price of fuel for the 'poor'." The "poor" in the project area did not use fuelwood, but sold it as their major source of cash. They used leftover sticks and leaves for their own consumption. The project had to change or it would undercut the local economy.

Most stove projects were evaluated on how many stoves were distributed. But research was needed on why some were adopted and others not. There was information on fuel saving in the laboratory, but not on the actual savings when the stoves were used by families. To understand and perhaps change the way families (mainly women) used fuel, and the way cooking was done, required understanding how people already used their resources (including time), what and how they cooked, and their relative concern over saving fuel.[3]

I heard women being asked by engineers if they would use a stove that required half as much wood, and of course, they answered, "Yes." But they were not being told the stove required added labor, because the wood had to be cut into small pieces. Some stoves did not accommodate leaves and sticks used locally. Solar stoves required standing in the sun and focusing the mirrors as the sun moved. This was uncomfortable for cooks who constantly stirred long-cooking millet mush, common in the hot climate of West Africa.

Researchers who asked adopters of efficient stoves why they used the stoves got answers including less smoke, children burned less often and even husbands who supported their use because they looked more modern. In India, women made and charged their friends for stoves with flues, and were pleased to have a trade that gave them status by bringing money to the family. Saving fuel was seldom mentioned. Stoves should, perhaps, be distributed on health and safety grounds instead of being introduced by foresters on the basis of fuel economy.

A field guide for planning, monitoring and evaluating stove programs (Joseph 1990) was developed by Stephen Joseph, an Australian anthropologist/engineer. It was built on years of work in rural areas, as well as the results of workshops and research, and included questions for stove design that cooks, usually women, and planners could consider together. Did the stoves cook the type of food people ate?

3 This issue continues to be examined, e.g. by recent Cornell PhD graduate in Economics, Andrew Simon.

Were they used by more than one cook at a time? Were they too big or too small for the space? Did they need to be mobile when the weather changed?

Tree-planting projects also had mixed results that research helped explain. Women who cooked in open kettles would not use a commonly introduced species which had acrid smoke that flavored the food. One team discussed a community tree-growing project only with the men. Seedlings were delivered without taking into consideration that women in that community tended and watered seedlings. The seedlings, which arrived during the busiest time for women's activities, dried up and died. The long growth period of trees handicapped people with minimal incomes who could not wait until trees matured for returns on their investment of time and labor.

Madhu Sarin, an Indian architect and activist for Indigenous land and women's rights, described an unusual project in India. Although the project planted trees in land used by women, it considered women's needs. Trees were planted close together so that they could be thinned and the thinning used for fuel. Later, lowest branches would be trimmed. However, when the trees grew tall, there was nothing left for women to use for fuel, and they had lost their traditional source of firewood.

In Haiti, all trees legally belonged to the government. Farming families wanted a few trees, but not enough that the government would notice and claim the land as government forest. In Honduras, many trees in a community plantation project died. Women told me that they jerked the seedlings so the hair roots would be cut, causing the trees to "mysteriously" die. Women wanted a few trees, but not a plantation. Since they were squatters, they feared a plantation would make the land more valuable and "city folk" would come to kick them off their garden plots. Groups without land security and migrating groups usually do not invest in planting or caring for trees, whereas resettled groups with assured land tenure more often do.

Another issue was the aspect of payments. In Nepal, reports from forestry nurseries recorded only men working, but one usually saw mostly women. The salary of women was less. Projects recorded the cost of men, but hired women instead, keeping the difference. This inequality needed to change.

A very different payment issue related to ownership of trees. Local people were often paid to plant foresters' trees. In projects of village woodlots, project staff often gave food or funds and materials to start up woodlots when villagers were facing hardships. Managers were sometimes startled when community members did not maintain the trees. Villagers assumed the trees would be owned by the project and the project assumed the trees belonged to the planter. Projects in which residents raised and planted their own seedling were sometimes stopped when participants learned a neighboring project was paying for the work. It was a difficult balance when participants were poor and labor was scarce.

Workshops on fuelwood focused on maximizing tree planting, until we introduced the concept that in no place was there a fuelwood scarcity, but plenty of food. The two scarcities had to be addressed together, not maximizing one at the expense of the other.

Only through this kind of information from the field could decision makers understand what was and was not working. This was a slow process, requiring new understanding, new goals (and attitudes) and new tools.

Seeing and planning with women

In order to reach policy makers and senior foresters with the importance of considering gender, we wrote a policy-level document showing the importance of forest products to both women and men, the difficulties men and especially women have in obtaining them, and what can be done to improve the situation. Through pictures and vignettes from our own field experience, as well as information from a number of authors, the book told of trees in the household economy, as food, fuel, fodder and construction (FAO 1987). It documented the long hours of women's work in different regions of the world being extended when forest products were scarce. It looked at the complementary and sometimes conflicting uses of tree products by men and women. Our publication was sent to forestry offices throughout the world to stimulate making women visible. It emphasized the potential for gender-sensitive forestry policies and activities not only to improve local livelihoods, but to make forestry activities themselves more successful.

New tools were also needed to assure FAO projects were designed with women's concerns integrated into planning and managing projects. Mary Rojas, then-Director of the Women and Development office at Virginia Tech University, wrote a field guide for people designing and managing forestry projects (Rojas 1993). It focused on critical points in forestry activities determining outcomes for women. These guides were given to all teams designing FAO forestry projects and were requested by many other organizations.

All our efforts were gender sensitive and many subjects were of special interest to women. However, in the 1990s, we decided to initiate a major effort on a multimedia and comprehensive training guide, which could be used for including gender considerations in all work (Wilde and Vainio-Mattila 1995). We started in Asia with case studies in seven countries. A two-person team was selected in each country, one from a local research group and the other from a local training institution. The purpose was to leave the information where it could have the greatest impact. In each country, the highest-level forester we could identify who was willing to be the forestry advisor to the project, was asked to help. This was extremely useful in some countries where the forester became aware of gender issues during the research and made changes in the country's forestry program.

The case studies showed that one could not generalize about men's and women's roles. In one community, people planted trees. In a neighboring community in the same country, land tenure or migration was different and no one planted trees. In one community in Thailand, women felt projects helped raise their status, while in Bhutan, women said they had more equality before outside projects favoring men were initiated. In most communities, women collected wood; in others, men did.

The case studies were accompanied by videos. Six manuals were included: participatory rapid appraisal, organizing and running participatory workshops, making case studies, methods for training, analyzing gender issues and applying findings. The approach was based on the understanding that everywhere in the world men and women had differing roles to play in their families, including in relation to forestry.

There was an Asian regional forestry meeting being held as the materials were finished. When I asked to show the film at that meeting I was told that the Asian foresters would walk out of a program on gender. But, if I really wanted to show it, they would let me take the risk. I insisted, and we showed the film. No one walked out. At the end, the Asian forestry officials (all men) stood and clapped. They thought the approach respected their cultures and men's and women's roles within families. A parallel set of case studies and materials was then developed from South and Central America and published in Spanish.

Forestry schools and in-service training

In non-industrial countries, most of the foresters came from an urban background. It was high status to enter forestry school to become a forester. A great deal was needed in the training to help these young people become participatory and to focus on local as well as forester priorities.

A training manual and video entitled *What is a tree?* were developed and sent to forestry training institutes. *What is a tree?* focused on why people in rural communities want various species of trees. It tried to expand the vision of foresters beyond fast-growing pulp wood and valuable timber species. At first, students of forestry found the title almost insulting, but they soon realized that it was talking about what foresters versus local people visualize as the end use of various trees.

A popular set of training materials was developed by Nancy Peluso, Matt Turner and Louise Fortmann (Peluso et al. 1996) for forestry schools as well as for in-service training. It included a pre-test which asked such questions as how women used trees differently than men. This approach allowed the training staff to know whether the trainee had observed women and men and their activities or if more gender training would be needed.

A field manual called *The Community Tool Box* was written by D'Arcy Davis-Case (1989). It was a new methodology for participation. It was not like methods of the time, which limited participation to identifying local terms, answering outsiders' questions, or involving some community members in the research team. This was a method that helped local men and women identify their own questions and help find ways to answer them. As with all our field manuals, it was tried out in the field before being completed. In this case, extension agents and local women discussed perfecting the methods in a CARE forestry project. Both the extension agents and the women commented on how much they had gained from the process; the end-product was enriched by their participation.

Also very popular was a concept note on rapid appraisal methods by Augusta Molnar (1990).

A whole set of information was collected on food that came from the forest (Falconer and Arnold 1991). A project guideline was then written for project designers and staff, starting with ways forestry could avoid competing with food production (Ogden 1993). It also included a methodology for foresters together with health workers to identify nutritional needs that might be addressed with careful tree selection. An example of one activity was working with the World Blind Union to identify where vitamin A-deficiency blindness was a major problem. Primary school children in selected regions planted seeds of locally appreciated food trees with vitamin A-rich fruit or leaves. Children took home the seedlings and planted them, and cared for others on communal land available to those who had less land and were at greatest risk for malnutrition.

A series of comic books was created at the request of primary and secondary school teachers and was also popular for adult literacy courses. One on nutrition was entitled *I am so hungry I could eat a tree*. All comic books were translated into numerous local languages and were also used by several environmental groups for children in the US.[4]

We did research and published concept documents and manuals on many other topics. John Bruce, a lawyer specializing in land tenure, wrote about how countries can make laws to help men and women participate and benefit from community forestry (Bruce 1999).[5]

Conflict management especially affected women, who often lacked power to express or obtain their objectives. Since we found little in the literature specifically focused on forestry conflicts in in less-industrialized countries, we held a four-month virtual conference coordinated by Dr. Garry Thomas. The documented inputs from around the world were taken to the World Forestry Congress held in Turkey in 1997 (Chandrasekhran 2000).

Forest products cannot be managed in small plots as the plants may be scattered over large areas; they need to be managed more like pasture land. Research on communal management of forest-lands from different parts of the world was based on such different foci it was not possible to draw conclusions. We contacted Elinor Ostrom (see Ostrom 1990)[6] and her team at Indiana University, which had become well known for their work on communal management of fish and water resources. We asked them to help us by integrating trees in their methodology. Through collaboration, an original and innovative research methodology for

---

4 For example, CIFOR translated them into Indonesian.
5 See Holmgren and Arora-Jonsson or Bandiaky-Badji et al. this volume, for evidence of this enduring issue.
6 Ostrom was the first woman to receive the Nobel Memorial Prize in Economic Science for her work on understanding that communal management can work, including within forestry, and understanding the rules that make it function. The International Forestry Resources and Institutions (IFRI) methodology and materials were developed by her group with collaboration of the Community Forestry Unit of the FAO.

forests was devised (now called the International Forestry Resources and Institutions, IFRI). It measured the resource, both trees and understory, at identified points and recorded various local laws, markets, users, rules, and so on. Through this method, it was possible to compare different sites with similar or different situations, and also the effects of changes in the same place over time. As more information was collected, it was possible to identify the rules, laws, markets and the organization of users which influenced forest sustainability and its relation to the well-being of forest-dependent people.

Although the method did not focus on gender as such, it identified inequalities in laws, members of user groups and other data invaluable in identifying handicaps for participation. It showed whether men or women were involved and what species were needed by which users. The program, working with additional funders, established training centers in Africa, Central and South America, Asia and Europe, so the methodology would be improved and available to all regions.

Sharing successes

A number of case studies and articles were also published and incorporated in talks, journals, workshops and an FTP Newsletter, to give examples of approaches which might be useful. For example, in 1984, Wangari Maathai's[7] program for tree planting in Kenya demonstrated that middle-class urban women could support rural women by helping grow seedlings and making information on planting trees available (Maathai 2003). Another example was an NGO project in East Africa where a traditional type of humorous play demonstrated to the community that rules against women cutting fuelwood caused the men to have cold tea. The discussions which followed helped create new customs on tenure and what species women could cut for fuel.

Case studies were published on many projects. One was about women's participation in a very strict Muslim community in Sudan (Faidutti and Webb 1991). The women had found a way to be involved and appreciated, as well as profiting by expanding on a windbreak project the men had started. This proved to be of great interest to project managers and women in other Muslim communities.

One problem a number of projects faced was increasing the value of local products women traditionally processed and sold. When marketing was improved, businessmen from outside the community came and threatened the sustainability of the resource as well as the income of traditional processors. Case studies from Thailand and India (Campbell 1991) gave examples where the women were protected by organizing and setting rules on harvesting *before* marketing was improved.

Changes from FLCD to FTP

When the FLCD was first started, most activities were studies and support for project design and implementation. A second phase, the FTP, first focused on four

---

7 Maathai was the first African woman to receive a Nobel Peace Prize.

projects to be designed by communities and carried out with their input. Donors, the FAO and governments withheld approval and funding for the planning phase, finding it too difficult to work with the flexibility required for this kind of project. All four were closed and we decided to have no more projects.

We then designed an entirely new approach. We started a program with existing national training or research groups already focused on community development or community forestry and willing to contribute a facilitator and an office. The FTP brought funds to better carry out their work and to create joint activities with other groups and organizations. It also gave facilitators and their activities status with government officials and other national groups since this was an FAO program. As a learning program, Swedish, Swiss, Norwegian and Dutch donor representatives followed activities and gave helpful advice. Collaborating FTP organizations had freedom to identify and work with compatible people and organizations forming national and international networks.[8]

Every year a meeting was held in a different country where the host facilitator took other facilitators and donor representatives to see activities, exchange information on what was going on in their various countries, and their successes and issues. Each facilitator presented what had been designed for the year, what had been accomplished, and why there was a difference. Since it was a learning program there was no blame, only discussion of ideas for learning and future success. Facilitators told how they had planned budget use and activities for the coming year, with what groups they were collaborating, the outcomes they expected.

Outcomes

Looking back over the changes during these early years, there are three issues on which we did not make sufficient progress, and which I believe still need attention. They are: (1) support for participant design and evaluation of community activities with attention to strengthening social capital and legal arrangements to withstand pressures from outside; (2) better evaluation of training, especially when the goal is changing attitudes and developing appreciation for local perspectives, and (3) the "joy of letting go."

Support for participant design and evaluation of community activities

The first phase of FTP failed to start projects with actual community control. It takes confidence in people to allow them to identify their problems and use their knowledge in planning and evaluating local activities. It also takes focus on social capital and legal arrangements to protect the resource from growing outside

---

8 FTPP facilitators exchanged experiences through a monthly journal coordinated and edited by Daphne Thuvesson.

pressure. One activity requested by an indigenous group in Bolivia was gaining land rights. The FTP facilitator adapted the IFRI inventory to include local people and a forester to document historical use and make a management plan. The people had a great celebration when they obtained title. Some years later, however, as in many countries, companies are moving in with large amounts of money and/or force to take over the resource. There often is not enough attention paid to building social protection and legal arrangements to help community members maintain progress and benefits from their resources (cf. Elmhirst et al., this volume).

Better evaluation of training

Training has often used evaluation forms, asking trainees if they found the room comfortable, the trainer interesting and the material useful. Applied training evaluations ask if the participant remembered some fact mentioned during a class. These really do not tell you if the trainee will listen with respect to the women and men, if training will be used and if the community will benefit. More pre- and post-evaluations are needed with follow-up to measure actual long-term changes trained participants are able to make.

The "joy of letting go"

"The joy of letting go" was the most difficult. When we started the FLCD program most decisions were made in the FAO with the community forestry advisory committee. Then, decisions moved to host governments, donors and the forestry project management. Finally, the decisions were largely given to the facilitators with groups and communities in the participating countries. Each move left former collaborators, often people of good will and concern, with less or no role to play. They felt the program was no longer participatory; *they* were no longer participating. Once I saw a statue of a bird breaking out of its shell with the title "The Triumph of the Egg." The egg triumphed only when it was no longer needed. We need to capture that spirit, so former decision makers and persons handing over authority feel the joy of accomplishment, of "letting go." Often NGOs and project managers fear local people taking over, fearing they will lose funding if they are no longer needed. We need better monitoring methods that track the skills community members gain and reward outsiders letting go.

Conclusions

It is difficult to know outcomes from the various efforts. Reports were made of the hundreds of foresters trained in new approaches. One knows that thousands of documents were requested and many people reported that they were useful. Even when it was not obvious, new ideas sometimes were put to use. One Chilean forester told us that he smuggled in our documents to read during the time when his country was under an authoritarian government forbidding talk of community activities. He said the documents were very useful as he was ready to start when the government changed.

If I were asked to identify the most lasting impact of the program, I would say it was an educational impact on the facilitators, the interns and others who worked in it most intimately. They had experience and support and the opportunity to develop and work with emerging topics, creative activities, and the top-level thinkers of the time. They worked with people of other countries where they visited, discussed and contrasted programs, and attended international meetings. The network they formed is still functioning, sharing ideas; some have cross-continent collaboration. Many of these individuals have moved up into high-level national or international positions, are doing cutting-edge research, or are at universities teaching the next generation that gender awareness and forestry are both important.

The photo of the tired, pregnant woman watering saplings has long ago faded, but I still think of her and wonder how she is doing. A lot has been done to help forestry build on a stronger understanding of local realities and methods to work with rural women and their families. Has it made the lives of forest–dependent women both visible and improved?

When I first went to Africa, I had a copy of *The Ugly American* (Burdick and Lederer 1958) under my arm. I wanted to be like the man who introduced long-handled brooms to women who were suffering back pain from bending over short-handled brooms. One day, in a village, I asked women I saw using short-handled hoes if bending over like that hurt their backs. "Of course," they answered. "Well," I said, "the women in the next village use longer handled hoes. Would that be a good idea?" They gave me an overly patient look and asked if I had seen the soils. The other village had soft soils—this one had rocks. I learned quickly that things may not be as they seem and outside answers are not necessarily useful.

What I really want to know about the woman in the photo is whether people come to her better prepared than I was when I started? Do they begin with a phase of learning her concerns and questions and how to incorporate her knowledge with what has been learned elsewhere to work together toward viable solutions? Is her life better?

References

Bruce J. 1999. *Legal bases for the management of forest resources as common property.* Rome, Italy: Community Forestry Note 5. Community Forestry Unit, FAO.

Burdick E and Lederer W. 1958. *The ugly American.* New York: WW Norton and Company.

Campbell J. 1991. *Case studies in forest-based small scale enterprises in Asia.* Community Forestry Unit Case Study 4. Rome, Italy: FAO.

Chandrasekhran D. 2000. *Addressing natural resources conflicts through community forestry.* Rome, Italy: Community Forestry Unit, FAO.

Davis-Case D. 1989. *The community's tool box.* Manual 2. Rome, Italy: Community Forestry Unit, FAO.

Eckholm EP. 1975. *The other energy crisis, firewood.* Washington, DC: Worldwatch Institute.

Faidutti R and Webb C. 1991. *Women and community forestry in Sudan*. Community Forestry Unit Audio Visuals and Booklets. FAO.

Falconer J and Arnold JEM. 1991. *Household food security and forestry*. Community Forestry Unit Note 1. Rome, Italy: FAO.

FAO (Food and Agriculture Organization). 1978. *Forestry for local community development*. Forestry Paper 7. Rome, Italy: FAO.

——. 1987. *Restoring the balance: Women and forest resources*. Rome, Italy: Community Forestry Unit FAO.

Hoskins MW. n.d. *Women in forestry for local community development*. Washington DC: USAID.

Joseph S. 1990. *Guidelines for planning monitoring and evaluating cookstove programmes*. Community Forestry Field Manual 1. Rome, Italy: FAO.

Maathai W. 2003. *The green belt movement: Sharing the approach and the experience*. New York: Lantern Books.

Molnar A. 1990. *Community forestry rapid appraisal*. Community Forestry Unit Note 3. Rome, Italy: FAO.

Ogden C. 1993. *Guidelines for integrating nutritional concerns into forestry projects*. Community Forestry Unit Field Manual 3. Rome, Italy: FAO.

Ostrom E. 1990. *Governing the commons: The evolution of institutions for collective action*. Cambridge: Cambridge University Press.

Peluso N, Turner M and Fortmann L. 1996. *Introducing community forestry*. Community Forestry Unit. Rome, Italy: FAO.

Rojas M. 1993. *Integrating gender considerations into FAO forestry projects*. Community Forestry Unit Guidelines. Rome, Italy: FAO.

Scott JC. 1998. *Seeing like a state*. Binghamton, NY: Vail-Ballou Press.

Wilde V and Vainio-Mattila A. 1995. *Gender analysis and forestry*. Community Forestry Unit Training Package. Rome, Italy: FAO.

PART II
Gender and climate change

3

LIVING CONSERVATION VALUES

Women and conservation easement protection in central New York

Virginia Kennedy

> Do women have a different attitude towards the right use of land? I don't tend to like generalizations based around sexual divides unless they are, say, about childbirth, but, perhaps, in general, they do, or perhaps they may have. That's better . . . My hope for my gift is to give the land back to itself, and to the other creatures of the earth which have learned to live on, and with it, nourishing it in turn as they have been nurtured. May it renew us, as we help to preserve it.
>
> *(Dr. Mary Anne Whelan, January 2015)*

In the summer of 2013, Dr. Mary Anne Whelan, a retired pediatric neurologist living in Cooperstown, NY, decided to make a gift of over seventy acres[1] of farm and forest lands to Otsego Land Trust. She allowed that Otsego Land Trust could sell the land and use the funds generated for its land protection mission. She designated that as a condition of the sale, the land be protected by a conservation easement. Dr. Whelan is one of eight women landowners who have sought to protect land with conservation easements since my arrival as Executive Director at Otsego Land Trust in 2013. In fact, during my tenure at the land trust, ten of twelve active easement projects have involved women landowners. My purpose in what follows is not to claim that these women represent a statistical trend toward greater numbers of women landowners protecting their land than men and/or in increasingly greater numbers overall. However, a comprehensive study of statistical trends regarding women landowners and easement protection, both in terms of their numbers and motivations, may indicate that women do protect land in greater numbers than men, and women landowners who protect their land may be increasing. It is a study worth undertaking.

Currently, there is a paucity of data focused on women landowners in the United States, and in New York specifically. There is, however, general agreement regarding the need to educate and become educated about women forest and farm landowners in terms of their needs and their desires for their lands

---

1 Editors' note: 2.471 acres = 1 hectare.

(Pinchot Institute 2006; American Farmland Trust 2013). The Women Owning Woodlands Network reports:

> ... across the country, more women are taking the reins of forest management, as owners, as stewards, and as family members. These women are a historically underserved demographic who are making or influencing management decisions on millions of acres of working forests.
>
> *(Muth et al. 2013)*

According to the USDA National Woodland Owner Survey, in 2011, 24 percent of woodland owners were women, up from 19 percent in 2006 (Wilmot 2013).

In what follows, I focus on what encouraged this particular group of women—the eight women with whom I have worked directly and six additional women landowners who have granted easements to Otsego Land Trust—to embrace the conservation easement as a tool for sustaining the conservation values of their lands. I communicate their motivations, their relationships with their lands, and their attitudes toward generating (or not) financial resources from their lands. Because the majority of these women have been involved to one degree or another in resisting the introduction of horizontal gas drilling ("fracking") in New York State, I also discuss possible connections between this resistance, concerns with climate change, and their desire to protect their lands with conservation easements.

Redmore et al. have made the point that "while there is a plethora of information on women and land management in developing countries . . . little was known about women in private forest management in the US" (2011, 75). It is, of course, essential to explore the evolving conditions of women in relationship to forests and land conservation in the Global South. However, in the context of globalization and the pervasive spread of capitalism's patriarchal structures of power and resource exploitation, a global analysis, including North America, of women's relationships to land conservation may prove essential. Such a comprehensive analysis would offer the possibility for defining paths forward that may contribute to addressing our mounting ecological challenges, especially climate change. The motivations and goals of this group of women are both revelatory and provocative in terms of potential frames and paths of inquiry for future analysis of women landowners and conservation easement protection in general in the United States.

Defining the conservation easement

According to US law, a conservation easement is a binding agreement between a property owner and a land trust (or other non-profit organization or government entity) that restricts development rights on the property in order to protect conservation values, which are the natural ecological and environmental attributes of the property.[2] This protection does not preclude farming, forestry, or other

2 Conservation and preservation easements are also used to protect historic sites, buildings and structures. For my purposes here, I refer solely to conservation easements protecting ecological and natural attributes.

working or recreational uses of the land, and in some cases may even encourage such use. Easements will, however, limit the specific manner of these uses in ways that ensure conservation values of a property will be sustained into the future or "perpetuity," the term used in most easement agreements. The land trust or entity holding the easement is responsible for monitoring adherence to its restrictions.

In many cases, the landowner giving the easement, or "grantor," will be qualified for a tax deduction because voluntary conservation easements are understood to contribute to the public good by conserving land, protecting water, flora, fauna and habitats, and encouraging the sustainable use of land when it is used for agriculture and forestry. Landowners are motivated to place their land under a conservation easement for a variety of reasons: deductions from federal and state taxes, preempting the threat of government land-acquisition or land-use regulations, protection of the natural attributes of the land from fragmentation and development, personal attachment to a specific property or place, or contribution to the community or public good (Farmer et al. 2011; Halpenny 2006). Some analysis demonstrates that the financial considerations of tax deductions are the least cited motivation for conservation easement protection (Farmer et al. 2011; Halpenny 2006). As a land trust professional, I hasten to add that though the tax deductions may not, in many cases, be the primary motivation for land protection, these financial benefits often make land protection possible for landowners who cannot afford the financial hardship of losing development value in the land. The majority of the women I interviewed for this chapter have taken advantage of tax deductions, though, as their perspectives communicated below will demonstrate, these deductions were ancillary to their decisions to protect their lands with conservation easements. In what follows, I elaborate the motivations of this particular group of women landowners for protecting their lands with conservation easements. In doing so, I hope to offer provocative paths forward for analysis that will lead to a greater understanding of women landowners who seek this kind of protection for their lands; and thus, how to most effectively engage women landowners in conservation practices that can support them and their families while protecting the lands and waters that sustain life.

Introducing the participants

The women landowners of this group are all over fifty years of age, Caucasian and middle to upper middle class economically. Coordinator of the US Forest Service's National Woodland Survey, Brett Butler, explains that "nationally, more than half of the 766 million acres of forest land is owned privately by [non-industrial] proprietors whose average age is 62.5" (Ring 2014). All the women in this conversation are close to that national average. Eleven of the fourteen participants are single, meaning divorced, widowed, or never married. The length of property ownership on their protected properties ranges mostly from fifteen to forty years. One participant has owned her land for only four years, but her parcel is adjacent to family lands owned by her parents for forty-five years. Of the fourteen participants, all but five live full time on their properties, though three of the five grew up on their properties, return to them regularly, and spend a significant amount of time there annually. Six of the fourteen women have children; although none of the women raised considerations of their own children in our conversations regarding easement protection. Rather, they spoke more generally about future generations.

The racial homogeneity of this group reflects the geographic location of the Otsego Land Trust and our service range for land protection. The counties of central New York State—Herkimer, Delaware, Otsego and Schoharie—where these protected properties are located, are not racially or ethnically diverse. For all four of these counties, the US Census Bureau reports a significant majority Caucasian population, between 94 and 97 percent (US Census Bureau 2013). All four counties report roughly 50 percent female population (US Census Bureau 2013). This homogeneity as well as the socio-economic bracket within which the grantors sit raise important questions about the demographics of land ownership and the possibility of conservation easement protection in general. Butler and Leatherberry in their 2002–03 survey of forest landowners focus on age and other variables that do not include race, ethnicity, or gender (Butler and Leatherberry 2004, 4–14). In his 2006 survey, Butler makes explicit that the majority of family forest landowners and decision makers are white and male (2006, 32). However, as the opening paragraph of this chapter suggested, these demographics are shifting and so attention to new demographic trends like greater numbers of women landowners will be important to analyze.

Among the women in this particular group, there was a great amount of experiential diversity. From Marion Karl,[3] raised by missionaries overseas, to Sandra Freckelton, an artist who plans to create a nature and artists' retreat on her property, to Marianne Younkheere, a financial services advisor in California with deep family ties to and love for Schoharie County, New York, these women represent a diverse variety of life experiences that have shaped their views on land and land protection. Like the similarity in their demographics, the variety of their lives' paths provokes interesting questions regarding relationships between specific experiences and a desire to conserve land, though my particular focus is the influence of gender on their interest in conservation.

The troubling question of gender

According to both oral and written responses to the question of gender motivations in their decision to protect their lands, the entire group agreed that considerations of gender did not specifically or consciously influence their actions. One participant, Connie Young, captured the sentiment for the majority of the group:

3 All women's names used in the chapter are the actual names of participants who have given permission for use of their names in the chapter. Participants include Susan Burdsall, Sandra Freckelton, Pat Gambitta, Cat Gareth, Susan Huxtable, Marion Karl, Marjorie Kellogg, Barbara Newman, Gail Sondergaard, Lynn Tanner, Connie Young, Rebecca Young, Mary Anne Whelan, and Marianne Younkheere.

I think that my relationship with nature is gender neutral. I don't know that being a woman causes me to have a stronger or weaker connection to nature. I don't think that the fact that I am female has any identifiable impact on my decision to protect my land.[4]

Another participant, Susan Burdsall, complicated the "gender neutral" characterization. In her written response, Susan stated:

...my late husband and I were unable to have children, so the generative part of being female had to be expressed in a different way—not that I can create nature, but that I can be a help in preserving it. Mostly, though, I do not think of it in terms [of] my being female so much as I think of it as being human.

Susan went on to articulate her resistance to the gendered term "mother earth:" "'Earth' is a breathing, living thing. It is as much father as mother, for after all, both give us life. I think the term 'mother earth' often tends to reinforce gender stereotypes." Like Susan, other participants' responses at once denied conscious gender considerations but invoked them nevertheless. Gail Sondergaard, for example, explained that her late husband loved the land as she loves the land, but "differently as a man, as a farmer, farming the land, what it produced." Susan Huxtable asserted that it was the women in her family that made the protection of the land possible; her mother insisted upon it and her sisters agreed:

We knew we had to protect the farm. We didn't want to go into farming. But, we knew the land had to be protected. I give my mother the credit. I applaud strong women who work to protect history and tradition in the land.

As Dr. Whelan communicates in the quote that opens the chapter, "perhaps" gender—or "sexual divides" to use Dr. Whelan's term—may matter in these women's decisions to protect their land, though they themselves did not originally see their decisions in those terms.

In her influential work, *Gender Trouble: Feminism and the Subversion of Identity*, Judith Butler argues that gender is performative (1999). That is to say there is no natural opposition between masculinity and femininity. Rather, masculine and feminine roles are defined through the hegemony of heteronormative[5] power

---

4 All direct quotations from participants are from email responses to a series of questions, personal conversations and single interviews, and an organized group meeting at Otsego Land Trust on January 22, 2015.

5 Heteronormativity posits that the binaries defined by Western culture—male/female, and heterosexual/homosexual—are the only "natural" or "normal" categories into which people fall. The former of each binary—male and heterosexual—are privileged and considered superior while the latter—female and homosexual along with any other alternative constructions of sex and gender—are marginalized. Heteronormativity also considers the sexual relationship between a heterosexual couple "normal"; any non-heterosexual relationship is not so considered.

that insists on essentialized definitions and divisions. Butler writes that "gender is the repeated stylization of the body, a set of repeated acts within a highly rigid regulatory frame that congeals over time to produce the appearance of substance, a natural sort of being," but the "compulsory frames" of these "natural sorts of beings" are "set by the various forces that police the social appearance of gender" (1999, 43–4). To borrow Boucher's summation, Butler's "theory of identity rejects the essentialist conception of gender as a substantial difference expressing an underlying natural sexual division" (2006, 115). Where Dr. Whelan's quote at the opening of the chapter conflates gender and sex, Butler separates them only to join them in her conclusion that both gender *and* sex are socially constructed within the frame of heteronormative power (Butler 1993).

In the introduction to *The Gender Box: A framework for analyzing gender roles in forest management*, Colfer, in seeming opposition to Butler, expresses a separation that has been naturalized between male and female knowledge and goals in the context of forestry (2013). She writes:

> Foresters are recognizing (a) that they need to do a more thorough job of taking women's knowledge, roles, interests and goals (and in many places, men's as well) into account, and (b) that they don't really know how to do it.
> *(2013, 1)*

In a bulleted list elaborating suggestions on better management of forests, Colfer acknowledges women's roles as "the bearers and usually enculturators of the next generation," implying both a naturalized sexual (biological) function of women as child bearers and a culturally performative role of women as the teachers and progenitors of cultural norms and traditions (2013, 1). Where Butler focuses on the hegemony of heteronormativity, Colfer refers to "a hegemonic masculinity, which is seen as creating and/or reinforcing gender hierarchies" of patriarchy where men have a disproportionately large share of power, resources and choices in communities (2013, 9).

Butler explores the potential for resisting the hegemony of heteronormativity in the marginalized spaces beyond or outside of its regulating structures, structures that regulate not just gendered identities, but constructed identities of the "other," the "alternate" in any given cultural formation.[6] Colfer recognizes her own immersion in these structures. She writes: "We are all enmeshed in

---

6 There is not space here to elaborate Butler's considerably expansive body of work and its exploration of the construction, politics and ethics of the self and the other, or to explore the body of criticism that interrogates Butler's theories. I understand that Butler's conception of heteronormativity would subsume masculine hegemony and gender ideals as themselves performances within the hegemony of heteronormativity. My point in juxtapositioning *Gender Trouble* with *The Gender Box* is to first articulate and then accept the limitations of gendered roles within the theoretical constructions of "masculine hegemony" and "gender ideals."

gender systems of our own. Although I make every effort to approach this research topic as objectively as I can, none of us can fully escape our own experience and resulting biases" (2013, 3). While recognizing this challenge, Colfer argues for the necessity of identifying "gender ideals" (2013, 14). She asserts that in "writing about such ideals one is in danger of reinforcing essentialist stereotypes; yet there is little doubt that such notions, in fact such stereotypes, cultural as they are, have real impacts on the world's forests" (2013, 14). This statement indicates the conceptual-theoretical box inside which the "Gender Box" is contained. Colfer recognizes the conditions of masculine hegemony and communicates the gendered identities—essentialized male and female identities—it constructs. I agree with the necessity Colfer insists upon to address essentialized gender ideals as they are perpetuated globally, especially in the context of the homogenizing forces of capitalism and their influence on unsustainable land-use practices and resource exploitation. While I suggest that continued studies in forests and gender struggle to trouble the conceptual framework of masculine hegemony and gender ideals—Butler's theory of the hegemony of heteronormativity and critiques of it[7] are fertile ground for such struggle—I agree with Colfer that "women's knowledge, roles, interests and goals" need to be taken much more into account even as they are defined within the confines of gendered constructions that may be considered stereotypical. The institutions of colonialism and Western capitalism have perpetuated masculine hegemony of patriarchal structures that determine gendered hierarchies (inclusive of structures that elevate white males above *othered* populations, male and female) where they may not have previously existed and exacerbated them where they have.[8] Considerations of these structures and resistance to them from their margins *and* from within them is a necessary political strategy. In the context of global ecological emergencies like climate change and rapid, rampant biodiversity decline, embracing more sustainable approaches to land use and human relationships with the land that resist the tendency of masculine hegemony to exploit the land[9] could have important impacts on the quality of human survival and the survival of the living beings with whom humans share the planet.

Patriarchy perpetuated

Women in the United States, though they have made progress since gaining full citizenship with the vote in 1920, still lag behind men in acquiring equal wealth and access. Along with land ownership, economic and political power and access to corporate power and governance are still mainly in the hands of men in the United States. In 2013, women's fulltime earnings were 82 percent of men's, up

7 See e.g. Prosser 1998; Boucher 2006; Hawkesworth 2006; Schwartzman 2007.
8 See e.g. Cheyfitz 1997; Pagden 1998; Mies and Shiva 2014.
9 See Mies and Shiva 2014; Jahanbegloo and Shiva 2013; Kolodny 1975 and Merchant 1990.

from 62 percent in 1979, but still largely shy of equal.[10] In government, women hold 17 percent of the hundred Senate seats and 16.8 percent or 73 seats of the 435 in the House of Representatives (Catalyst 2015). Across all strata of publicly elected offices in the United States, women represent less than, sometimes significantly less than, 25 percent of office holders; for example, only 6 percent of current state governors are women (Catalyst 2015). These statistics indicate that though women's presence in positions of power and authority may be improving, they are still considerably underrepresented in terms of decision-making power in political and economic arenas.

A number of participants in this women landowners group, without specific reference to these statistical realities, nevertheless referenced their implicit impact in the context of their decision to protect their lands. In answer to my question "what impact did the nature of your personal relationship with nature have on your decision to protect your land with a conservation easement?", financial advisor Marianne Younkheere answered, "As a woman who's worked in a man's world for 35 plus years, perhaps it's my instinct to protect, because I have the same strong instinct with my clients." Rather than focus on her relationship with nature, although she freely discusses this relationship in other places, the question evoked for Marianne her subject position in the gendered-male world of finance. She understands her position and experience in this world to have provoked what she interpreted as her implicitly feminine instinct to protect herself, her clients and her land. Another participant, Rebecca Young, referred specifically to a lack of respect by male professionals who did not take her seriously when she questioned them about various issues of land management. Rebecca indicated the necessity for women to be especially strong in a world of men who denigrate women's intelligence and capabilities. She stated, "We were encouraged in our family to be strong women, so whenever men made me feel like I asked too many questions, I didn't care." For Cat Gareth, this conversation raised a question related to both male power and the desire to protect the land: "We all want to protect the land from what might happen to it," she stated, "but what is it about being female that creates the concern that something could be taken away if we don't protect it?" Sandra Freckelton agreed with the sense of fear that Cat expressed, but was quick to specify that the fear and the resulting desire to protect were not about the desire for a kind of stasis, about wanting the land to always remain the same in the face of a changing world. She emphasized the dynamism of the land and the inherent drama in the fact that the land is always changing, both on its own and as a result of people's engagement with it. The desire for protection, most of the women

10 The Bureau of Labor Statistics Report, *Women in the labor force: A databook* reports that in 2013, women who worked full time in wage and salary jobs had median weekly earnings of $706, which represented 82 percent of men's median weekly earnings ($860). Among women, earnings were higher for Asians ($819) and Whites ($722) than for Blacks ($606) and Hispanics ($541). Women's-to-men's earnings ratios were higher for Blacks and Hispanics (91 percent for each group) than for Whites (82 percent) and Asians (77 percent). The 82 percent statistic best represents the demographics of these women landowners.

agreed, arises from a fear of specific kinds of engagement with the land that they view as destructive to it and thus to people and, they unanimously emphasized, all the living beings with whom people share the biosphere.

Mary Anne Whelan expressed what I would characterize as the most consciously gendered response of the group to the question of differences in male and female approaches to the land. She explained that, "some years ago, as an assignment, I had to write haiku[11] in, I think, 15 minutes. I based two of them on nineteenth-century prints that were hanging in the Otesaga,[12] where the class was held." The haiku Mary Anne composed are as follows:

> I. In these scenes, always,
> Arms of men are in action,
> Raised to chop, thresh, drive

> II. In these scenes, always,
> Arms of women encircle
> Holding bread, fruit, child

Mary Anne went on to interpret the paintings that inspired her haiku:

> So perhaps we might infer (from these pictures at least) that men saw the new land as something to conquer, something to wrest a living from by force. But the women looked further ahead, to see its fruits, to see what it might yield. Their postures embrace and enfold, rather than subdue and alter. It is a broader and longer view, and a gentler one. It is protective.

Mary Anne's invocation of the conquering of the American continent in the nineteenth century in gendered terms, raises the specter of Manifest Destiny, the belief held by European Americans that they possessed a God-given right and authority to press across the continent and subdue the land and the indigenous peoples who had been living from it for millennia. Mary Ann's comments recall the scholarship of writers like Caroline Merchant and Annette Kolodny, characterized as "ecofeminists." Ecofeminism is a movement in academia and in activism to understand connections between ecology and feminism and more specifically between the degradation of the earth and the oppression of women. Merchant and Kolodny have theorized an entwined patriarchal impulse toward the subjugation of women, "othered" populations like African Americans and Native Americans, and the land (Merchant 1980; Kolodny 1975). Ecofeminist theory seeks to make visible the interconnections between the subjugation of these groups and the exploitation of the biosphere within a paradigm of "mastering" (Kennedy 2012).

---

11 A haiku is a form of Japanese poetry and is a short verse of seventeen syllables, written in three lines, the first and last being five syllables and the second being seven. Haiku usually focus on images of nature.
12 The Otesaga is a hotel and conference center in Cooperstown, NY.

Mary Anne tapped into the tensions regarding the land that were pivotal in the nineteenth century when America was attempting to form a national identity independent from Europe. In the first half of the century, Kolodny explains, a significant number of American writers, for example, James Fenimore Cooper and John James Audubon, communicated a struggle between a need to preserve wilderness as a pastoral utopia and to acknowledge that the growth of the nation in which they themselves were participating would inevitably exploit what they perceived should be saved (Kolodny 2007). Kolodny defines the contradiction as the "pastoral impulse dangerously confused with the myth of progress" (2007, 67).

By the second half of the nineteenth century, the rapid felling of forests, usurpation of water, overuse of pasturelands, and so on encouraged the beginnings of a nationwide movement toward ecological sustainability most famously embodied by John Muir and Gifford Pinchot. Muir demanded that America's vast natural spaces he defined as "wilderness" be preserved and kept safe from people. Pinchot believed nature's resources existed for human use. At the turn of the nineteenth into the twentieth century, Muir battled with Pinchot over the latter's determination to dam what Muir believed to be the pristine landscape[13] of Hetch Hetchy Valley in Yosemite National Park to bring more water to a rapidly growing San Francisco. Muir's preservationists lost the battle. The concept of preservation of "untouched" wilderness was subsumed within Pinchot's ethics of conservation, which promoted what Pinchot believed was a wise and sustainable use of "nature" defined as "resources" for the benefit of people in the present and in the future (Kennedy 2012). Pinchot unapologetically envisioned humans as masters of the land, but insisted upon a wise and sustainable use of land that would ensure the availability of resources for future generations. America's body of environmental law is the culmination of both Muir's desire for untouched wilderness and Pinchot's demand for sustainable use of resources. It is, as I elaborate more fully in my conclusion below, a body of law caught in a contradiction with a notion of progress that demands more and more resources to feed a consumerism that must necessarily grow to keep the system running (Kennedy 2012).

These women landowners seem to resist this paradigm of mastery with their focus on protection and also through an expression of resistance to concepts of land ownership that prioritizes its resource value. They express a humility that subverts Pinchot's expression of superiority and Muir's demand for people-free nature; a humility that comprehends a shared place for people on the land rather than a dominant one—a philosophy more in line with American Indian philosophical orientations to the land or with Aldo Leopold's "land ethics."[14]

---

13 Indigenous peoples, specifically the Miwok, had lived in and from the Hetch Hetchy Valley for millennia. In the earlier part of his life, Muir chose either not to see Indigenous influences on the land, or to see them and criticize them. This point of view did transform somewhat over time. See e.g. Mark Dowie 2011; Julie Cruikshank 2005, Chapter 5; and Mark David Spence 1999.

14 See e.g. Leopold 1966; LaDuke 2002; Kimmerer 2003; Nelson 2008; Kennedy 2012.

Marjorie Kellogg related an interesting anecdote in relationship to the notion of mastering the land. She explained that the husband of her good friend, Barbara Newman (another participant in this dialog) with whom she owns her land, is a city planner. "When Barbara's husband comes," she explained, "he walks the boundaries as if he has an instinct to count the land, to know it as a commodity." Marjorie suspected that this desire for marking the boundaries might have just as much to do with a professional background of city planning as in any particularly male inclination. In any case, the notion of marking boundaries and defining the land as a "commodity" provoked a discussion about property ownership—seeing land solely as something to be owned and dominated. Marianne Younkheere directly questioned the ability of people to *own* the land. "We are only temporary stewards of land," she asserted, "like Indians of long ago, the concept of 'owning' land seems presumptuous to me. How can we own deer, water, and fields?" Susan Bursdall echoed Marianne's sentiments. She stated, "I consider myself to be a steward of the land. Although I have a document that says I own it; that is simply hubris." Pat Gambitta also resisted the notion of land ownership. In the spirit of Mary Anne Whelan's formation of "giving the land back to itself," Pat explained that "I've tried to really stay focused on what the land seems to want to be and then try to support and encourage that process." Cat Gareth expressed the same sentiment in a desire to accept the land and "pay attention to what it is" rather than focus on "owning" it. In the statement of "owner's intention" that she wrote for the baseline documentation[15] of her conservation easement, Cat summarized her perspectives as follows:

> I am too close to it to speak more formally about what this conservation easement will mean to this land's ecology, its water, its wildlife, the sustainability of this fragment of the natural world, but I do hope that it survives as what I have known it to be—a habitat for the human spirit.

These arguments were not arguing for Muir-like preservation that leaves the land untouched by human interventions. Of this group of fourteen women, twelve currently or have in the past farmed their fields themselves or leased them for use and/or did commercial timber harvests of their forestlands. Their conservation easements allow that these activities can occur. In fact, Otsego Land Trust encourages "working lands" or engagement with the land in a way that allows landowners to live from their lands as long as the methods they undertake ensure protection of the lands' "conservation values." However, as opposed to Pinchot who saw the land solely as exploitable resource, the subject of the "natural resources" as interpreted by this group of women reveals an ethic of land use focused on long-term

---

15 Baseline documentation is a recorded description of the conservation values and other characteristics of a property at the time a conservation easement is signed and recorded. The baseline is then used as the description of record in annual monitoring of the easement. Otsego Land Trust requires a "statement of landowner's intentions" so that those original intentions can be clearly communicated for future reference.

sustainability and what I would characterize as *reciprocity*. In addition to what the land produces in a purely economic context, all of these women understand the land, in exchange for their good stewardship and protection, gives something back to them that cannot be valued in the economic terms of a capitalist market.

To elaborate, Lynne Tanner offered a particularly provocative story. She spoke of a path behind her house, "a magical place; a wetland" with a view of the valley. She related that the day after 9/11/2001,[16] she took a walk on that path and watched the trucks on the interstate highway that she could see down in the valley. "I had a sense of comfort," she explained, "a sense that people were still moving on the roads and the land was still here." She asserted that the land felt like a "spiritual center," a place that healed her in that moment. She went on to explain that "farming is good, the land producing is good; it's pragmatic. But we have known it's more than just that."

Connie Young, who defines herself as a "capitalist," gave a nuanced perspective of "natural resources" that expanded the capitalist definition of natural resources with which she began her explanation of the term and indicated the concept of reciprocity to which I referred:

> The technical definition of "Natural Resources" are things of the earth (minerals, timber, etc.) that have monetary value in our modern economy. This definition, unfortunately, does not provide for placing value—monetary or otherwise—on the Resource that is Nature. There is value in the Resources in Nature in that they restore our spirit, provide habitat for non-human creatures, and keep the globe from becoming an overly toxic environment. The negative connotation of the conventional definition of "natural resources" is that it can involve stripping away resources for the benefit of mankind's bank accounts thereby damaging or destroying habitat. The potential positive connotation for treating Nature as a resource is that it contributes to planetary health not just physically but spiritually.

Gail Sondergaard offered that to her, natural resources are things "people think of using for themselves, for 'progress.'" Gail viewed "progress" negatively, seeing it as a movement forward toward greater and greater exploitation of land solely for economic reasons. Like Lynn Tanner, Gail asserted that "the land is more than that." Susan Bursdall argued that "'natural resources' is neither a positive or negative term. While plants, animals, water, etc. are 'natural,' so are oil, gas, coal, tornados, hurricanes, fires, and other things," but, she continued, "just because something exists 'naturally' doesn't mean that we can do whatever we want with it."

---

16 On September 11, 2001, three passenger jets flown by members of the terrorist group Al Qaeda were intentionally flown into buildings, the Twin Towers in New York City and the Pentagon in Washington DC. A fourth plane meant also for a building in Washington DC crashed in a field in Pennsylvania. The death toll was over three thousand people.

These varied approaches to their perception of land ownership and natural resources did not prevent a consensus on a critique of a capitalist economy that demands, once again referring to Connie Young's words, a "stripping away of resources for the benefit of mankind's bank accounts thereby damaging or destroying habitat." Though Connie confidently claimed, "I am a capitalist" who is committed to working the land for what it might yield in terms of profit, she and the other women of the group embraced the limitations of easement protection "to preserve habitat for the multitude of species that live with us."

Antifracking, climate change, and the commitment to perpetuity

The privilege of ownership carries with it a serious obligation to the land itself, to maintain it in its natural state so that it can regenerate indefinitely. Establishing a Conservation Easement through the Trust enables me to meet this obligation now and for the future.

(Barbara Newman, December 2014)

Barbara Newman wrote these words as part of her "owner's statement" for the baseline documentation for her conservation easement. They reflect the strong feeling among all fourteen women participants of a duty and responsibility toward future generations, as well as the future of the land itself. Marion Karl expressed the relationship between a commitment to perpetuity and to the land in explicit terms. She asserted with great emotion that she protected her land because to her it was "home." She had been raised by missionaries and had never had a home, a place to settle and to come to know intimately. For Marion, her easement protects a place that will always be home for her, her children and grandchildren, and she hopes, for their children into the future. Marion, who is in her eighties, has been a staunch advocate for a permanent moratorium on fracking[17] in New York State, which she connected with her strong sentiment toward a home-place. "I participated in protecting my home," she explained, "but not just my home here, the earth too; everybody's home." Marion, Cat Gareth and Marjorie Kellogg, along with actively fighting against the introduction of fracking into New York State, marched in the September 21, 2014 climate march in New York City, a march that demanded quick and serious action globally on climate change. All three women related responsibility to their own land to responsibility to fight for protection of the earth overall. As Marjorie explained, "in the face of such a huge thing like global warming, the easement was one small thing I could do."

---

17 Hydraulic fracturing, commonly called fracking, is a drilling technique used for extracting oil or natural gas from rock formations deep underground through the injection of fluid made from water and chemicals. Most major energy companies claim the practice is safe; however, many scientists and environmentalists agree that overwhelming evidence exists of its hazards to drinking water, human health and the atmosphere, especially through the release of greenhouse gases that contribute to climate change.

Susan Burdsall also related her commitment to perpetuity to a legacy for future generations beyond the borders of her own local space. Susan explained that she considered her conservation easement protected land, her "legacy to future generations, not only the local community, but to all who appreciate open spaces." Susan, who also committed to keeping fracking from New York, did not explicitly consider climate change in her decision to protect her land, but she asserted that "it is irrefutable," and "we are responsible for doing something about it." Lynn Tanner stated that climate change is "at the forefront of my mind all the time. I'm not just worried about my own land," she explained, "I'm very sad about what we are doing to the whole earth."

Pat Gambitta, like Susan Burdsall, did not explicitly consider climate change when she decided to protect her land. But, perpetuity was paramount in her mind:

> I have no family who would be interested in this property, so there would be no one who knows or understands my love for this land, therefore perhaps no one who would fight to preserve it. Perpetuity [in the conservation easement] will protect what I've tried to do.

Since protecting her land, Pat, who would "not allow fracking even if it were permitted," has gotten solar panels and begun a reforestation project that will contribute to carbon sequestration. Like Marjorie, she sees her small acts in the local space of her land as a contribution to something bigger, an impact magnified by other small local acts.

For Connie and Rebecca Young, the idea of protecting their land into perpetuity was "absolutely vital." Connie elaborated that she is connected to her land through her family, a "history of place," and a "personal history and time spent on the land." She envisions her land in the future as it has been for her family for generations. But, she relates this vision of her land to a connection to "the universe at large." Connie described a part of this larger connection through water: "My property is situated on the cusp between the Mohawk/Hudson drainage and the Susquehanna drainage. I love to think about the water flowing in two directions into these two major river courses and then to the ocean."

She went on to connect her concern for water to climate change: "On a very small scale I think my desire to protect water resources is linked to climate change, as water resources in the northeast may be impacted by climate change." She also asserted that there are a "'hecka' lotta' [sic] trees on my land doing their part to make oxygen and to sequester carbon." Ironically, Connie, who adamantly claimed her own characterization as a capitalist, explained that she is "looking for ways to support education of the American Public that motivate people to modify their behavior as it relates to consumerism—both energy and stuff." This statement captures a sense of not just literal but ethical limitation on unchecked resource exploitation that conservation easement protection embodied for this group of women. Along with a sense of humility and acknowledgment of other species who share the lands they chose to protect, their easement protection represented an opportunity to connect to something larger in terms of both time and place: The act of protection of their land was connected to protection of the earth itself, for now and for future generations.

In conclusion: legal land protection and living conservation values

The conservation easement is a legal method of land protection developed by people within a system of land ownership that prioritizes short-term economic value over all other values (cf. Holmgren and Arora-Jonsson, this volume). The efficacy of this legal protection is only as sound as the actual land ethics upon which it is based. Easements can and have been constructed to rationalize activities like fracking, oil extraction and toxic agricultural practices as allowable and "safe" for the conservation values of the land. Can anyone rightly claim that easements are true and perpetual protection in a world where short-term economic gain is the goal of so many and such powerful people and corporations?

As I draw to a close, I relate a story relevant to alternate conceptualizations of engagement with the land that is illustrative of the weaknesses in America's structure of legally based environmental protections like the conservation easement. Oak Flat is a part of Tonto National Forest and land that is considered sacred to the Apache[18] of the nearby San Carlos Apache reservation. By special order of President Eisenhower, the area has been under federal protection from mining since 1955. The earth under Oak Flat is also filled with copper ore, and Resolution Copper Mining, a subsidiary of British–Australian mining conglomerate Rio Tinto, has sought ownership of the land for a decade, lobbying Congress to enact special legislation on its behalf more than a dozen times since 2005 (Fang and May 2015).

The Apache and communities close to Oak Flat have been fighting just as long against such legislation. In December, the legislation passed as part of the 2015 National Defense Authorization Act; a "land swap" will occur in which "Resolution Copper will swap roughly 7.8 square miles of land scattered across Arizona for roughly 3.8 square miles of Tonto National Forest, which includes Oak Flat" (Fang and May 2015). The company plans to use a particularly invasive kind of mining to extract the copper called "block cave mining" that will leave

> . . . a crater two miles wide and up to 1,000 feet deep, destroying the surface of the land. Block cave mining will also generate nearly a cubic mile of mine waste, which the company proposes to leave on a parcel of Forest Service land, just outside the town of Superior.
>
> *(Fang and May 2015)*

The arguments over whether or not to allow the mine to go forward are the usual. The copper is purportedly needed for national defense and the jobs that the

18 The Apache are an American Indian tribe.

mining industry can bring are purportedly needed for the region's depressed economy. Local residents have fought the project based on its potential for damage to the forest and surrounding homes and towns and to the health of the region's people. In addition to their concerns for the health of land and people, the Apache have fought because the land is meaningful and essential to their cultural identities and engagement with their traditions. Since the end of the nineteenth century, the federal government has captured more and more of Apache land to offer to mining companies. Apache tribal council member, Wendsler Nosie Sr., who has been fighting against the mine for years, opined, "Look at all the mines here already. Look at all the things they took from us. Seventy percent unemployment, that already tells you where it's going. It's not going to benefit anybody here. What's going to be left is contamination" (Fang and May 2015).

Oak Flat is publicly owned, legally protected land covered by a plethora of legal environmental and American Indian cultural protections like the National Environmental Policy Act, the American Indian Religious Freedom Act, the National Forest Management Act which governs Forest Service land and the Federal Land Policy and Management Act. Still, other legislation was passed: the 2015 National Defense Authorization Act that undermined those protections and traded Oak Flat to an international corporation whose priority goal for the land is economic profit defined in capitalistic terms. I do not mean that statement to be particularly pedantic or judgmental. I do mean to point out that in a global economic system where engagement with the land is based on greater and greater exploitation of resources extracted from it for monetary profit, legislation alone cannot protect it for the long term, certainly not for *perpetuity*.

When I give presentations on conservation easements to landowners, I always make clear that easements are *legal* protection. If this protection is not sustained by a set of ethics that provoke a sense of responsibility and an understanding of human beings' reciprocal relationship with the land and water that literally sustain us, this legal protection will not withstand the forward march of a system based on a consumerism that must keep expanding to be successful in its essential goal—the accumulation of more and more profit.

As my engagement with this group of women landowners revealed, the type of protection this particular group of women defines, in its emphasis on nurturing, reciprocity and an understanding of intrinsic value in the land outside of profit in capitalistic terms, forms the type of ethics to which I refer. The motivations for the legal protection of their conservation easements can be interpreted as both an act of love and of resistance to patriarchal norms, which would see their lands only in terms of their monetary value. The conservation easement seems to be for this group of women not just a legal encumbrance to exploitation, but an expression of their commitment to a sustainably reciprocal relationship with the land and a responsibility to the future of people and the living beings with whom we share the planet. To once again quote the words of landowner, Marjorie Kellogg, "a conservation easement is one small gesture that one small person can make to help stave off the reckless and raging destruction of species and habitats going on worldwide. Change has to start somewhere."

Acknowledgments

I am honored by and grateful to the women landowners who participated in this exploration: Susan Burdsall, Sandra Freckelton, Pat Gambitta, Cat Gareth, Susan Huxtable, Marion Karl, Marjorie Kellogg, Barbara Newman, Gail Sondergaard, Lynn Tanner, Connie Young, Rebecca Young, Mary Anne Whelan and Marianne Younkheere. I am also grateful to Marcie Foster and Sarah Scheeren for their participation in these dialogues and their passionate dedication to land protection at Otsego Land Trust. Thanks also to Bethany Keene of the Delaware Highlands Conservancy and Emily Monahan of Otsego Land Trust for their support and assistance. Thank you to the anonymous reviewers whose comments and suggestions improved this chapter considerably. Finally, special thanks to Amanda Subjin of the Delaware Highlands Conservancy, whose groundbreaking work for women landowners in Pennsylvania provided the original inspiration for this chapter.

References

American Farmland Trust. 2013. Washington, DC. Empowering women landowners to become conservation leaders. Accessed January 29, 2015. http://www.farmland.org/programs/protection/Empowering-Women-Landowners.asp.

Boucher G. 2006. The politics of performativity: A critique of Judith Butler. *Parrhesia: A Journal of Critical Philosophy* 1: 112–41. Accessed January 20, 2015. http://www.parrhesiajournal.org/past.html.

Bureau of Labor Statistics. 2013. *Women in the labor force: A databook.* Accessed January 31, 2015. www.bls.gov/cps/wlf-databook-2012.pdf.

Butler J. 1993. *Bodies that matter: On the discursive limits of "sex."* London and New York: Routledge.

——. 1999. *Gender trouble: Feminism and the subversion of identity.* New York and London: Routledge.

Butler BJ. 2006. Family forest owners of the United States: A technical document supporting the Forest Service 2010 RPA Assessment. Accessed June 19, 2015. www.fs.fed.us/cooperativeforestry/frcc/brett_butler_frcc.pdf.

—— and Leatherberry EC. 2004. America's family forest owners. *Journal of Forestry* 102(7): 4–14.

Catalyst. 2015. Knowledge center. Accessed January 31, 2015. http://www.catalyst.org/knowledge/.

Cheyfitz E. 1997. *The poetics of imperialism: Translation and colonization from* The Tempest *to* Tarzan. New York: Oxford University Press.

Colfer CJP. 2013. *The gender box: A framework for analyzing gender roles in forest management. Occasional Paper* 82. Bogor, Indonesia: Center for International Forestry Research. Accessed January 20, 2015. http://www.cifor.org/library/4303/introducing-the-gender-box-a-framework-for-analysing-gender-roles-in-forest-management/?pub=4303.

Cruikshank J. 2005. *Do glaciers listen? Local knowledge, colonial encounters, and social imagination.* Seattle: University of Washington Press.

Dowie M. 2009. *Conservation refugees: The hundred-year conflict between global conservation and native peoples.* Cambridge, MA: MIT Press.

Fang S and May A. 2015. Apache sacred land handed to mining conglomerate in Arizona. *Al Jazeera America* at ReaderSupportedNews.org. Accessed February 21, 2015. http://readersupportednews.org/news-section2/318-66/28695-apache-sacred-land-handed-to-mining-conglomerate-in-arizona.

Farmer JR, Knapp D, Meretsky VJ, Chancellor C, and Fischer BC. 2011. Motivations influencing the adoption of conservation easements. *Conservation Biology* 25(4): 827–34.

Halpenny E. 2006. Examining the relationship of place attachment with pro-environmental intentions. *Proceedings of the 2006 Northeastern Recreation Research Symposium*. Accessed January 20, 2015. www.nrs.fs.fed.us/pubs/gtr/gtr_nrs-p-14/9-halpenny-p-14.pdf.

Hawkesworth M. 2006. *Feminist inquiry: From political conviction to methodological innovation.* Piscataway, NJ: Rutgers University Press.

Jahanbegloo R and Shiva V. 2013. *Talking environment: Vandana Shiva in conversation with Rahmin Jahanbegloo.* New York: Oxford University Press.

Kennedy V. 2012. *The Nature of nature: Environmental ethics in American and American Indian literatures from the 17th century to the present.* Dissertation. Available at: ecommons.library. cornell.edu/bitstream/1813/33797/1/vmk6.pdf.

Kimmerer R. 2003. *Gathering moss: A natural and cultural history of mosses.* Portland: Oregon State University Press.

Kolodny A. 1975. *The lay of the land: Metaphor as experience and history in American life and letters.* Chapel Hill: University of North Carolina Press.

LaDuke W. 2002. *The Winona LaDuke reader: Collection of essential writings.* Stillwater, OK: Voyageur.

Leopold A. 1966. *A sand county almanac.* New York: Oxford University Press.

Merchant C. 1990. *The death of nature: Women, ecology, and the scientific revolution.* New York: Harper Collins.

Mies M and Shiva V. 2014. *Ecofeminism.* New York: Zedbooks, Ltd.

Muth A, Subjin A, Sagor SN, and Walkingstick T. 2013. Growing your peer learning network: Tools and tips from the Women Owning Woodlands network. Accessed January 22, 2015. https://catalog.extension.oregonstate.edu/em9064.

Nelson MK, ed. 2008. *Original instructions: Indigenous teachings for a sustainable future.* Rochester, VT: Bear and Company.

Pagden A. 1998. *Lords of all the world: Ideologies of empire in Spain, Britain and France c. 1500–1800.* New Haven, CT: Yale University Press.

Pinchot Institute for Conservation. 2006. Understanding the role of women in forestry: A general overview and a closer look at female forest landowners in the US. Accessed January 7, 2016. www.pinchot.org/?module=uploads&func=download&fileId=68.

Prosser J. 1998. *Second skins.* New York: Columbia University Press.

Redmore LE, Tynon JF, and Strong NA. 2011. Women owning woodlands: A case study from the U.S. In Aguilar L, Daniel D and Quesada-Aguilar A (eds.) *Forests and gender.* Gland, Switzerland in collaboration with WEDO, New York: IUCN. Accessed January 7, 2016. http://www.wedo.org/library/new-publication-forests-and-gender.

Ring W. 2014. Experts fear for long-term health of US forests. *Washington Post.* Accessed January 7, 2016. http://www.washingtonpost.com/politics/experts-fear-for-long-term-health-of-us-forests/2014/12/07/0f37abf4-7e4b-11e4-81fd-8c4814dfa9d7_story.html.

Schwartzman L. 2006. *Challenging liberalism: Feminism as political critique.* University Park, PA: Penn State University Press.

Spence MD. 1999. *Dispossessing the wilderness: Indian removal and the making of the national parks.* New York: Oxford University Press.

US Census Bureau. 2013. Quick facts beta. Accessed February 11, 2105. http://www.census.gov/quickfacts/table/PST045214/00.

Wilmot S. 2013. Women and their woods program educates and motivates woodland owners. *The Forestry Source* 8(11): 1.

4

THE FOREST KINGDOM AND VALUES

Climate change and gender equality in a contested forest policy context[1]

Sara Holmgren and Seema Arora-Jonsson

Introduction

'The Forest Kingdom – with values for the world' is the Swedish government's vision for the forest sector, launched in 2011 by the ministry responsible for forest policy in Sweden. The vision has been propounded in a setting where the conventional boundaries of the forest sector are challenged, as forest issues are increasingly intertwined with global climate change and related politics, energy systems and broader land use issues (Beland Lindahl and Westholm 2010). The explicit ambition with the Forest Kingdom is to find new and innovative ways of relating to and generating income from forest resources; to stimulate entrepreneurship in rural areas, and the valuing of multiple services that forests provide. In this chapter, we explore the Forest Kingdom as a positioning of the Swedish state in an increasingly contested forest policy setting. The study is an attempt to demonstrate implicit values guiding Swedish forest policy making, enabling critical reflection on 'the taken for granted'. We analyse the Forest Kingdom as discursive practice, i.e. as the means through which some statements but not others are made credible and consequential (Barad 2003, 818–21), and focus on the representation of two fairly new but seemingly disparate policy issues: climate change (CC) and gender equality. By analysing these two issues, we aim to contribute to a broader and deeper picture of the Forest Kingdom as a recent forest policy initiative on the one hand, *and* bring attention to implicit values underpinning Swedish forest governance.

CC and gender here serve as examples of how previously disconnected policy arenas are increasingly being brought into the forest policy domain, and provide an opportunity to study how 'old structures' deal with 'new issues'. CC is often

1 This chapter is a revised and shortened form of Holmgren S and Arora-Jonsson S. 2015. The Forest Kingdom – with what values for the world? Climate change and gender equality in a contested forest policy context. Scandinavian Journal of Forest Research 30(3): 235–245. Reprinted material by permission of the publishers Taylor & Francis Ltd (www.tandfonline.com).

considered to be one of the most prominent global forest governance issues, affecting all human systems, requiring adjusted land use, adapted forest management practices and altered forest policies (Storch and Winkel 2013). Moreover, while CC can be connected to environmental considerations and a will to 'do good', gender equality is about core values of democracy and justice – questions and values that have been problematised in recent forestry-related debates (Zaremba 2012; Nilsson 2012). Consequently, the inclusion of CC and gender on the Swedish forest policy agenda and the meaning assigned these concepts deserve analytical attention.

In the following, we begin by conceptualising forest governance and values, which we link with discourse analysis. We then describe Swedish forest governance and previous inclusions of gender equality and CC. Following this, we describe our material and methods, and we present our analysis. In the concluding discussion, we highlight the effects of the policy discourse in terms of depoliticisation, the discrepancy between the values underpinning the Forest Kingdom and those at which the 1993 forest policy aimed.

Governance, values and discourse

We approach *values* and *discourse* as central dimensions of governance, i.e. the steering of the public sphere towards targets that are often socially and politically contested (Peters 2006). Drawing on sociological reasoning, Kooiman and Jentoft (2009) suggest that societies consist of large numbers of interacting governance actors (individual and collective, public and private) whose agency is enabled or constrained by structures (material and technical prospects, law, culture) that delimit or expand their potential for action. As cultural structures, values guide those who govern in how to think and make judgements about how to act in specific situations as well as providing more general ideas about how 'the world works' (Kooiman and Jentoft 2009). Theoretically, we thus conceptualise values as implicit and stable guiding mechanisms mediated through discourse, which unconsciously influence the planning of action, give policies a particular purpose and seek to reform behaviour (Schwartz 1999). We further define discourse as not only what is said, but as that which constrains and enables what can be thought, said, or written about a certain object or practice (Foucault 1979). The values explicitly referred to in official documents are thereby not our focus. Instead, we direct attention to the implicit values shaping how CC and gender equality are represented in the documents.

The Swedish social science literature on forestry has used the language of 'values' to explain attitudes, motivations and objectives in forest management among small-scale forest owners (Hugosson and Ingemarsson 2004; Nordlund and Westin 2010); to illustrate the changing roles of environmental public administration in their shift from authoritative to facilitating roles (Appelstrand 2012), and to illustrate the development of attitudes towards intensive forestry among actors in Swedish forest policy – expressing ecological, aesthetic and recreational values of forests (Lindkvist et al. 2011). Some research has also stressed the importance of surfacing and explicitly discussing different democratic values in Swedish forest policy processes (Sundström 2010). We add to this literature by enabling reflection on the values underlying the taken for granted lines of action in Swedish forest governance.

When it comes to gender, we use the concept in two ways: (i) gender equality as a policy issue whose meaning requires analysis, (ii) gender as an analytical category that demonstrates assumptions of what is appropriate for women and men to do (Arora-Jonsson 2013). By using gender as an analytical category, it is possible to illustrate the 'making' of women and men, i.e. their subject positions (e.g. Reed 2008; Lidestav and Sjölander 2007, in forestry contexts), make visible relations of power (Arora-Jonsson 2013) as well as the pre-existing values drawn upon. In sum, we approach the Forest Kingdom as discursive practices performed by the government that produce particular problem representations, subject positions and functions, while drawing on pre-existing values. More specifically, we ask: *how are the issues of climate change and gender (in)equality represented? What subject positions are produced? What values underpin these representations and subject positions?*

Analysis and material

We draw on Bacchi's (2009) discourse analytic framework, 'What's the Problem Represented to be?' and Keller's (2011) Sociology of Knowledge approach to discourse, both inspired by Foucault. In their analytical frameworks, these scholars draw attention to (i) *problematisations,* i.e. *how* policy problems are represented and how the representations make people think about the policy 'issue', themselves and others. (ii) How *political subjects* (identities) are shaped, i.e. how social actors are attributed with certain qualities and related to the 'issue' at hand. Problematisations and subject positions altogether affect what can be thought, said and done about a given policy issue, e.g. CC or gender equality, and ascribe responsibilities to social actors.

In our analysis, we have applied the questions of the analytical frameworks to a limited number of texts. These questions have helped us structure the analysis and 'locate' the problem representations of CC and gender (in)equality also when they are not explicitly articulated; that is, when the texts offer solutions without referring to a problem. During the interpretive process, the texts were read repeatedly and 'answers' to the questions were marked in the margins. When no new answers were appearing, the answers were brought into a table, summarised and categorised. From there, we returned to the research questions, and structured the presentation of the analysis accordingly. In order to make the analysis more transparent we also use quotes, which were marked and selected during the interpretive process.

Our in-depth analysis is based on twenty public policy documents directly linked to the Forest Kingdom, and to the issues of gender equality and CC. Unlike the issue of gender equality, CC does not have its own strategy in the Forest Kingdom. CC is more of a background factor influencing the use of forests. We have thus complemented Forest Kingdom documents with the section on forestry and land use in the Climate Bill (Regeringen 2009), a Swedish Forest Agency

(SFA)[2] report contributing to the Climate Bill (SFA 2007), and the Forest Policy Bill from 2008 (Regeringen 2008). The majority of the texts are produced by or for the Ministry for Rural Affairs (MRA) and are available on the ministry's webpage. Documents appearing under the heading of the Forest Kingdom are what we refer to as 'key' documents. These include *The Action Plan,* the *Gender Strategy* (a short and a long version), *Declaration of International Forest Policy*, and *Swedish Forestry*. The *Gender Strategy* has generated a number of reports and interviews that have also been included in our analysis. A few of these reports differ from the texts produced by the Ministry for Rural Affairs. They have been commissioned from outside authors and are more explicit in pointing out problems, whereas ministerial documents have a promotional character and tend to depict an overall positive picture of Swedish forestry as world-leading.

Swedish forest governance

Around two-thirds of Sweden is covered with forests and during the last century Swedish forests have primarily been valued for their timber. In 2012, the export of forest products accounted for 10 per cent (SEK 122 billion = US$14 billion) of the total Swedish export value; forests thus play an important role in the Swedish economy. Fifty per cent of the forest land is owned by around 328,000 individual forest owners (38 per cent females, 61 per cent males), 25 per cent by private companies and 14 per cent by state-owned companies (SFA 2013). The centre of attention in forest policy and research has been on the individual forest owners, and policy making has involved a relatively small set of interests and actors: the state, private forest owners and the forest industry (Appelstrand 2012; Hysing and Olsson 2008; Schlyter and Stjernqvist 2010). Throughout the 1900s, reciprocal relationships evolved between these actors, facilitated by a common perception of knowledge and expertise that relied on natural science theories and methods. The homogenous assembly of actors – white, middle-aged or older men – had in general studied forestry in the same location. Whether in the official policy arenas (Sundström 2010) or at the local level (Arora-Jonsson 2004), the governance of Swedish forests has thus traditionally been carried out by men and male-dominated organisations, and Swedish forest governance is dominated by a narrow set of collective actors: *public authorities, private forest owners, the forest industry* and, more recently, environmental non-governmental organisations (Hysing and Olsson 2008). Research on the ground illustrates how local perspectives and participation have no clear role in Swedish forest policy making (Lindahl 2008). Male dominance, again shown by research, has implied that women in some locations choose to organise separately, creating spaces of local forest management that simultaneously reproduce and challenge gendered stereotypes. These are seldom acknowledged in forest policy studies (Arora-Jonsson 2013). Taken together, these local studies illustrate a marginalisation of local perspectives on forests, of

---

2 The SFA is the national authority responsible for forest-related issues and forest policy achievement.

indigenous people and rights, and of women, that contrasts with Sweden's forest policy goals and international commitments.

The Forest Kingdom advocates sustainable use of forests and confirms the coequal policy goal of environment and production adopted in 1993 (MRA 2011a). The state's intention with the 1993 Forestry Act was to achieve a more diverse forestry after decades of intensive production-oriented forestry that had resulted in large clear-felled areas. The 1993 governmental bill called explicitly for new knowledge and values in policy formulation and application (Regeringen 1992, 226), and largely followed the global discourses on sustainable development and local participation prominent in the early 1990s. It promoted a broadened conception of forests, forest use and forest products to meet the challenges associated with global issues such as climate change, which at the time was primarily represented as a threat due to spreading of insects, diseases and changing forest ecosystems (Regeringen 1992, 226). The role of the state was to be focused on protection of public interests in forests, i.e. 'natural values', wood production would be left to the market (Finansdepartementet 1991). Combined with extensive deregulation and greater voluntary responsibility for forest owners, i.e. 'freedom with responsibility', the co-equal policy goal adopted in 1993 is often referred to as the 'Swedish forestry model' (KSLA 2009).

However, the 'Swedish foresty model' has been criticised for not achieving its policy goal in terms of production and protection (SFA 1998, 2001; Regeringen 1998). The Environmental Quality Objective 'Sustainable forests' is not considered achievable by 2020 (SEPA 2012, 408). According to an official government report, forest owners have not taken the responsibility expected of them either in protection or production and the SFA has not fulfilled its task as a supervisory authority (SOU 2005, 39). Others have argued that the problem of decreasing legitimacy and insufficient environmental considerations is a consequence of the 'culture' of the forest sector that is linked to timber production and resistant to change (personal communication, SFA official, 9 October 2013).

Climate change and gender equality

Swedish forestry was affected by climate policy for the first time in 1991, when a carbon dioxide tax on fossil fuels incentivised increased bioenergy production from forests and forest products. The long-term sector goals, adopted in 2004, further state that forests should contribute to reducing climate impacts, primarily through production of renewable energy, wood and other products that do not produce net emissions of greenhouse gases (SFA 2007). Additionally, CC has reinforced the polarisation between production and conservation interests, which the adoption of the coequal policy goal of production and environment in 1993 aimed to balance (Appelstrand 2012). The difference stems from production advocates arguing for increased production as a mitigation strategy due to the carbon-sequestering capacity of *growing* forests, and environmental advocates arguing that old boreal *standing* forests are ideal for carbon sequestration (Lindahl and Westholm 2011; Arora-Jonsson 2013).

Whereas CC policy started to intersect in Swedish forest policy in the early 1990s, gender as an 'issue' has been less evident in domestic forest policy. Paradoxically, while Sweden has spearheaded popular participation and gender equality in forestry abroad through international forums and development aid since the early 1990s, these issues did not have a corresponding space domestically (Arora-Jonsson 2013). A decade after the adoption of the 1993 Forestry Act, the governmental report, 'Slow to advance . . . Gender Equality in the Agricultural and Forestry Sector', concluded that the Swedish forest sector lagged behind other policy domains when it came to gender equality and recommended the government formulate a special gender-equality plan (Näringsdepartementet 2004). In 2011, the Ministry for Rural Affairs finally produced a gender strategy, 'Competitiveness requires gender equality' (MRA 2011b), which now has a prominent position in the Forest Kingdom. In the following section, we analyse the Forest Kingdom and how the representation of CC and gender equality reflect particular 'Swedish' values.

The Forest Kingdom – a policy initiative

Initially, the Forest Kingdom was mainly an action plan for rural development. Two years after its introduction, the government expanded the Forest Kingdom's agenda. A range of activities have been added to its framework, such as the development of regional action plans, strategies for increased export of wood products, a plan for coordinated Swedish participation in international negotiations, and a pilot study on a National Forest Programme[3] (MRA 2013; SOU 2013).

With what values for the world?

The Forest Kingdom presents the reader with a profusion of values, and with goals meant to protect those values. In addition to references to social and environmental values, there are *three core values* that the government actively intends to convey in international forest policy settings: sustainability, responsibility and competitiveness (MRA 2012a). While these are the explicit and articulated values, our analysis demonstrates three other values that guide the representations of CC and gender equality, and Swedish forest governance more generally: economic growth, individualism and faith in markets. In practice, they result in a privileging of (i) forest production, (ii) competitiveness (as also explicitly articulated), and (iii) profitability. Below, we describe how these dominant values are reflected in the representations of climate change and gender, and discuss some of their powerful effects.

---

3 NFPs (National Forest Programmes) were introduced in the mid-1990s by the Intergovernmental Panel on Forests, and considered as an important tool for sustainable forestry. Until recently, Sweden has claimed to already fulfil the aim of NFPs. Developing a programme would merely imply duplicate work.

Forest production

The *Action Plan* states that Sweden has sound forestry expertise, that Sweden's views are listened to in international contexts and that by spreading knowledge about the Swedish model and sustainable forestry, Sweden has the potential to contribute to increased poverty reduction and the fight against global warming (MRA 2011a). From such an 'international' perspective, CC is represented as a threat or a battle that needs to be fought. However, on a domestic level CC is instead represented as an opportunity, as illustrated in the following.

Climate change – an opportunity for increased forest production

In the documents, the broad range of services forests deliver, such as biodiversity that creates resilient ecosystems, food and mitigation of environmental pollutants are explicitly acknowledged (MRA 2012b). Despite the recognition of forests' multifunctional role, there is a consistent privileging of forest production and forest products, which is particularly evident in relation to CC. Carbon sequestration is given a prominent role in relation to wood production and products. The Ministry for Rural Affairs states:

> In 2010, the Swedish forest sink was about 40 million tonnes of carbon dioxide. This represents well over half of Sweden's total emissions. In addition, the harvested wood products created a net sink of about 5 million tonnes of carbon dioxide.
>
> *(MRA 2012b, 11)*

Major growth in Swedish forests helps to combat climate change. This is expressed in a significant production of bioenergy which replaces fossil fuels, and forest products that store large amounts of carbon while replacing other more fossil-fuel intensive materials and a large-scale forest sink (MRA 2012b, 11).

Growing forests and production of wood products as the most efficient climate mitigation strategy were also expressed in the *Governmental Bill* from 2007, where the government stated that high and stable growth are a prerequisite for climate mitigation (Regeringen 2008). CC mitigation thereby encouraged more intensive forest management practices that involved fertilisation, improved plant material and fast-growing species aimed at producing highly productive forests. A similar focus is apparent in relation to CC *adaptation*. In its contribution to the Climate Bill from 2007, SFA concludes that possibilities for forest owners to spread the risks associated with forestry need to increase, with continued high and valuable wood production that has limited effects on the environment. This involves more variation when it comes to tree species, thinning and rotation periods, as well as felling regimes (SFA 2007). Thus, in the Forest Kingdom, CC mitigation and adaptation measures are turned into a problem of finding proper forest management practices, requiring further research – all in order to maintain or increase the production of biomass and to safeguard the interests of the Swedish forest industry. In that sense, CC is not really represented as a problem per se, but rather as an opportunity for increased production.

However, *protected* forest lands are also referred to as carbon sinks (MRA 2012b). Nevertheless, protected forest lands are not presented as an opportunity or efficient long-term climate mitigation strategy such as growing forests. Here, there is no mention of research investments, no incentives for active carbon sequestration in standing forests. Neither are protected forests linked to the fulfilment of the environmental goal of the forest policy. Forest protection as climate mitigation is not mentioned as a potential source of income for forest owners or entrepreneurs, not linked to innovation, business, or as a contribution to economic development in rural areas, and consequently not a resource to be exploited.

Active female forest owners

The focus on forest production apparent in the representation of CC is also evident in the Gender Strategy. *The female forest owner* is mainly linked in these documents to the key concern in the Forest Kingdom, i.e. long-term wood supply (MRA 2012b). In a report proposing interventions to increase gender equality among forest owners, SFA (SFA 2011a) refers to Nordlund and Westin (2010). In a quantitative study on forest values and forest management attitudes among Swedish private forest owners that take the forest owners' place of residence into account, Nordlund and Westin (2010) argue that female forest owners value the forests for recreational and ecological purposes to a greater extent than men, and hold more environmental and human-centred forest management attitudes. Consequently, the authors suggest that as the number of female forest owners is increasing, a change in forest management can be expected (Nordlund and Westin 2010). Rather than approaching this assumed development towards a more heterogeneous forest ownership as an opportunity for more diverse forest use, making female forest owners *active* is a central component of the Gender Strategy:

> A way to a more active forest ownership, independently of whether male, female, distant or nearby, can be to increase one's knowledge and become more aware of one's own objectives regarding one's forest ownership. Especially women have in different settings expressed the lack of knowledge as an obstacle for becoming active in their forest ownership. It is therefore likely that increased efforts to develop the competence of especially female forest owners can lead to a more gender equal forestry.
>
> *(SFA 2011a, 16)*

What it means to be an *active* forest owner is not stated explicitly. In one report, the SFA interprets *active* forest owners as those who 'take the economic and overall responsibility for the management of forests, or actively order the services necessary for the management of forests' (SFA 2011a, 4). In addition, 'In order to be an active forest owner one does not have to carry out operational work but one requires sufficient knowledge in order to be a good buyer of necessary services' (SFA 2011a, 1).

The report does not state what the 'necessary services' are, but looking at an information leaflet produced by the SFA (2010), it becomes clear that they imply services connected to timber production and felling. The competence development presented as necessary for increased gender equality is not related to what the SFA describes as female forest owner concerns, such as 'tourism' and 'health' (SFA 2011a, 8), but instead to conventional large-scale forestry. In that way, women's assumed lack of interest in conventional forestry becomes a 'problem' to be solved rather than making their 'unconventional' interests in forests an opportunity to be developed.

Competitiveness, profitability and employment

Although the Forest Kingdom includes many forms of activities and actors of which the majority is referred to as 'industries' (MRA 2011a), it is primarily in relation to paper, pulp and sawn timber industries that climate change and gender equality are constructed. It is in relation to these large-scale, forest production-dependent industries that competitiveness, profitability and employment are articulated.

Building wooden houses – fighting climate change

In the Forest Kingdom *Action Plan*, it is stated that 'using wood in construction is climate-smart and contributes to a sustainable society' (MRA 2011a, 8). In a press release from July 2013, the Ministry for Rural Affairs stated that 'industrial wood construction is one of the focus areas of the governmental investments in the Forest Kingdom' (Regeringen 2013). In a report from a wood construction site available on the Forest Kingdom webpage, a male building contractor describes the environmental benefits of wood construction and how it reduces CO_2 emissions. He states: 'If Sweden and Stockholm are to achieve their environmental objectives, we have to start building much more in wood' (Regeringen 2012). This focus on wood construction in relation to CC mitigation can be traced back to the Climate Bill that describes how large quantities of carbon are sequestered in wood and paper products (Regeringen 2009). The Bill states that as a small country with a large forest industry, the inclusion of carbon stored in forest products in a future Climate Change agreement becomes a central issue for the Swedish forest industry and its competitiveness, and for cost-efficient climate policy achievement. Consequently, the Swedish government argues in the Bill that carbon sequestered in forest products should be included in a future international climate agreement (Regeringen 2009). However, related to the production and use of forest products is the question of employment opportunities, which in our analysis become intimately linked with gender equality.

The problems of gender inequality

In an opening speech for a workshop connected to the development of the forest sector's Gender Strategy, the Minister of Rural Affairs declared that:

> Forests are important for the climate, biodiversity and for economic growth. Human diversity also generates diversity in forests. It is positive that the labour market within this industry looks so bright. It is therefore important that there are enough competent personnel for the future. To enable such a development we need . . . to recruit employees from the whole population that is fit to work. The work with gender equality within the forestry sector is an important and decisive question for the future of the industry and for the future of Sweden.
>
> *(Erlandsson 2010)*

The role of forestry in the Swedish economy is described as pre-eminent (MRA 2011a) and although the labour market, according to the minister, looks bright (Erlandsson 2010), increased profit is needed: 'It is important that the profitability of forestry increases so that the potential for increased production and more jobs is unlocked' (MRA 2011a, 5). The documents further state that:

> Like other primary industries the forest industry has difficulties to interest, recruit and keep women at all levels of positions. An increased number of women have graduated from forestry educations and the industry has to a certain extent begun to request and recruit women with non-forestry academic backgrounds. Despite this the rate of change is low.
>
> *(MRA 2011b, 9)*

In these quotes, we see two problem representations. First, considering the current situation and the emphasis on profitability, it is apparent that forestry is not considered profitable enough. Second, the industry appears to have recruitment problems that also threaten the industry's competitiveness:

> Sweden must provide equal opportunities for women, men and people with foreign backgrounds to work in the forest sector, and an integration project is under way in the green industries. We need a gender equal forest sector that also attracts young people.
>
> *(MRA 2011a, 6)*

This quote suggests that there is a problem with social homogeneity in the sector, not only when it comes to gender, but also in reference to age and origin. It is however gender that is in focus in the Forest Kingdom, and gender equality is linked to an economic rationale rather than to the consequences of social homogeneity. This is captured in a statement by the director-general of the Swedish Forest Agency:

> It is important that the strategy is hands-on and a positive reception of the forest industry sector is crucial. It needs to be clarified that a gender equal forest sector means a lot for the work and the economic results in the longer run.
>
> *(MRA 2011c, 10)*

In the Forest Kingdom, gender equality is a strategy for meeting the needs of the forest industry, rather than a questioning of the practices and culture that generate gender inequalities. Repeated references to the 'industry' further illustrate a tendency to conflate the forest industry with the forest sector as a whole. The concepts are used synonymously, reflecting a bias towards private interests in Swedish forest policy discourse, as the interests and needs of industry are seen to be one with those of society. Interestingly, in the Ministry Publication Series (Finansdepartementet 1991) in the early 1990s, the ministry responsible for forest policy at the time represented public interests as the protection of natural values. As the interests of the 'public' or of 'society' in the Forest Kingdom are represented foremost as securing the needs of the industry, current policy discourse in some ways has returned to the production values dominating Swedish forest policy making before 1993, when forestry and environmental goals were first put on an equal footing. The economic rationale is demonstrated in the formulation of the objective of the forest sector's Gender Strategy. The overall objective of Swedish *gender policy* is that men and women should have the same power to shape society and their own lives and an equal division of power and influence (Regeringen 2011). In the context of the forest sector, the interim target that women and men should have the same power is translated as 'women and men should have the same opportunities to own and manage forests, to work in or to do business within forest management' (MRA 2011b, 3). Apart from a narrow delineation of the forest sector to business interests, this translation implies that the issues of power, influence and citizens' rights expressed in the overall gender policy are eliminated. Gender equality is turned into an issue of individual opportunities in forest management and business, attention is thereby directed at individual women rather than structural power relations. In the Forest Kingdom, the problem of gender inequality is constructed as 'a lack of women' rather than male domination. It is in this context that the 'female employee' is conceptually produced.

The female employee

Throughout the Gender Strategy, there is a focus on *changing* women rather than changing what makes them 'underrepresented' or 'inactive'. Paradoxically, while *female employees* (like female forest owners) are attributed with a natural difference that is perceived as a resource that needs to be tapped, they still need to be reconstructed. The Swedish Forest Agency has been tasked to develop interventions aimed at increasing interest in forestry among girls, with the overall objective of having more women applying for forestry education. The forest agency approached the task through cooperating with 'Forests in School', a national programme among schools and forestry stakeholders aimed at raising a general interest in forests among teachers and students in elementary schools. This involves '. . . comprehensive and balanced explanations of ecological, economic, social and cultural contexts in order to increase the understanding of forests as an environmentally friendly resource that has great climate benefits' (SFA 2011b, 5). The substance of the knowledge produced through 'Forests in School' is tied to the future need of industrial employers. The forest agency does not mention alternative partners in relation to the task of attracting more women to forestry education and work. The narrow choice of partners illustrates how the meaning of forestry work is still primarily related to large-scale industries and production, and is focused on individuals.

Taken together, CC, gender inequality and the making of the female employee mainly reflect values that instil profitability, competitiveness and the need to secure an adequate work force for the forestry industry. Just as climate change is tied to production forests and wood construction industries, the 'change' that the Gender Strategy aims to generate is connected to the goal of increased timber production rather than the coequal environmental goal or to gender equality per se. Hence, neither gender inequality nor CC is articulated as a political issue that might require profound changes in the system itself.

Discussion

Overall, there is a difference between the values reflected in the Forest Kingdom and those that were explicitly articulated in 1993. The latter involved a more holistic view of forests, including not only rational economic dimensions, but also environmental and social perspectives (Regeringen 1992). In the Forest Kingdom, underpinning values systematically promote production goals above the coequal environmental goal formalised in the Forestry Act (Regeringen 2008). The Forest Kingdom thus veers towards a pre-1993 productionist model, promoting intensified forestry and increased use of forest products to sequester carbon dioxide; and the education and activation of female forest owners so that they may enrol in the cause of these narrowly defined forest policy goals. The explicit ambition to find new and innovative ways of relating to and generating income from forest resources; to stimulate entrepreneurship in rural areas, and the valuing of multiple services that forests provide is repeatedly contradicted by promotion of industrial forest production and a delineation of the forest sector that is more or less interchangeable with 'industry'. Here the core meaning of forests is reduced to a source of employment and economic growth all over Sweden.

Accordingly, CC is turned into a business opportunity and a means to secure growth and employment throughout Sweden. Women are reconstituted as employees and active forest owners in order to meet the needs of the industry, rather than as active citizens involved in forest policy making. Gender equality does not involve equal participation among men and women in forest management, public decision making, or advisory committees. Gender equality does not have value in itself but is instrumental, a means to achieve something else: increased production of biomass, increased export of technical know-how and forest products, economic gains, rural development, and the workforce for the forest industry. The problems of CC and gender inequality are thereby displaced from the political to an economic sphere, linked to industrial needs, private forest ownership and profit, rather than to public and collective decision making. Furthermore, as 'society' and societal needs are not articulated, and the needs and interests of the industry are linked to the future of Sweden; industrial interests become identical to public interests. This is an interesting change from 1993, when public interests primarily were linked to 'nature values' and wood production was to be left to the market (Finansdepartementet 1991, 31).

Kooiman and Jentoft (2009) note how governance ultimately means choosing between values. In this chapter, we have made values underpinning Swedish forest policy explicit in order to enable a discussion of their implications. In the Forest Kingdom, dominant values lead to the privileging of forest production at the cost of environmental concerns, and a problematising of individual women rather than of power relationships between men and women as groups. Previous research shows how gendered differences are present in the ways that forests are managed and governed both at the national and the local level (Lidestav and Sjölander 2007; Sundström 2010; Arora-Jonsson 2013). These are differences that do not necessarily adhere to gendered forest stereotypes, presuming that women are concerned with environmental protection and non-timber products and men with timber production and economic gains. The difference rather lies in the space for agency that men and women as social groups have in relation to the masculine culture that characterises the Swedish forest sector – a culture that not only circumscribes the possibility of many women to take action and engage (Arora-Jonsson 2013), but also marginalises the representation of diverse local perspectives on forests in public forest policy processes (Lindahl 2008).

In order to address gender equality and climate change in more diverse ways, we recommend that forest policy makers and stakeholders reflect on the values guiding their thought and action, and ask themselves which values have priority and why. A more concrete bit of advice is to acknowledge alternative actors and forms of political organising as a way to broaden the types of voices heard in Swedish forest policy, not only in order to fulfil national and international policy goals and commitments, but simply because this has democratic value.

Acknowledgements

We warmly thank Daniela Kleinschmit and Peter Edwards for commenting on a draft version of this chapter, and two anonymous reviewers for their constructive comments.

References

Appelstrand M. 2012. Developments in Swedish forest policy and administration – from a 'policy of restriction' toward a 'policy of cooperation'. *Scandinavian Journal of Forest Research* 27: 186–99.

Arora-Jonsson S. 2004. Relational dynamics and strategies: Men and women in a forest community in Sweden. *Agriculture and Human Values* 21: 355–65.

——. 2013. *Gender, development and environmental governance: Theorizing connections.* New York: Routledge.

Bacchi C. 2009. *Analysing policy: What's the problem represented to be?* Frenchs Forest, NSW: Pearson.

Barad K. 2003. Posthumanist performativity: Toward an understanding of how matter comes to matter. *Signs* 28: 801–31.

Beland Lindahl K and Westholm E. 2010. Food, paper, wood, or energy? Global trends and future Swedish forest use. *Forests* 2: 51–65.

Erlandsson E. 2010. *Inledningstal – workshop för en jämställdhetsstrategi inom skogsbrukssektorn.* Accessed 16 September 2013. http://www.regeringen.se.

Finansdepartementet. 1991. Skogspolitik för ett nytt sekel – sammanfattning. Expertgruppen för Studier i Offentlig Ekonomi. Ds 1991: 31.

Foucault M. 1979. *Discipline and punish: The birth of the prison.* New York: Vintage Books.

Holmgren S and Arora-Jonsson S. 2015. The Forest Kingdom – with what values for the world? Climate change and gender equality in a contested forest policy context. *Scandinavian Journal of Forest Research* 30(3): 235–45.

Hugosson M and Ingemarson F. 2004. Objectives and motivations of small-scale forest owners: Theoretical modelling and qualitative assessment. *Silva Fennica* 38.

Hysing E and Olsson J. 2008. Contextualising the advocacy coalition framework: Theorising change in Swedish forest policy. *Environmental Politics* 17: 730–48.

Keller R. 2011. The sociology of knowledge approach to discourse (SKAD). *Human Studies* 34: 43–65.

Kooiman J and Jentoft S. 2009. Meta-governance: Values, norms and principles, and the making of hard choices. *Public Administration* 87: 818–36.

KSLA. 2009. *The Swedish forestry model.* Stockholm: The Royal Swedish Academy of Agriculture and Forestry.

Lidestav G and Sjölander AE. 2007. Gender and forestry: A critical discourse analysis of forestry professions in Sweden. *Scandinavian Journal of Forest Research* 22: 351–62.

Lindahl KB. 2008. *Frame analysis, place perceptions and the politics of natural resource management. Exploring a forest policy controversy in Sweden.* Doctoral Thesis No. 2008: 60, Swedish University of Agricultural Sciences. Uppsala: SLU Service/Repro.

—— and Westholm E. 2011. Future forests: Perceptions and strategies of key actors. *Scandinavian Journal of Forest Research* 27: 154–63.

Lindkvist A et al. 2011. Attitudes on intensive forestry. An investigation into perceptions of increased production requirements in Swedish forestry. *Scandinavian Journal of Forest Research* 27: 438–48.

MRA (Ministry for Rural Affairs). 2011a. The Forest Kingdom – with values for the world. Action Plan. Stockholm: XGS grafisk service.

——. 2011b. Konkurrenskraft kräver jämställdhet. Jämställdhetsstrategi för skogsbrukssektorn. Linköping: LTAB.

——. 2011c. Konkurrenskraft kräver jämställdhet. Kort om jämställdhetsstrategin för skogsbrukssektorn. Linköping: LTAB.

——. 2012a. Programförklaring för Sveriges internationella skogspolitik. Hållbarhet, ansvar och konkurrenskraft.

——. 2012b. *Swedish forestry.* Stockholm: Edita Västra Aros.

——. 2013. Skogsriket – med värden för världen. Faktablad. Government Offices of Sweden.

Näringsdepartementet. 2004. Det går långsamt fram . . . Jämställdheten inom jord- och skogsbrukssektorn. Ds 2004: 39.

Nilsson S. 2012. Future Swedish forest policy process. Report prepared for the All Party Committee on Environmental Objectives. Accessed 10 August 2015. http//www.sou.gov.se.

Nordlund A and Westin K. 2010. Forest values and forest management attitudes among private forest owners in Sweden. *Forests* 2: 30–50.

Peters BG. 2006. *Institutional theory in political science: The 'New Institutionalism'.* 2nd edn, London: Continuum.

Reed MG. 2008. Reproducing the gender order in Canadian forestry: The role of statistical representation. *Scandinavian Journal of Forest Research* 23: 78–91.

Regeringen. 1992. Regeringens proposition om en ny skogspolitik. *Prop. 1992/93:226.*

———. 1998. Uppföljning av skogspolitiken. *Prop. 1997/1998: 158.*

———. 2008. En skogspolitik i takt med tiden. *Prop. 2007/08: 108.*

———. 2009. En sammanhållen klimat- och energipolitik – Klimat. *Prop. 2008/2009: 162.*

———. 2011. Jämställdhetspolitikens inriktning 2011–2014. Stockholm: Ministry of Education.

———. 2012. *Klimatsmart på höjden.* Accessed 9 January 2014. http://www.regeringen.se.

———. 2013. Trästad – regeringen satsar på klimatsmart byggande. Accessed 18 July 2013. http://government.se.

Schlyter P and Stjernquist I. 2010. Regulatory challenges and forest governance in Sweden. *In* Bäckstrand K, Khan J, Kronsell A and Lövbrand E. eds. *Environmental politics and deliberative democracy. Examining the promise of new modes of governance.* Cheltenham: Edward Elgar. 180–196.

Schwartz SH. 1999. A theory of cultural values and some implications for work. *Applied Psychology* 48: 23–47.

SEPA (Swedish Environmental Protection Agency). 2012. Steg på Vägen. Fördjupad utvärdering av miljömålen 2012.

SFA (Swedish Forest Agency). 1998. Skogsvårdsorganisationens utvärdering av skogspolitiken, SUS.

———. 2001. Skogsvårdsorganisationens utvärdering av skogspolitikens effekter – SUS 2001. Borås.

———. 2007. Svenskt skogsbruk möter klimatförändringarna. Jönköping.

———. 2010. Beställa skogsbrukstjänster. Accessed 16 September 2013. http://www.skogsstyrelsen.se

———. 2011a. Förslag på åtgärder för att skapa förutsättningar för ökad jämställdhet bland skogsägare.

———. 2011b. Utveckla insatser inom sSIS för att särskilt öka flickors intresse för det skogliga området.

———. 2013. Skogsstatistisk årsbok 2013. Mölnlycke: Elanders Sverige AB.

SOU. 2005. Skog till nytta för alla. Stockholm. 39.

———. 2013. Långsiktigt hållbar markanvändning – del 1. Delbetänkande av Miljömålsberedningen. 43.

Storch S and Winkel G. 2013. Coupling climate change and forest policy: A multiple streams analysis of two German case studies. *Forest Policy and Economics* 36: 14–26.

Sundström G. 2010 In search of democracy: The process behind the Swedish forest-sector objectives. In Sundström G, Soneryd L and Furusten S. eds. *Organizing democracy: The construction of agency in practice.* Cheltenham: Edward Elgar.

Zaremba M. 2012. Skogen vi ärvde. *Dagens nyheter.*

5

GENDER GAPS IN REDD+

Women's participation is not enough[1]

*Anne M. Larson, Therese Dokken, Amy E. Duchelle,
Stibniati Atmadja, Ida Ayu Pradnja Resosudarmo,
Peter Cronkleton, Marina Cromberg, William
Sunderlin, Abdon Awono and Galia Selaya*

Introduction

Concern over the contribution of forest clearing and degradation to climate change
has led to the promotion of strategies for Reducing Emissions from Deforestation
and Forest Degradation (REDD+). REDD+ is intended to be a performance-
based mechanism whereby forest stakeholders at multiple scales (from national to
household levels) could be rewarded for protecting or enhancing the carbon sequestration capacity of forests (Angelsen 2009). Since the initial proposal by the coalition
of rainforest countries, led by Papua New Guinea and Costa Rica, at the UNFCCC
11th Conference of Parties (COP) in 2005 in Montreal, REDD+ has been placed
firmly on the global climate change agenda.

REDD+ has been controversial, however. Objections have been raised by business interests intent on converting forests to other land uses. In addition, indigenous
peoples and other local communities have demanded that REDD+ policies, programs and projects fully guarantee respect for their land and forest rights as well as
ensuring their participation in related decision-making arenas. Such challenges have
been discussed elsewhere (Griffiths 2008; Gomes et al. 2010; Sikor et al. 2010; Larson
et al. 2013), but the specific problems for women as members of these communities
have so far received much less attention. Given the tendency to see 'communities' as
undifferentiated, even REDD+ policy makers and proponents (organizations or entities designing and implementing subnational REDD+ programs and projects) that
are sensitive to the needs of forest-based peoples may fail to understand or address the
specific needs of women related to forest and REDD+ policies. As a result, women

---

1 This chapter is an edited and shortened version of Larson et al. (2015), published in the
International Forestry Review and available free online at http://www.ingentaconnect.com/
content/cfa/ifr/2015/00000017/00000001/art00005.

must struggle on two fronts: Even if 'communities' are taken into account and have opportunities to benefit from REDD+, women may still be left out.

The data presented here are based on the Center for International Forestry Research Global Comparative Study on REDD+ (GCS-REDD), from focus group interviews in 77 villages participating in 20 subnational REDD+ initiatives in six countries: Brazil, Cameroon, Indonesia, Peru, Tanzania and Vietnam (Sunderlin et al. 2010). These data were collected early in the planning stages of REDD+ initiatives and before activities were fully underway (2011–12). The data demonstrate that women-only focus groups were less informed about the REDD+ initiatives in comparison to mixed-gender (mostly male) focus groups in the same villages. We explore several hypotheses to explain differences. We hypothesized four conditions that should contribute to more equal participation of women in REDD+ relative to men: (1) women have a strong voice in village decision making; (2) women have a strong role in forest rule making; (3) women use forest resources as much or more than men, or (4) initiatives take an explicit, gendered approach to REDD+. We find that, overall, women's groups were less knowledgeable even when three of these four key variables suggest that women should participate more fully. This chapter uses the research findings to argue that 'participation' is only a partial solution to addressing women's strategic needs in ways that could strengthen their position in REDD+ (see also Khadka et al. 2014; cf. Harris-Fry and Grijalva-Eternod, this volume).

Forests, REDD+ and women

Concerns about the risks of REDD+ for poor people living in and near forests have been brought to international attention primarily by social movements, particularly international indigenous organizations, and NGOs working with these groups. The effect of REDD+ policies, programmes and projects could be either positive or negative for local people, or could pass them by altogether (Larson 2011). Benefits could arise from policies that secure tenure rights, secure borders from unwanted outsiders, or provide new sources of income or other advantages for poor local communities. Harm could come from the usurpation of land rights by outsiders or elites, or from new limits on forest livelihoods without consent or compensation (Sunderlin et al. 2009; Sikor et al. 2010; Corbera et al. 2011; Larson 2011; Beymer-Farris and Bassett 2012). Communities that already conserve forests may be left out if benefits only accrue to those who *stop* deforesting or degrading forests.

Concerns about women's participation and role in REDD+ stem from substantial field evidence that women tend to have less voice than men in forest communities and participate less in decision making, particularly with regard to forests and forest resources (e.g. Rocheleau et al. 1996; Saigal 2000; Agarwal 2001; Gupte 2004; Benjamin 2010; Jackson and Chattopadhyay 2010; Sunam and McCarthy 2010). Research findings regarding women's participation, however, are not always straightforward (Agarwal 2001, 2009, 2010a).

For example, Agarwal (2001) describes a variety of types of participation ranging from the least effective, which she calls 'nominal participation' (physical presence), to the most effective, called 'interactive and empowering participation' (taking initiative and exercising influence). She argues that many studies focus on the numerical strength of women rather than their 'ability to participate better in the very process of decision-making' (Agarwal 2010a, 8). Physical presence on boards, committees or at meetings, for example, is far from guaranteeing voice and influence (see Agarwal 2001). Even with specific efforts at inclusion, women often lack the experience, confidence and skills to engage in the public sphere (Mai et al. 2011).

At the same time, studies from India also show that a greater presence of women on decision-making bodies tends to lead to better forest conservation and regeneration outcomes (Agarwal 2009), and having a critical mass of women, rather than a single woman or a small number, can make a significant difference (Agarwal 2009, 2010a, b). In a study comparing data from Kenya, Uganda, Bolivia and Mexico, however, forest user groups with more women than men perform less well than more equally mixed or male-dominated groups in adopting forest-enhancing behaviour (Mwangi et al. 2011).

'Participation' can also be a burden, particularly for women who are almost always the ones in charge of the household in addition to other responsibilities, and women (and men) may believe that 'women's place is not in the forest' (Bolaños and Schmink 2005). But even if women do not personally harvest forest resources, they may be concerned about the supply of those resources, access to land in forests, water supplies, climate variation, forest conservation, or the cultural value of forests (see e.g. Mairena and Cunningham 2011). It should not therefore be assumed that women's failure to participate, or even desire to participate, is necessarily related to less use of or interest in forests. Understanding women's participation requires in-depth knowledge of specific local social norms and gender dynamics by those promoting forest management and conservation policies and measures, such as REDD+.

Women also use forests differently than men, and this is often poorly understood or simply not acknowledged (den Besten 2011). There is broad-based evidence that men tend to be more cash oriented and women more tied to subsistence uses, and that men use more high-value and processed resources such as timber, and women, more non-timber or unprocessed forest products (Awono et al. 2002; Fu et al. 2009; Awono et al. 2010; Shackleton et al. 2011). Nevertheless, there are important differences between world regions (Sunderland et al. 2014), and above all gender findings tend to be context specific. Also, there is no particular reason to assume that men in communities would effectively represent women's interests; they may not even understand how women's specific criteria or priorities regarding forest goods and services may vary from their own (Cruz-Garcia, personal communication).

Women's rights to land and forests are often not as secure as men's. Whether land is individual or collective, women may not be permitted to control their own plots, to be included on land titles, or to inherit land. The specific arrangements of women's relationship to men (e.g. husbands, partners, fathers, brothers) and their social position (e.g. single, married, widowed) form a complex matrix of factors that often affect women's rights and, hence, their dependence on men for their livelihoods. Even where the law guarantees women's rights, in practice women are sometimes forced to obtain access to land and natural resources through husbands and sons

(Lastarria-Cornhiel 2011). Past interventions in forests and natural resource management or conservation have often failed to recognize gender differences, resulting in greater hardships for women (Schroeder 1999; Colfer 2005; den Besten 2011).

If these gender differences are not recognized and taken into account in REDD+, policies and actions that are assumed to be gender neutral could have detrimental effects on women and on women's contribution to household income and well-being. The design of appropriate policies and interventions, and avoiding unwanted outcomes, requires research on 'people in nested and overlapping constituencies that reflect the multiple roles, identities and interests of men and women across class, location, occupation and other points of difference and affinity' (Rocheleau and Edmunds 1997, 1368). Yet integrating gender into forestry research is constrained by the broad perception that forestry is a male-dominated profession, as well as a lack of clarity among researchers about the concept of gender, and a lack of technical skills, interest and/or awareness of gender (Mai et al. 2011).

While women are still peripheral to REDD+ debates, a number of organizations have recently called for greater attention to women and gender (Gurung and Quesada 2009; UN-REDD 2011; Peach Brown 2011; UN-REDD 2013); relatively, at least at the level of discourse, the situation is improving. For example, the World Bank's Forest Investment Program (FIP), which provides funding to support REDD+ in developing countries (such as REDD-readiness and pilot activities), refer to women in a footnote in the 2009 FIP Design Document (Climate Investment Funds 2009: 4), whereas the more recent FIP document on the grant mechanism for indigenous peoples and local communities refers to an overarching principle that includes gender equality and twice mentions ensuring the participation of women (Climate Investment Fund 2011). Maginnis et al. (2011: 7) report that the climate change 'negotiation documents went from zero [previously] to eleven mentions of gender in Cancun.'

Finally, the second version of the REDD+ Social and Environmental Standards (SES) safeguards has addressed gender issues much more prominently than its previous version and more than other safeguard standards (Mackenzie 2012; REDD+ SES 2012). This attention suggests a small, yet growing consensus on the importance of addressing women's particular interests and concerns in REDD+. The rest of this chapter uses an exploration of early field research findings from sub-national REDD+ sites in six countries to contribute to the debate on how that should be done.

Methods

CIFOR's GCS-REDD+ is a multi-year research project (2009-2016) that aims to provide policy and technical guidance to REDD+ stakeholders (see Figure 5.1, http://www.cifor.org/gcs/).

In each of the 77 villages, a village survey was used to gather secondary data from key informants and to guide focus-group interviews. We asked village leaders to invite focus-group participants with an aim of 15 participants and a mix of both men and women. On average, 17 villagers participated in the village focus groups with greater male participation (66 per cent) overall. A women's survey

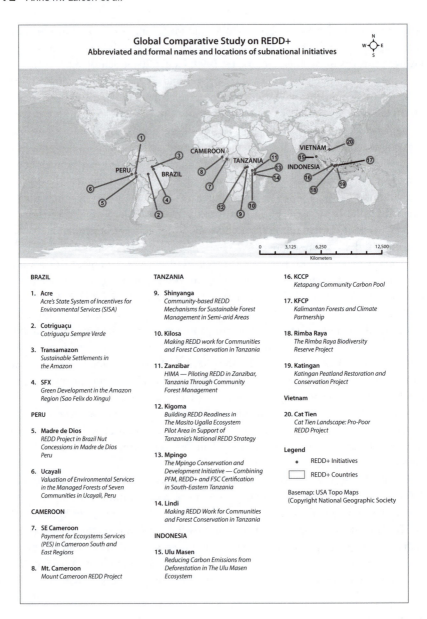

FIGURE 5.1 Map of GCS-REDD+ project sites [Global Comparative Study on REDD+ (GCS-REDD)]

was implemented by holding a separate focus-group interview with women only; there were 13 participants on average. Both groups were asked the same questions about their knowledge of and participation in REDD+. The women's survey focused on perceptions of participation in community decision making, as well as on how men and women use the forest. For most questions, the group was asked to agree on a response; for the questions regarding perceptions of representation and participation, the women were asked to vote, and responses were reported in shares. This chapter also draws from data on a survey with proponents of REDD+ initiatives to assess plans to address gender issues. These proponents were most often NGOs, but also include some private-sector initiatives and those of regional or municipal governments (for full list, see http://www.cifor.org/gcs/modules/redd-subnational-initiatives/redd-initiatives-study/).

On knowledge of REDD+, both village and women's groups were asked whether they had heard about (1) REDD+ in general, and (2) the local REDD+ initiative. If the group answered affirmatively to at least one of these questions, they were asked to provide a short explanation. Based on the answers, the researchers facilitating the meetings used standard criteria to evaluate whether the group demonstrated a basic understanding. The surveys were done prior to the full implementation of the local REDD+ initiatives, so we did not expect many respondents to have heard of REDD+. Nevertheless, during that same early phase, a comparison between the village and women's surveys in the same village would illustrate the differences between the two groups.

During the focus group interviews, participants were asked to list all the main decision-making bodies in the village and agree on the most important. The women's focus groups reported the total number of members of the body and how many were women. Membership in decision-making bodies is a common measure of participation but, as discussed previously, does not necessarily reflect women's ability to influence decisions.

The participants of the women's focus groups were also asked to vote on a series of statements regarding their perception of participation and influence in village and household decisions, including forest rule making (see Appendix). Each statement was read aloud to the group, and the number of people that agreed or disagreed noted.[2] To correct for the differences in group size, the share of participants in each group that agreed/disagreed to each statement was calculated. Based on these shares, the weighted average at site, region and country level was calculated; each village is given equal weight.

With regard to (perceptions of) forest use by men and women, the women's group was asked to agree on a single answer. We did not ask the village group about forest use, meaning that we use data where women report men's use of the forest, hence we are not able to triangulate the information to see if the answers by men would differ.

Results

The presentation of the results follows the logic of our analysis. The first section compares knowledge of REDD+ between the mixed village and women's focus groups. As explained above, the results demonstrate that on average, the women's groups were less knowledgeable than the village groups. The following sections explore four factors related primarily to women's participation to help understand those results and their significance for REDD+.

---

2 The participants could also choose not to respond, or state that they did not know.

Knowledge of REDD+

Across all villages, awareness of REDD+ among the women's focus groups was lower than among the village focus groups (Table 5.1). More than half of the women's focus groups (58 per cent) had heard about the local REDD+ initiative, demonstrating that quite a few of the proponents were able to make their initiatives known at their respective sites, even in the early stages. Overall, 41 per cent of the women's focus groups were considered to have a basic understanding of REDD+ and/or the local REDD+ initiative. When the same questions were posed to the village focus groups, we found consistently higher numbers: 71 per cent had heard about the local REDD+ initiative, while 67 per cent demonstrated a basic understanding. In Brazil, the women's focus groups demonstrated a basic understanding of REDD+ in all the villages where the village focus groups demonstrated understanding. But in Peru, Cameroon, Tanzania and Indonesia, the number of villages where the women's groups demonstrated a basic understanding was lower than the number of villages where the village focus groups demonstrated this understanding. In Vietnam, none of the women's or village focus groups had heard of REDD+ in general or the local initiative.

Women's voice in village decision making

The main decision-making bodies identified by the focus groups varied; they included elected village governments/associations in Tanzania and Brazil and in a few sites in Indonesia and Peru; the assembly of all (adult) villagers in one site in Peru; traditional councils in Cameroon, and the party committee in Vietnam. In the decision-making body identified as the most important by the women's group, the average share of women was 24 per cent, while 17 of the 76 did not have any women (Table 5.2). The share of women in these bodies varied considerably across countries from 48 per cent in Peru,[3] 30 per cent in Tanzania to only 4 per cent in Indonesia.

Across the twenty sites, we found that 64 per cent of the women's focus group participants agreed that women were sufficiently represented in important village decision-making bodies, 65 per cent agreed that they were usually able to influence village decisions, and 79 per cent agreed that women participated actively in meetings (Table 5.3). Again, there was substantial variety across countries and sites, and even villages within sites. In two of the sites in Indonesia, Rimba Raya and Katingan, the overall share of women in the main decision-making body was low, and in most of the villages there were no women represented at all. Yet, in both sites, all women in the focus groups in all villages agreed to all three statements regarding their participation in community decision making; they perceived that women in the village were sufficiently represented, were usually able to influence village decisions when they wanted to, and participated actively in village meetings. In Brazil and Tanzania, the situation was the opposite. In Acre, Brazil, although all

---

3 Nevertheless, in all four villages in one site, the women identified organizations of particular interest to women rather than village government.

TABLE 5.1 Knowledge of REDD+ in women's and village focus groups (in number of villages)

Country (sites)	Women's groups' knowledge of REDD+			Village groups' knowledge of REDD+			Total villages
	REDD in general	Local REDD initiative	Basic understanding	REDD in general	Local REDD initiative	Basic understanding	
Brazil (3)[a]	2 (17%)	8 (67%)	7 (58%)	2 (16%)	8 (67%)	7 (58%)	12
Peru (2)	5 (63%)	4 (50%)	4 (50%)	7 (88%)	6 (75%)	6 (75%)	8
Cameroon (2)	3 (50%)	5 (83%)	3 (50%)	5 (83%)	6 (100%)	6 (100%)	6
Tanzania (6)	16 (70%)	13 (57%)	10 (43%)	21 (91%)	18 (78%)	17 (74%)	23
Indonesia (5)	2 (10%)	12 (60%)	6 (30%)	9 (45%)	14 (70%)	13 (65%)	20
Vietnam (1)	0	0	0	0	0	0	4
Total average	**28 (38%)**	**42 (58%)**	**30 (41%)**	**44 (60%)**	**52 (71%)**	**49 (67%)**	**73**

Source: Field research data, CIFOR, 2010–12.

a There are four sites total in Brazil; in one site, focus groups were not asked about REDD+.

Note: Data is presented based on the number of villages (in focus-group discussion) demonstrating knowledge.

TABLE 5.2 Share of women in the main decision-making body (identified by women)

Country (total number of villages)	Share of women, in % (# of villages with women in decision-making body)
Brazil[a,b] (16)	13[a] (14)[b]
Peru (8)	48 (8)
Cameroon (6)	15 (5)
Tanzania (23)	30 (23)
Indonesia (20)	4 (6)
Vietnam (4)	26 (4)
TOTAL (77)	**24 (60)**

Source: GCS-REDD field research, CIFOR, 2010–12.

a Observations missing from two villages in one site. In one, the respondents did not know the total number of members in the body, and in the other they did not know the number of women.

b One observation missing due to non-existing decision making body.

Note: Data in % is calculated based on the share of women in each village reported by the focus group, averaged across the villages in each REDD+ initiative site, then averaged across the total sites in each country.

villages had at least one woman in the main decision-making body, a majority of the women disagreed to at least one of the three statements. In Tanzania, the share of women in the main decision body was the highest, and there was at least one woman represented in every village; in most villages, women perceived themselves to be sufficiently represented, but in 13 of 23 villages, the majority still disagreed that women were able to influence decisions, and in less than half of the villages the majority of the participants in each group agreed to all three statements jointly. The results demonstrate a disconnect between presence of women on village committees and the perception of participation and influence in the village.

TABLE 5.3 Perception of women's participation in village decision making (in shares)[a]

Country	Sufficiently represented (%)		Usually able to influence (%)		Participate actively in meetings (%)	
	Agree	Disagree	Agree	Disagree	Agree	Disagree
Brazil	31	66	65	31	67	27
Peru	88	12	78	22	83	17
Cameroon	67	33	83	17	83	17
Tanzania	79	08	55	33	93	01
Indonesia	60	40	65	30	70	24
Vietnam	73	28	75	25	100	0
TOTAL	*64*	*32*	*65*	*29*	*79*	*15*

Source: GCS-REDD field research, CIFOR 2010–12.

a The remaining responses are 'The respondent does not know' and 'The respondent chooses not to respond.'

Note: Data is presented based on the share of women agreeing and disagreeing in each village focus group, averaged across the villages in each REDD+ initiative site, then averaged across the total sites in each country.

Women's participation in forest decisions

Overall, women were more involved in forest-related decisions at the household level than at the village level: in all but four sites, more participants in the women's groups agreed that women participated actively in decisions about forest use in the household compared to the village level (Table 5.4). The weighted average indicates that less than half (47 per cent) of the participants agreed that women actively participated in making village rules for forest resource use while 73 per cent agreed that women played an active role in household decisions about land and forest use.

There are variations across and within countries, making it hard to generalize regarding women's influence on forest decisions. Study villages in Cameroon had the highest weighted percentage of respondents who agreed that women actively participated in making village rules for forest resource use (82 per cent).[4] In contrast, only 6 per cent of women respondents in the Vietnam study villages thought so. Analogous values in Peru, Brazil, Indonesia and Tanzania ranged from 38 to 50 per cent. Yet within Tanzania, for example, the range of values across REDD+ initiative sites is large, from 25 to 95 per cent.

Women's perception of their participation in forest decisions was lower than their perceived participation in community decision making in general. The share of women's focus-group participants that agreed to actively participating in

TABLE 5.4 Women's participation in resource use decisions at village and household level (in shares[a])

Country	Village forest resource use rule making (%)		Forest resource monitoring (%)		Household's cash income (%)		Household's land and forest use (%)	
	Agree	Disagree	Agree	Disagree	Agree	Disagree	Agree	Disagree
Brazil	40	48	62	32	85	15	46	50
Peru	38	62	69	31	75	25	82	18
Cameroon	82	18	65	35	99	01	99	01
Tanzania	49	36	44	45	69	18	73	18
Indonesia	50	24	36	45	85	15	85	15
Vietnam	06	31	06	56	50	50	71	29
TOTAL	*47*	*37*	*48*	*40*	*64*	*31*	*73*	*23*

Source: GCS REDD field research, CIFOR, 2010–12.

The remaining responses are 'The respondent does not know' and 'The respondent chooses not to respond.'
a Data is presented based on the share of women agreeing and disagreeing in each village focus group, averaged across the villages in each REDD+ initiative site, then averaged across the total sites in each country.

4 Women control the management of non-timber forest products. Moreover, they are involved in decision bodies such as forest management institutions both in SE Cameroon and Mount Cameroon sites.

making rules for forest resource use in the village was lower in all but three of the sites when compared to the share of women that agreed they were usually able to influence village decisions (Kigoma and Mpingo, Tanzania; Ulu Masen, Indonesia), and lower in all sites when compared to whether they participated actively in meetings, except in Acre, Brazil.

Finally, the results indicate that in villages where women were more involved in forest resource-use decisions and monitoring, they were also more likely to demonstrate a basic understanding of REDD+ similar to the mixed village group. Overall, the share of women that agree women are actively participating in making village rules for forest resource use is significantly higher (30 per cent) in the villages where the women's and mixed village focus groups demonstrate the same basic understanding of REDD+.

Women's use of the forest

The results showed that men went to the forest more often than women in 56 per cent of all the villages in our sample (Table 5.5); there was no difference between women and men in 33 per cent of villages, while in the remaining 11 per cent, women went into the forest more often. Again, there were differences between countries. Women and men went equally often into the forest in Indonesia in 60 per cent of the villages and in all the villages in the Vietnam site. Men went into the forest more often on average in Brazil, Peru and Tanzania, while all of the villages where women went into the forest more often were in Cameroon or Tanzania.

TABLE 5.5 Who goes more often to the forest, and how far do they go inside the forest?

Country (total # of villages)	Who are more often in the forest? # of villages			Mean walking distance inside the forest, in minutes	
	Women	Men	Same	Women	Men
Brazil (16)	0 (0%)	13 (81%)	3 (19%)	51	158
Peru (8)	0 (0%)	7 (88%)	1 (12%)	240	300
Cameroon (5ᵃ)	2 (40%)	0 (0%)	3 (60%)	131	161
Tanzania (23)	6 (26%)	14 (61%)	3 (13%)	23	35
Indonesia (20)	0 (0%)	8 (40%)	12 (60%)	89	188
Vietnam (3ᵇ)	0	0	3	65	125
Total (75)	*8 (11%)*	*42 (56%)*	*25 (33%)*	*78*	*141*

Source: GCS-REDD field research, CIFOR, 2010–12.

a One observation missing.

b In one village, the respondents do not know who goes more often.

Note: The mean walking distance is based on the minutes agreed by each focus group, averaged across the villages in each REDD+ initiative site, then averaged across the total sites in each country.

There were also differences regarding how far[5] women and men went from the forest edge into the forest. On average across the sites, men walked almost twice as far as women, but there were variations. For example, in the Cotriguaçu site in Brazil, SE Cameroon and Kigoma in Tanzania, women walked further into the forest, while in some other sites the difference between men and women was small.

Women carried out a wide range of activities. There were clear differences in forest use: women's main activities included collecting firewood, fruits and vegetables, while hunting and collecting poles were the top two activities of men. There were also differences across countries. In Vietnam, both men and women carried out a relatively small variety of activities in the forest compared to all the other countries.

We found that in the small portion of villages where women were the ones who most often went to the forest, a higher share of women respondents agreed that they participated in forest monitoring. But otherwise, we did not find a clear relationship between women's use of the forest and their participation in forest decisions. The data suggest that women were not necessarily included in forest decision making even when they were the ones that went into the forest as much as or more often than men, with the exception of Cameroon. In Cameroon, a high proportion of women agreed that women actively participate in making rules for forest resource use (82 per cent). Also, a weighted average of 99 per cent agreed that women participated in decisions about land and forest use at the household level.

No relation was found between women's use of the forest and their knowledge of REDD+. In 13 of the 19 villages where the village focus group demonstrated a basic understanding of REDD+ while the women's focus group did not, women went into the forest at least once a week. In the remaining villages, they went only a few times a year, but so did the men in the same villages. Hence women's relative lack of a basic understanding of REDD+ cannot be explained by their lack of use of the forest.

REDD+ initiative commitments to women

Data from the proponent appraisal interviews were examined to see whether or not proponents were planning to take women's specific needs into account in the design and implementation of REDD+ initiatives, and if they were proposing concrete actions to do so. At the time of these early interviews, none of the proponents[6] listed women as a stakeholder group, though five proponents stated fair benefits to women as an equity goal (SE Cameroon, Ucayali in Peru, Cat Tien in Vietnam, and Zanzibar and Mpingo in Tanzania).

Nevertheless, at only one of those sites – SE Cameroon – did we find a similar basic understanding of the local REDD+ initiative between women and village

---

5 The question asked was about time rather than distance, so this should be seen as a proxy. In any case, the point is the relative time/distance between men and women and not the absolute numbers.
6 The proponents in Tanzania had not finalized this exercise yet at the time of the research.

groups. At the time of the research, the SE Cameroon initiative had already taken a gendered approach, helping individual women who had been involved in NTFP collection to organize and obtain rights to a community forest in the village under a village umbrella association that the initiative also helped form. Stakeholder activities such as Payments for Environmental Services (PES) and organizing the group sale of produce were coordinated through the association, which was being led by a woman when this chapter was written.

In the two Tanzanian and one Peruvian initiatives with explicit gender-equity goals, however, mixed village groups showed greater basic understanding relative to the women-only groups. (As mentioned earlier, in the Cat Tien site in Vietnam, neither the women's nor the mixed village groups showed knowledge of REDD+, due to timing: Workshops had not yet been conducted in the villages.)

Discussion

Overall, the data comparing the results of women-only and male-dominated mixed focus groups suggest that women were less familiar with REDD+ or local REDD+ initiatives than men. This result corresponds with the findings of other recent, more qualitative research on REDD+ (e.g. Gurung and Setyowati 2012) and raises concerns regarding the future of REDD+. Of the four aspects of decision making that were explored further, only participation in forest rule making and monitoring was correlated with greater knowledge of REDD+. Each factor is considered briefly in turn, before turning to the implications for REDD+.

In terms of women's participation and perceived influence in village decision making, results show that having women on the main decision-making bodies was not correlated with the perception of influence over village decisions. This is not particularly surprising. In most cases, there is only one woman on these bodies, and, as discussed above, past research suggests that such arrangements often fail to result in effective representation (Agarwal 2010b). Overall, however, almost two-thirds of women believed that they influenced village decision making, whether or not they were represented on formal decision-making bodies.[7] Interestingly, women at the two sites in Cameroon showed higher influence in both the communal and household arena when compared to sites in the other countries. Perceptions of influence, of course, cannot be assumed to represent influence in practice, as women in a certain culture or community may simply be more demanding or have higher expectations.[8] In any case, the perception of influence was not correlated

---

7 One reviewer asked if this may be because women perceive that, though they do not participate directly, their husbands represent their viewpoint as well. This is probably true for some women. Overall, about 45 per cent agreed and about 46 per cent disagreed with the statement that their primary influence was through their husbands.

8 For example, this may be the case in Brazil, where women are least content with their representation on official bodies and have the lowest share agreeing that they actively participate in meetings, but in practice they have equal knowledge of REDD+ as the mixed village groups.

with greater knowledge of REDD+ initiatives. That is, perceived participation in village decision making did not guarantee participation in REDD+.

For women's participation in forest rule making and monitoring, perception of influence relative to overall village decision making dropped, with just under half of women stating that they participated in forest rule making. At the household level, with the exception of Brazil, the majority of women believed that they influenced household decisions on land and forest use. Notably, in the villages where the women's and mixed village focus groups demonstrated the same basic understanding of REDD+, a higher portion of women participated in forest rule making and, to a lesser degree, forest monitoring. Of the four factors studied, this is the only one that appears to be associated with knowledge of REDD+, though it does not hold across all of the villages.

In terms of women's use of forests and forest resources, we anticipated finding similar knowledge of REDD+ between the two types of focus groups where women use forests as much or more than men. The data demonstrate that there is great variation across countries and sites regarding women's use of forests and relative uses by women and men. In almost half the villages (44 per cent), women reported going into the forest as much as or more than men. Nevertheless, there was no correlation between this data and women's participation in forest rule making, nor were these women as informed about REDD+. Conversely, less knowledge of REDD+ is not associated with less forest use.

With regard to proponents' explicit attention to gender equity in REDD+, while a similar number of women and village groups showed a basic understanding of REDD+ at several sites, with one exception these were not the places where proponents stated in early interviews that they were concerned with equity. This finding suggests that other initiatives may have taken this on without necessarily planning to, while those that planned to did not do so, at least early on in the initiative, with much success.

The results raise a number of questions. Importantly, it is somewhat reassuring that women involved in forest rule making appear to be better placed in relation to the provision of information on local REDD+ initiatives. Also, some REDD+ proponents had clearly managed to provide equal information for women and men in the early stages, even when this was not among their apparent goals. However, there is substantial evidence, from the study and from the literature, that this is far from enough to ensure a sustainable and equitable REDD+.

The results of the study confirm that understanding women's participation, representation and influence is not at all straightforward. The women interviewed often believed they are 'sufficiently represented' in village decision-making institutions, but this may be because they did not see existing governance bodies as particularly effective institutions through which to exercise influence. The majority also believed they were able to exercise influence when they wanted to, apparently through other means. This could be interpreted to mean that REDD+ proponents do not need to be concerned, but the relative lack of information on REDD+ suggests otherwise.

TABLE 5.6 2013 gender-related rankings, by country

UNDP[a] (2014) Gender Inequality Index		WEF[b] (2013) Global Gender Gap		IUCN[c] (2013) Environment and Gender Index[d]	
Country	Rank (of 152)	Country	Rank (of 136)	Country	Rank (of 72)
Vietnam	58	Brazil	62	Brazil	24
Peru	77	Tanzania	66	Vietnam	28
Brazil	85	Vietnam	73	Indonesia	33
Indonesia	103	Peru	80	Tanzania	44
Tanzania	124	Indonesia	95	Cameroon	63
Cameroon	138	Cameroon	100		

a United Nations Development Programme
b World Economic Forum
c International Union for the Conservation of Nature
d Peru not ranked.

The evidence that women using forests as much as or more than men do not participate more in rule making and are not equally informed about REDD+ is particularly worrisome. REDD+ policies, programmes and projects most certainly involve changing the rules for forest use – otherwise there would be no change in behaviour to reduce emissions. Not all such rules necessarily would affect community uses – such as a project that simply secured the borders of a community forest from outside incursion. Nevertheless, among the cases studied, all of the initiatives contemplate some sort of norms and regulations regarding the community's forest use.

In addition, REDD+ may have unexpected consequences for women and for households. Even though most women believed they could influence decisions that they cared about, they may not understand the ways in which REDD+ could affect them until too late: for example, after contracts have been signed (cf. Tiani et al., this volume). Even in a hypothetical situation in which women make no use of and have no interest in forests, the impacts of a REDD+ intervention could have an important effect on families, such as in cases where men lose access to forests in return for cash payments. This requires that women be fully informed.

The data demonstrate important inequities but also that the results are highly context specific. Though the number of villages studied in each country is varied and clearly not representative at the national level, it may be useful to examine country-level differences in this light. Table 5.6 presents the rankings of the six countries on gender inequalities using three different indices. Though there is substantial variation even among the different indices, Vietnam and Brazil are consistently in the top half (among our study countries), and Cameroon is consistently on the bottom. Contrary to the expectations that might be set by national rankings, and perhaps demonstrating even further the importance of the specific context, the villages in Cameroon had the highest percentages in some key variables such as the ability to influence decisions and participation in forest rule making (cf. Colfer et al. 2015). More consistent with the national indices, Brazil is the only country in our study where women and village focus groups were equally informed about REDD+.

Finally, it is important to note that the key variable that we have used in this study – knowledge of REDD+ – to compare men and women is a very modest measure of difference. In fact, it is a measure that virtually requires only physical presence at meetings – the *weakest* form of participation. The complexities of gendered participation demonstrated here suggest, in fact, that more extensive participation in REDD+ processes is necessary but alone is insufficient. Rather, women's opportunity to participate in REDD+ will not necessarily bring a strong gendered understanding of forest resource access and control to the table, or result in influence over – and more equitable – outcomes.

Conclusions

Results across our study sites show great variation in the extent to which women participate, influence and are represented in village and household decision-making processes, but some broad patterns emerged. In many sites, women reported that they influenced household land-use choices and village decisions, used forests substantially and as much as or more than men, and were sufficiently represented on village decision-making bodies. Yet these indicators of participation and representation were not linked to the perception that they could influence forest decisions, and they were not correlated with their knowledge of REDD+ when compared to male-dominated village focus groups in the same villages.

We cannot fully explain whether women's lower level of participation in REDD+ processes comes from their free choice to *not* be engaged, or because they are prevented from being engaged due to a set of cultural, social and economic conditions. Nevertheless, other recent studies have also found that women are not being sufficiently included in REDD+ processes (Gurung and Setyowati 2012; Nhantumbo and Chiwona-Karltun 2012; WOCAN 2012; see other chapters, this book). As Khadka et al. (2014) argue based on similar research in Nepal, limiting attention to including women in meetings or even in payment mechanisms without addressing the underlying power issues behind inequity is insufficient.

At least two points should be taken into account in the search for an effective gendered approach to REDD+ policies, programmes and projects. First, given the variation across villages, sites and countries, there are no simple solutions for improving women's participation or a blueprint appropriate for all locations. Nevertheless, there are strategies for improving participation that have had some success, such as adaptive collaborative management and other deeply participatory strategies for engagement (Colfer 2005; McDougall et al. 2013; Evans et al. 2014). Women should be involved in all aspects of REDD+ design, decisions, capacity building and benefits (see also Gurung and Setyowati 2012).

Second, promoting women's participation alone is insufficient. This is true on the one hand because of social and cultural norms, discrimination and lack of experience, confidence and skills (Mai et al. 2011) and power relations (Khadka et al. 2014) that may limit women's voices in the public sphere; and on the other hand because of the limited analysis and understanding of gendered forest uses and community and household relations that may be affected by interventions. Although REDD+ SES has made important progress by focusing on women and gender in a number of principles and criteria, the overwhelming emphasis is on promoting women's participation rather than accompanying this with relevant gendered data and analysis.

Thus, REDD+ initiatives should integrate gender into design, monitoring and evaluation (Gurung and Setyowati 2012), to explore the ways in which men and women interact and differ with regard to key processes related to REDD+ implementation in their respective sites. This includes household and village decision making, rights to and management of land and natural resources, and information dissemination. Gender-responsive analysis will be crucial to understanding real and perceived gender differences in interests and needs, and to anticipating threats or risks – to ensure that REDD+ implementation on the ground can lead to the effective engagement of rural men and women, encourage greater awareness and understanding of gender and forests, and lay the groundwork for community empowerment and informed participation in REDD+. Most importantly, interventions that do not seek to address gender inequities at the outset may be doomed to perpetuate them.

Acknowledgements

We thank the numerous people involved in the collection of these data in the field, our partner proponent organizations, and the women and men of the study villages, for graciously participating in our research efforts. GCS-REDD is supported by the Norwegian Agency for Development Cooperation (Norad), Australian Aid, the Department for International Development (DFID) of the United Kingdom, the European Commission (EC), and the CGIAR Research Program on Forests, Trees and Agroforestry. We also thank four anonymous reviewers for comments on two previous versions of this chapter.

Appendix: Data description – list of variables and sources

Variable name	Question asked	Survey
The main decision making body in the village	What are the main village decision-making bodies? Identify the most important	Women Village
Female members of main decision making body	How many women are on this body? Number of women and number of all members	Women
Selection of leaders of the body	How are leaders of this decision-making body selected?	Village
Selection of female members of the body	If there are one or more women on this body, how were the women members chosen?	Women

Perception of participation
(voting question: recorded number of people who agree/disagree/do not know/choose not to respond)

Sufficiently represented	Women are sufficiently represented on important village decision-making bodies	Women
Usually able to influence	Women are usually able to influence village decisions when they want to	Women
Participate actively in meetings	Women participate actively in village meetings	Women
Village forest resource use	Women actively participate in making rules for forest resource use in the village	Women
Forest resource monitoring	Women actively participate in monitoring forest use (for example, as park guards, observers, reporting on infractions)	Women
Household's land and forest resource use	In most households, women play an active role in decisions about land and forest use (e.g. what products to grow, collect, from where, how much, where to clear, etc.)	Women

Perception of forest use by men and women

Frequency of use	On average over the year, how often do women/men go to the forest?	Women
Distance	On average, how far do women/men go from the forest edge inside the forest, in terms of walking time?	Women
Activities	What do women/men do when they are inside the forest?	Women

Knowledge of REDD+

Knowledge of REDD+	Have you heard of REDD+ prior to this interview?	Women Village
Knowledge of local initiative	Have you heard of (name of the local REDD+ project) prior to this interview?	Women Village
Basic understanding of REDD+	Do the respondents show a basic understanding of what REDD or the local REDD+ project are, in the sense of stating knowledge of at least one of their attributes? *(Evaluation by the researcher)*	Women Village

Project strategies and goals

Stakeholders	Please list all of the major groups targeted by the project (e.g. small farmers, indigenous people, logging firms, concession holders, private companies, women's organization, particular ethnic groups, herders/pastoralists etc.) who currently use the forest	Proponent
Equity goals	Does your project have a specific plan for equitable distribution of project costs and benefits with project stakeholders? If yes, what are the equity goals?	Proponent

References

Agarwal B. 2010a. *Gender and green governance: The political economy of women's presence within and beyond community forestry.* Oxford: Oxford University Press.

——. 2010b. Does women's proportional strength affect their participation? Governing local forests in South Asia. *World Development* 38(1): 98–112.

——. 2009. Gender and forest conservation: The impact of women's participation in community forest governance. *Ecological Economics* 68(11): 2785–99.

——. 2001. Participatory exclusions, community forestry, and gender: An analysis for South Asia and a conceptual framework. *World Development* 29(10): 1623–48.

Angelsen A. 2009. Introduction. In Angelsen A, ed. *Realising REDD+: National strategy and policy options.* Bogor, Indonesia: CIFOR. 1–9.

Awono A, Ndoye O and Preece L. 2010. Empowering women's capacity for improved livelihoods in non-timber forest product trade in Cameroon. *International Journal of Social Forestry* 3(2): 151–63.

——, Ndoye O, Schreckenberg K, Tabuna H and Temple L. 2002. Production and marketing of safou (*dacryodes edulis*) in Cameroon and internationally: Market development issues. *Forests, Trees and Livelihoods* 12(1–2): 125–47.

Benjamin AE. 2010. Women in community forestry organizations: An empirical study in Thailand. *Scandinavian Journal of Forest Research* 25(S9): 62–8.

Beymer-Farris BA and Bassett TJ. 2012. The REDD menace: Resurgent protectionism in Tanzania's mangrove forests. *Global Environmental Change* 22(2): 332–41.

Bolaños O and Schmink M. 2005. Women's place is not in the forest. In Colfer CJP, ed. *The equitable forest: Diversity, community and resource management.* Washington, DC: Resources for the Future/CIFOR.

Climate Investment Funds. 2009. *FIP design document for the Forest Investment Program, a targeted program under the SCF Trust Fund.* Accessed 5 June 2015. http://www.climateinvestmentfunds.org/cif/node/111.

——. 2011. *Design for the dedicated grant mechanism for indigenous peoples and local communities to be established under the Forest Investment Program.* Accessed 5 June 2015. http://www.climateinvestmentfunds.org/cif/node/5707.

Colfer CJP. 2005. *The equitable forest: Diversity, community and resource management.* Washington, DC: Resources for the Future/CIFOR.

——, Achdiawan R, Roshetko JM, Mulyoutami E, Yuliani EL, Mulyana A, Moeliono M, Adnan H and Erni. 2015. The balance of power in household decision-making: Encouraging news on gender in southern Sulawesi. *World Development* 76: 147–64.

Corbera E, Estrada M, May P, Navarro G and Pacheco P. 2011. Rights to land, forests and carbon in REDD+: Insights from Mexico, Brazil and Costa Rica. *Forests* 2(1): 301–42.

Den Besten JW. 2011. Women in REDD. *Arborvitae. The IUCN Forest Conservation Programme Newsletter* (43): 14.

Evans K, Larson AM, Mwangi E, Cronkleton P, Maravanyika T, Hernandez X, Muller P, Pikitle A, Marchena R, Mukasa C, et al. 2014. *Field guide to adaptive collaborative management and improving women's participation.* Bogor, Indonesia: CIFOR.

Fu Y, Chen J, Guo H, Hu H, Chen A and Cui J. 2009. Rain forest dwellers' livelihoods: Income generation, household wealth and NTFP sales, a case study from Xishuangbanna, SW China. *International Journal of Sustainable Development and World Ecology* 16(5): 332–8.

Gomes R, Bone S, Cunha M, Nahur A, Moreira P, Meneses-Filho L, Voivodic M, Bonfante T and Moutinho P. 2010. Exploring the bottom-up generation of REDD+ policy by forest-dependent peoples. *Policy Matters* 17: 161–8.

Griffiths T. 2008. *Seeing 'REDD'? Forests, climate change mitigation and the rights of indigenous peoples and local communities. Update for Poznan (UNFCCC COP 14).* Moreton-in-Marsh: Forest Peoples Programme.

Gupte M. 2004. Participation in a gendered environment: The case of community forestry in India. *Human Ecology* 32(3): 365–82.

Gurung JD and Quesada A. 2009. *Gender-differentiated impacts of REDD to be addressed in REDD social standards.* Arlington, VA: CARE International and the Climate, Community and Biodiversity Alliance.

—— and Setyowati AB. 2012. *Re-envisioning REDD+: Gender, forest governance and REDD+ in Asia.* Brief 4 of 4. Washington, DC: Rights and Resources Initiative.

IUCN 2013. *The Environment and Gender Index (EGI) 2013 Pilot.* Washington, DC: IUCN. Accessed 5 June 2015. http://cmsdata.iucn.org/downloads/reportthe_environment_and_gender_index___2013_pilot___final.pdf.

Jackson C and Chattopadhyay M. 2010. Nature in a South Bihar village. In Agrawal A and Sivaramakrishnan K, eds. *Social nature: Resources, representations, and rule in India.* New Delhi: Oxford University Press.

Khadka M, Karki S, Karky BS, Kotru R and Darjee KB. 2014. Gender equality challenges to the REDD+ initiative in Nepal. *Mountain Research and Development* 34(3): 197–207.

Larson AM. 2011. Forest tenure reform in the age of climate change: Lessons for REDD+. *Global Environmental Change* 21(2): 540–49.

——, Brockhaus M, Sunderlin WD, Duchelle A, Babon A, Dokken T, Pham TT, Resosudarmo IAP, Selaya G, Awono A and Hyynh TB. 2013. Land tenure and REDD+: The good, the bad and the ugly. *Global Environmental Change* 23(3): 678–689.

——, Dokken T, Duchelle A, Atmadja S, Resosudarmo I, Cronkleton P, Cromberg M, Sunderlin W, Awono A, and Selaya G. 2015. The role of women in early REDD+ implementation: Lessons for future engagement. *International Forestry Review* 17: 43–65.

Lastarria-Cornhiel S. 2011. Las mujeres y el acceso a la tierra communal en América Latina. In Deere CD, Lastrria-Cornhiel S and Ranaboldo C, eds. *Tierra de Mujeres: Reflexiones sobre el acceso de las mujeres rurales a la Tierra en América Latina.* La Paz, Bolivia: International Land Coalition. 19–38.

Mackenzie C. 2012. *REDD+ social safeguards and standards review.* Burlington, VT: Forest Carbon, Markets and Communities Program.

Maginnis S, Aguilar L and Quesada-Aguilar A. 2011. Introduction. In Aguilar L, Shaw DMP and Quesada-Aguilar A, eds. *Forests and gender.* Gland, Switzerland: IUCN, and New York: WEDO.

Mai YH, Mwangi E and Wan M. 2011. Gender analysis in forestry research: Looking back and thinking ahead. *International Forestry Review* 13(2): 245–58.

Mairena D and Cunningham M, eds. 2011. *Conocimientos tradicionales, mujeres indigenas y bosques: Estudios de caso en la Costa Caribe de Nicaragua.* Puerto Cabezas, Nicaragua: CADPI.

McDougall C, Leeuwis C, Bhattarai T, Maharjan MR and Jiggins J. 2013. Engaging women and the poor: Adaptive collaborative governance of community forests in Nepal. *Agriculture and Human Values* 30: 569–85.

Mwangi E, Meinzen-Dick R and Sun Y. 2011. Gender and sustainable forest management in East Africa and Latin America. *Ecology and Society* 16(1): 17.

Nhantumbo I and Chiwona-Karltun L. 2012. *His REDD+, her REDD+: How integrating gender can improve readiness.* IIEC Briefing. United Kingdom: IIEC.

Peach Brown C. 2011. Gender, climate change and REDD+ in the Congo Basin forests of Central Africa. *International Forestry Review* 13(2): 163–76.

REDD+ SES 2012. *REDD+ SES Social and Environmental Standards Version 2. September 10, 2012.* Accessed 5 June 2015. http://www.redd-standards.org.

Rocheleau D and Edmunds D. 1997. Women, men and trees: Gender, power and property in forest and agrarian landscapes. *World Development* 25(8): 1351–71.

——, Thomas-Slayter B and Wangari E, eds. 1996. *Feminist political ecology: Global issues and local experience*. London and New York: Routledge.

Saigal S. 2000. Beyond experimentation: Emerging issues in the institutionalization of joint forest management in India. *Environmental Management* 26(3): 269–81.

Schroeder R. 1999. *Shady practices: Agroforestry and gender politics in the Gambia*. Berkeley: University of California Press.

Shackleton S, Paumgarten F, Kassa H and Zida M. 2011. Opportunities for enhancing poor women's socioeconomic empowerment in the value chains of three African non-timber forest products (NTFPs). *International Forestry Review* 13(2): 136–51.

Sikor T, Stahl J, Enters T, Ribot JC, Singh NM, Sunderlin WD and Wollenberg E. 2010. REDD-plus, forest people's rights and nested climate governance. *Global Environmental Change* 20(3): 423–5.

Sunam RK and McCarthy J. 2010. Advancing equity in community forestry: Recognition of the poor matters. *International Forestry Review* 12(4): 370–82.

Sunderland T, Achdiawan R, Angelsen A, Babigumira R, Ickowitz A, Paumgarten F and Reyes-Garcia V. 2014. Challenging perceptions about men, women, and forest product use: A global comparative study. *World Development* 64(S1): S56–S66.

Sunderlin WD, Larson AM and Cronkleton P. 2009. Forest tenure rights and REDD+: From inertia to policy solutions. In Angelsen A, ed. *Realising REDD+: National strategy and policy options*. Bogor, Indonesia: CIFOR. 139–49.

——, Larson AM, Duchelle A, Sills EO, Luttrell C, Jagger P, Pattanayak S, Cronkleton P and Ekaputri AD. 2010. *Technical guidelines for research on REDD+ project sites*. Bogor, Indonesia: CIFOR.

UNDP. 2014. Human Development Report. Human development statistical tables. Viewed at http://hdr.undp.org/en/content/table-4-gender-inequality-index.

UN-REDD Programme. 2011. *The business case for mainstreaming gender in REDD+*. Geneva, Switzerland: United Nations Collaborative Programme on Reducing Emissions from Deforestation and Forest Degradation in Developing Countries.

——. 2013. *Guidance note on gender sensitive REDD+*. Geneva: UN-REDD Programme. Accessed 17 October 2014. http://www. unredd.net/index.php?option=com_docman& task=cat_ view&gid=1044

WEF. 2013. *The global gender gap report 2013*. Geneva, Switzerland: World Economic Forum. Accessed 5 June 2015. http://www3.weforum.org/docs/WEF_GenderGap_Report_2013.pdf

WOCAN. 2012. *An assessment of gender and women's exclusion in REDD+ in Nepal*. WOCAN Case Study. Bangkok, Thailand: WOCAN.

6

FOREST CONSERVATION IN CENTRAL AND WEST AFRICA

Opportunities and risks for gender equity

Helen Harris-Fry and Carlos Grijalva-Eternod

Introduction

In recent years, the forestry sector has tried to mainstream gender[1] into legislation, policies and programs in order to promote "women's empowerment" (FAO 2007). This focus on women in gender mainstreaming has arisen because women have often been more excluded from forest conservation governance, employment and decision making than men (FAO 2007). Case studies have shown how the inclusion of women in forest conservation programs can empower women by, for example, increasing women's incomes or promoting female participation in forest management committees (Schroeder 1995; Yatchou 2011).

Similarly, the focus on women's empowerment has arisen because women often have less power than men (Kandiyoti 1988). Some believe that women's empowerment is inherently important (Kabeer 1999; Sen 1985); whereas, in the wider development discourse, women's empowerment has been justified by the potential value for achieving other development outcomes (Hickel 2014). For instance, it is widely perceived that when women have greater control over household resources, this greater control will be associated with more resources being spent on food and health care (Hopkins et al. 1994; Russell 1996) and in turn with better nutritional outcomes in the household (Malapit and Quisumbing 2015; John 2008).

Others have focused on women's empowerment from a conservation perspective (Agarwal 2009). Some forest conservation programs that excluded or did not

---

1 "Gender mainstreaming" is defined by the UN as "the process of assessing the implications for women and men of any planned action, including legislation, policies or programs, in all areas and at all levels" (UN 1997, 28).

benefit women have resulted in women sabotaging the programs, whereas many programs that included and benefited women have been reciprocally supported by women's labor, endorsement and better conservation outcomes (Schroeder 1999; Agarwal 2009). This labor may be especially valuable given the described gender-specific knowledge that women have about the value and distribution of various tree species (Rocheleau et al. 1996). On the other hand, simple gender main-streaming, without specific empowerment objectives and without attention to gender relations, has not always resulted in better conservation outcomes because women have had comparatively lower access to technology and time for additional non-domestic work (Mwangi et al. 2011).

Empowerment in the heavily forested region of Central and West Africa war-rants special attention because countries in this region have the lowest women's empowerment score in Africa, according to the Women's Empowerment in Agriculture Index (WEAI) (Malapit et al. 2014).[2] Conservation programs in Central and West Africa are now (or may be soon) supported by newly available multilateral funds that are designed to incentivize developing countries to protect their forests and reduce carbon emissions from deforestation. This climate change mitigation mechanism, REDD+ (Reducing Emissions from Deforestation and Forest Degradation), provides funds to promote forest conservation and presents a new opportunity to empower both women and men within forest conservation programs (Angelsen and McNeill 2012; Nartey 2014; Brown et al. 2008; Luttrell et al. 2012). Yet, despite their potential, national REDD+ programs have been described as limited and tokenistic in their efforts to mainstream gender (Peach Brown 2011; Larson et al., this volume).

Although there are many studies on gender and empowerment in community-based forest management,[3] research from Central and West Africa is notably sparse (Mai et al. 2011). Literature and recent policy documents (such as Cameroon's and Ghana's gender REDD+ roadmaps) have focused on gender bias within insti-tutions (FAO 2007) and on how programs may affect gender-specific roles and access to resources (Quesada-Aguilar et al. 2012). Yet, they have not questioned how programs may affect intra-household "empowerment and gender dynamics" (collectively termed "gendered power dynamics"). The Democratic Republic of Congo (DRC), the Republic of the Congo, Nigeria and Ivory Coast are receiv-ing REDD+ funds, and future recipients of REDD+ funding in this region (Cameroon, Central African Republic, and Ghana) are now developing their

---

2 WEAI is a measure of women's empowerment levels in the agriculture sector and is an aggregate of various measures of five different domains (e.g. control over income and time allocation) of women's empowerment (Malapit et al. 2014).
3 See, for example, Agarwal (2001, 2010) for descriptions of gendered participation in community forest user groups (2001) and the effect of gender composition in commu-nity forestry institutions (CFI) on the likelihood of women speaking up in CFI meetings (2010).

REDD+ Readiness Plan Idea Notes and their Readiness Preparation Proposals.[4] So this is a critical time to address this gap by reviewing evidence on the effects of previous community-based forestry programs on intra-household gendered power dynamics.

To our knowledge there are no reviews of the impacts of Central and West African community-based forest conservation programs on gendered power dynamics. Drawing on bargaining power theory by Sen (1987) and Agarwal (1997), we review the literature on Central and West African community-based forest conservation programs to describe how community-based forest conservation programs can affect intra-household gendered power dynamics, and learn lessons from programs that succeeded or failed to empower women and men, to understand how REDD+ programs could make gendered power dynamics more equitable.

Context and methods

Study setting and context

Here, we provide some examples of wider contextual factors and gender roles within forest-dwelling households in Central and West Africa. This is a considerable challenge given the sociocultural heterogeneity of such a large region, so we illustrate any trends with locally specific examples. We acknowledge the dangers of homogenizing or simplifying gender issues across different societies (Barry et al. 2010; Vansina 1990). But, we suggest a broad overview is a useful prerequisite to analyzing empowerment effects of conservation programs and we expect that the described trends may change.

A context of disempowered forest communities

First, we situate our study in the context of the widespread marginalization of forest communities in Central and West Africa. In colonial and postcolonial periods, state laws have often disempowered forest communities. For example, in the formation of national parks there, forest dwellers were forcefully evicted and their homes destroyed. Anthropological accounts describe how forest-dwelling Pygmy[5] men struggled to provide for their families or afford gifts for their wives due to

---

4 REDD+ Readiness Plan Idea Notes (R-PINs) are submitted by national governments to the Forest Carbon Partnership Facility (World Bank) to characterize the extent and causes of forest deforestation and degradation and current forestry policies and programs. These should outline a process for engaging stakeholders (including forest dwellers) and monitoring livelihood benefits of proposed REDD+ programs. If the R-PIN is approved, the country can then submit Readiness Preparation Proposals (R-PPs) which outline plans, budgets and timelines of their proposed REDD+ programs.

5 We acknowledge the necessity to use this academic term, which has been used in a pejorative manner in the region; but have no viable alternative, in using the relevant literature (see Lewis 2000 for a more detailed discussion).

new poaching laws and land use restrictions. Many turned to alcoholism as they struggled to fulfill their traditionally masculine duty of providing meat for their families (Lewis 2000).

Economic influences like market expansions and fluctuations, economic subsidies and unemployment from structural adjustment programs have also altered the relative value of men's and women's productive capacity (Meagher 2010; Richards 1998). Again, this raised the issue of men being unable to fulfill duties they considered central to their masculine identity: those of providing for their families or, in some areas, having enough money to pay a bride price for a wife (Richards 1998; Schroeder 1995; Whitehead and Tsikata 2003).

Many countries in the region have been affected by conflict (Richards 1998). Recruitment of men, women and children to army or rebel groups; frequent drug consumption, and forced use of rape as a war tactic have had long-term negative consequences on the empowerment status of men and women in these areas (Richards 1998). Now, many ex-combatants are unemployed, landless and unable to marry or return to their families (Barker and Ricardo 2005).

In summary, external influences have disempowered many forest-dwelling men and women. In terms of gender relations, male disempowerment in the region has been linked to increased likelihood of men engaging in violence, suffering from alcoholism (Barker and Ricardo 2005) and desiring more control in their domestic spheres (Colfer and Minarchek 2013).

Land, labor and resources in forest-dwelling households

Traditional land inheritance patterns have been highly heterogeneous across Central and West Africa (Davison 1988; Whitehead and Tsikata 2003). For instance, Yoruba men from Nigeria own land semi-permanently and divide it up for women to use (Sudarkasa 1973), whereas in the Gola forest in Sierra Leone, land is inherited by both men and women (Leach 1991) and in Cameroon, the Kom community follows historic matrilineal inheritance (Davison 1988; Tiayon 2011).

Long lists of ecosystem benefits, such as wild food, fodder, wood for construction, fuel and income, and non-timber forest products have been categorized as being "men's" or "women's" resources (Barry et al. 2010; Leach 1991; Tiayon 2011; Wan et al. 2011). However, others have debated the validity of these gendered categorizations (Meinzen-Dick et al. 2012), finding less clear distinctions between "men's" and "women's" crops (Doss 2002) and flexibility in divisions between genders (Quisumbing et al. 2001). Indeed, the concept of ownership is also contested, since both men and women may be involved at different points in the production process (Leach 1991; Rocheleau and Edmunds 1997; see Elias, this volume).

A common generalization (Ingram et al. 2014; Sunderland et al. 2014) is that men tend to be more involved in cash generation than women, whereas women tend to be more responsible for household subsistence. In cases where men and women jointly contribute towards the same income-generating system, women tend to gather materials or produce a commodity and men sell this at markets

TABLE 6.1 Search terms used for background research and analysis

Theme	Search term
Forest	Forest★, wood★, tree★
Conservation	Poli★, manag★, protect★, conserv★, reforest★, afforest★, "community based forest management," CBFM, "community forest," "integrated conservation development project," ICDP, REDD, access, use, ownership
Gender	Gender★, women, woman, feminin★, masculin★, "gender alliance," "gender dynamic," "gender relation," "gender roles"
Empowerment	"bargaining power," empowerment, power

Note: ★ were used as multiple word wildcard operators, where indicated in Table 6.1. For example, "forest★" would search for "forest," "forestry" and "forested."

(Rocheleau and Edmunds 1997; Tiayon 2011). Where market selling is done by women (examples from Cameroon, Nigeria, DRC and Central African Republic), men often exert control over the use of that generated income (Babalola and Dennis 1988; Flintan 2003; Ingram et al. 2014; Pérez et al. 2002; Tshombe et al. 2000).

With such heterogeneity in land ownership, roles and resource control, we conclude that gender is an important factor, but that program planners need local research for context-specific information on how their programs may have gender-specific impacts.

Literature search method

To acquire texts on the impact of community-based forest conservation programs on gendered power dynamics, we conducted a structured literature review using relevant key-word search terms (Table 6.1) that were used in conjunction with specific place names. Owing to the limited availability of studies, this process was combined with a "snowball" search method by manually searching the reference lists from papers that were identified using the search terms. Although this method increases the risk of selection bias, we considered it worthwhile because of the scarcity of literature.

The databases used (International Bibliography of Social Sciences, JSTOR, Web of Knowledge, Eldis and Scopus) were selected because of their multi-disciplinary coverage.

We included studies from any countries in UN-defined regions of "Central" and "West" Africa and selected all studies that referred to impacts of community-based forest conservation programs on gender dynamics or power dynamics within the household. This inclusion of any "community-based" program was done with anticipation of likely variation in the level of participation in such programs (Temudo 2012). We included both quantitative and qualitative studies, and non-peer-reviewed gray literature such as policy reports. The exclusion criteria included any studies: not written in English, not explicitly about forests, not referring to community-based programs, or with no reference to the intra-household

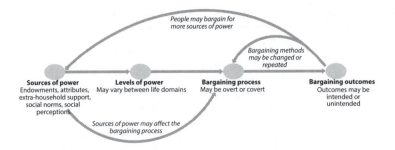

FIGURE 6.1 A framework of intra-household gendered power dynamics

gender or power effects of community-based programs. There was no exclusion criterion set for publication year.

Analytical framework

Kabeer defines empowerment as "the process by which those who have been denied the ability to make strategic life choices acquire such an ability" (1999: 435). Using bargaining power theory proposed by Sen (1987), and developed by Agarwal (1997) and Alsop et al. (2006), we expanded on this definition to produce a framework for analyzing the effects of forest conservation programs on intra-household gendered power dynamics (see Figure 6.1).

"*Sources of power*" are the points where household members access all support and resources beyond the household. Sources may include personal endowments or attributes, social norms, social perceptions and extra-household support (Agarwal 1997). Types of extra-household support may exist in formalized structures such as state support or conservation programs; they may be customary, such as kin groups, or ad hoc, through individual friendships (Rocheleau and Edmunds 1997). Here, REDD+ enters into the empowerment process, for example, through provision of subsidies or services by affecting "levels of power" and possibly "bargaining processes."

"*Levels of power*" represent the amount of influence and control a person has. Power levels may vary across different domains of a person's life. Power gained in one domain of someone's life does not necessarily result in a net increase in power, if they lose power in another domain.

Also termed "cooperative conflict" (Sen 1987), the *bargaining process* is the way that new or existing power is negotiated and utilized within the household. People may negotiate in ways that are covert (such as accepting the situation or ridiculing with friends), or overt methods (such as complaining or using physical violence; Colfer 2011), and this process may be affected by the power source.

Bargaining outcomes are the result of the "bargaining process." Program-provided sources of power may produce bargaining outcomes that were unintended by the program implementers. For example, if bargaining over program resources occurred aggressively, injury could be one unintended outcome (Colfer 2011; Mabsout and Van Staveren 2010). Alternatively, potentially disempowering assumptions that all women will use new-found power to make "altruistic" decisions that work in favor of programmatic intended objectives may not hold.

The theoretical framework proposes that empowerment processes are not linear; there is a *feedback loop*, whereby bargaining outcomes can determine sources of power and bargaining processes.

Results

A total of 14 results met the inclusion and exclusion criteria. Of these, publication dates ranged from 1988 to 2012. The 14 results (summarized in Table 6.2) often described just one or two components of the analytical framework. Most of the results described gendered differences in sources of power; very few described the levels of empowerment, bargaining process, outcomes, or feedback loops.

Effects of forest conservation programs on gendered power dynamics

Summarizing Table 6.2, we found that community-based forest conservation programs had mixed results for women's empowerment. Sources of power often came in the form of financial benefits from crop production (personal endowments and attributes section in Table 6.2). Without a deliberate gender- or women-based focus, by default, programs selectively provided men with sources of power in the form of grants for crop production (Sefa Dei 1994; Schroeder 1995) and support and employment for male groups (Flintan 2003; Abbot et al. 2000). This selective benefit for men was often also costly to women, due to reduced land access (Schoepf and Schoepf 1988; Leach 1991, 2000). Gender- or women-specific projects tended to engage women in income generation (Abbot et al. 2001) and crop production (Meinzen-Dick et al. 2012; Schroeder 1995).

Studies rarely described levels of power or bargaining process. One study described some public processes of bargaining, as well as a feeling of women being "emboldened" and men being displeased with these changes in power balances (Schroeder 1995).

Expected bargaining outcomes were described in more detail than unexpected outcomes. Most described how new income from programs was spent (Abbot et al. 2001; Yatchou 2011). Increased income did not consistently lead to increased household provisioning. One program that gave grants to landholding men led to men contributing less towards household subsistence and increasing the burden on women (Sefa Dei 1994). Another program that benefited women also led to men providing less towards the household (Schoepf and Schoepf 1988). One study from Nigeria described the negative outcome of women being excluded from a project—heavy time costs associated with the project caused the project to fail (Leach 1991).

Some studies described feedback loops. For instance, one study described how the simultaneous disempowerment of men and empowerment of women led to women negotiating marital obligations (Schroeder 1995, 1999). This may affect power levels and future bargaining processes. Another found that women's

TABLE 6.2 Summary of the effects of forest conservation programs on intra-household gendered power dynamics

Framework theme	Country	Reference
Sources of power		
(Personal endowments and attributes, extra-household support, social norms, social perceptions)		
Personal endowments and attributes		
Only landholding men were given state grants for tree-cropping cocoa and oil palm.	Ghana	Sefa Dei 1994
NGOs involved women in production of non-traditional crops, e.g. rice, onions, cotton and shea butter.	Upper East Zone, Ghana	Meinzen-Dick et al. 2012
Women were introduced to "male" activities such as beekeeping, supported through women's farmer groups.	Kilum-Ijim Forest, Cameroon	Abbot et al. 2001
Men's quinine cash cropping resulted in loss of high-quality land for women to use for food crops.	DRC	Schoepf and Schoepf 1988
The produce from a women's gardening project provided women with income at a time when men faced falling yields and removal of cash-crop subsidies. After this project, subsidies for orchards were given to male landholders; the orchards reduced women's time and land for agricultural production.	The Gambia	Schroeder 1995
Women's access to forest products and use of forests for grazing livestock was restricted but they gained skills from training on budgeting, resource management and forestry law.	Burkina Faso	FAO 2007, Coulibaly-Lingani et al. 2009
Coffee and cocoa production project reduced women's ability to intercrop land due to shade from the cash crops. Men often cleared trees valuable to women if the trees restricted growth of the cash crops.	Sierra Leone	Leach 1991
Extra-household support		
Recruitment of park guards using male brotherhoods strengthened male control over land.	Guinea	Flintan 2003
Project consulted male groups (KwiFon) but failed to involve female sacred societies (Fumbiem).	Kilum-Ijim forest, Cameroon	Abbot et al. 2000

Female representation in forest management committees increased women's control over finances.	Kilum–Ijim forest, Cameroon	Yatchou 2011
An integrated community development project registered women's groups with local governments to validate their claims to access land.	Gashaka Gumti National Park, Nigeria	Dunn et al. 2000
Social norms		
Coffee and cocoa cash cropping mainly benefited men because women traditionally do not clear land or participate in income-generating activities. Men adopted a wood collection role—traditionally a women's role—because of new market benefits.	Sierra Leone	Leach 1991
Women's access to conservation parks was restricted because they faced the threat of being bribed, beaten, or raped by park guards if they risked illegally gathering resources from restricted areas.	Guinea	Leach 2000
Social perceptions		
State transfer of power to local brotherhoods resulted in a new male identity and redefinition of female identity, where women's voice and sexuality were repressed.	Guinea	Leach 2000
Men lost respect because of their falling groundnut yields.	The Gambia	Schroeder 1995
Levels of power		
Project that encouraged women to plant shrubs and trees undermined men's control over management decisions. When women's income increased, men expected women to contribute more to household provisioning of food, clothing and ceremonial costs. Men viewed the increase in women's incomes and new-found confidence as an imbalance in power that needed to be rectified.	Gambia	Schroeder 1995
Committees provided a platform for women to change perceptions by demonstrating their knowledge and abilities. This was reported to be reducing the power imbalances.	Gashaka Gumti National Park, Nigeria	Dunn et al. 2000

(continued)

TABLE 6.2 (*continued*)

Framework theme	Country	Reference
Bargaining process		
In financial difficulties, women negotiated by changing marital obligations and allowing themselves a second husband. Resistance against men's allocation of tasks (threshing millet) to women was contested publicly with a group of many women. A focus of NGOs on "women's programs" generated resentment from men, but domestic violence was anecdotally reported to have decreased.	Gambia	Schroeder 1995, 1999
Bargaining outcomes		
Men, who were given grants, reduced their inputs in domestic food production thereby increasing the burden for women to meet household food demand. Men spent their new profits on leisure and personal items.	Ghana	Sefa Dei 1994
Project generated income for women. Women spent income on school fees and new clothes, but men had less pressure to provide for their wives.	Kilum–Ijim forest, Cameroon	Abbot et al. 2001
As a result of new benefits for women, men's financial contribution to the household decreased.	DRC	Schoepf and Schoepf 1988
Women were able to "buy" freedom from unhappy marriages. Women felt emboldened and gained confidence to find additional ways to make income.	The Gambia	Schroeder 1995
As a result of women's increased control over community group finances, the community built a water well, health center, community hall and electricity generator. Female participation was normalized.	Kilum–Ijim forest, Cameroon	Yatchou 2011
Community-based programs gave women more financial freedom and enabled them to change factors affecting levels of bargaining power, by making land management decisions and choosing which forest products to sell.	Burkina Faso	FAO 2007
Heavy time costs for women made a reforestation project unpopular and eventually unsuccessful.	Nigeria	Leach 1991

increased economic power enabled them to make land-management, product-marketing and sales decisions (FAO 2007), also potentially providing women with further sources of economic power.

Studies rarely described all components of the analytical framework, so we selected the most comprehensive case study to illustrate gendered bargaining processes (with reference to the framework in italicized parentheses, below).

A case from The Gambia (Schroeder 1995, 1999)

In 1991, the Gambian government responded to concern regarding high levels of undernutrition in women and children in the country's rural areas by implementing a woman-focused gardening project. The project was primarily meant to support household consumption of fruit and vegetables and the surplus was sold in local markets, thereby providing income for women (*sources of power*). The gardening project encouraged women to plant shrubs and trees and, in doing so, undermined men's control over land management decisions (*levels of power*). Combined with the falling yields and removal of subsidies for male-grown cash crops, women were able to negotiate overtly (*bargaining process*), and this resulted in women being able to "buy" freedom from unhappy marriages (*bargaining outcome*). Women were reported to feel emboldened (*levels of power*) and to have gained confidence to find additional ways to earn income (*sources of power*). Along with the sudden increase in female incomes and women's new-found confidence and falling groundnut yields, men viewed this imbalance in power relations as something that needed to be rectified (*bargaining process, not described*). Men started to expect women to contribute more to household provisioning of food, clothing and ceremonial costs (*bargaining outcome*).

However, in response to environmental concerns of soil and forest degradation, the government and the US Agency for International Development (USAID) supported a second project to promote reforestation with orchards in the same areas. This project took a different approach; unlike the focus on women in the earlier gardening scheme, subsidies for orchard planting were almost exclusively granted to male landholders. This reasserted male landholding privileges (*sources of power*) and women's new-found financial autonomy was undermined (*levels of power*). These forest programs accepted patriarchal structures and, in doing so, increased men's endowments whilst being costly to women (Schroeder 1995).

Opportunities and risks of conservation programs for women's empowerment

From the case study, we have seen the fluidity of gendered bargaining processes and the ease with which forest programs might affect power dynamics. Many programs have disempowered women and some that empowered women led to conflict between genders. Using the literature summarized in Table 6.2, we propose three reasons to explain why this may be, and we suggest lessons to be learned for REDD+ program implementers.

Sources of power were diverse and poorly understood

We found that many programs failed to promote equity in power dynamics because they did not fully understand how the program affected sources of power. Many failed to account for the time demands of program activities or fully understand how gender roles vary according to people's ages, ethnicity, the season and the location or agro-ecological zone (Leach 1991; Meinzen-Dick et al. 2012; Sefa Dei 1994). Some programs only focused on women's empowerment and ignored the possible program impacts on men. We found a number of examples where forestry programs were disempowering for men (Colfer 2011; Schroeder 1995).

Sources of power were not always received by the intended people

Other programs did not anticipate the fluidity of power and the ease with which existing gender imbalances in power levels could be maintained, despite changes in gender-specific power sources. For example, one wood collection program that was intended to benefit women (because it was a traditionally "female" role) was adopted by men instead, due to new cash benefits (Leach 1991).

Some programs accepted patriarchy—others challenged it

Some programs that accepted patriarchy aimed to benefit women by increasing women's access to or the value of traditionally female resources or roles, such as growing food (Abbot et al. 2001) and collecting fuelwood (Leach 1991). The former case was successful in increasing women's incomes; the latter had limited impact on women's empowerment. Other programs involved women in traditionally masculine domains. For example, in a successful project in Cameroon, women were introduced to the "male" income-generating activity of beekeeping (Leach 1991). Another program registered women's groups with local governments to legitimize their claims to access land within a traditionally "patrilineal" context (Dunn et al. 2000). Due to the lack of evidence, we were unable to assess whether programs that challenged patriarchy (possibly creating a more apparent loss of male power) posed risks to women by increasing tension and causing violent bargaining processes.

Implications for REDD+ programs

Before implementation, REDD+ programs should consult with communities to comprehensively characterize gender roles and identify gender-specific costs (including time costs) and benefits of proposed programs. This should include careful consideration of how programs may introduce benefits to both men and women.

Given the earlier described violence and alcohol-related problems sometimes associated with male disempowerment (Mabsout and Van Staveren 2010), REDD+ programs will ideally equalize power whilst collectively empowering forest-dwelling communities—in other words by reducing men's intra-household power levels but increasing their extra-household power. By including men in programs that aim to benefit women, a loss of male power within households may be perceived by men as more socially acceptable and could reduce the risk of a confrontational negotiation (Colfer and Minarchek 2013; Lwambo 2013).

It is difficult to know whether REDD+ programs that challenge patriarchy will maximize the potential for women's empowerment, or generate tensions in the household or animosity towards the program (Colfer and Minarchek 2013). An advantage for REDD+, which may promote community-level acceptance of women's empowerment in this way, is its inherent and monetary power from international and state actors. A disadvantage may exist if there is a poor reputation from previous conservation programs (Temudo 2012). Consultation with men and women may help to minimize potentially negative responses from men, especially if REDD+ programs are pre-acknowledged to empower women by engaging women in traditionally masculine domains (Schroeder 1995). Engagement of men and women in the formative stages of program design may create an enabling environment to implement such programs (Sefa Dei 1994; Yatchou 2011).

Conclusion

This chapter aimed to describe how conservation programs have affected intrahousehold power dynamics so that we can understand how future conservation development projects may affect power dynamics, broader development and livelihood outcomes. The chapter provided a new analytical framework to conceptualize the continuous construction and use of power. We found that forest conservation programs often failed to fully consider the costs and benefits of women's and men's participation (Coulibaly-Lingani et al. 2009; Leach 1991; Schoepf and Schoepf 1988). In particular, we identified the challenge that we do not have enough evidence on whether programs that operated against patriarchy posed risks for women's empowerment.

The conclusions of the study are limited by the publication dates. Impact evaluations are needed to monitor new programs and provide recent evidence on the links between conservation programs and empowerment.

A final note on REDD+ program design situates this chapter back into the realities of forest management in Central and West Africa. Without evidence of programmatic benefits of women's empowerment, the aim to make power dynamics more equitable may not be sufficient to make empowerment a priority in forest program design. Many countries in the study region are politically unstable, with limited resources, high levels of corruption, and weak governance contributing to a context that may not be conducive to the implementation of programs designed to increase women's empowerment (Baaz and Stern 2009; Lwambo 2013). REDD+ funds in this context may not necessarily be distributed transparently and equitably, and empowerment may not be a central focus of governments. However, the many non-state actors involved in forest governance, such as international conservation organizations and those who hold the REDD+ purse strings, can adopt these findings. Indeed, failure to do so would be a lost opportunity to improve the rights, prospects and freedoms of women and men in forest-dwelling communities.

Acknowledgments

Many thanks to Carol J. Pierce Colfer, Bimbika Sijapati Basnett, Marlene Elias, Tom Harrisson and two anonymous reviewers for their comments on this chapter. Thanks to Nora Fry for funding the study.

References

Abbot J et al. 2000. *Promoting partnerships: Managing wildlife resources in Central and West Africa. Evaluating Eden.* London: International Institute for Environment and Development.

Abbot JI, Thomas DH, Gardner AA, Neba SE and Khen MW. 2001. Understanding the links between conservation and development in the Bamenda Highlands, Cameroon. *World Development* 29(7): 1115–36.

Agarwal B. 1997. "Bargaining" and gender relations: Within and beyond the household. *Feminist economics* 3(1): 1–51.

——. 2001. Participatory exclusions, community forestry, and gender: An analysis for South Asia and a conceptual framework. *World development* 29(10): 1623–48.

——. 2009. Gender and forest conservation: The impact of women's participation in community forest governance. *Ecological Economics* 68(11): 2785–99.

——. 2010. Does women's proportional strength affect their participation? Governing local forests in South Asia? *World Development* 38(1): 98–112.

Alsop R, Bertelsen MF and Holland J. 2006. *Empowerment in practice: From analysis to implementation.* Washington, DC: World Bank.

Angelsen A and McNeill D. 2012. The evolution of REDD+. In Angelsen A, Brockhaus M, Sunderlin WD and Verchot LV, eds. *Analysing REDD+: Challenges and choices.* Bogor, Indonesia: Center for International Forestry Research.

Baaz ME and Stern M. 2009. Why do soldiers rape? Masculinity, violence, and sexuality in the armed forces in the Congo (DRC). *International Studies Quarterly* 53(2): 495–518.

Babalola S and Dennis C. 1988. Returns to women's labour in cash crop production: Tobacco in Igboho, Oyo State, Nigeria. In Davison J, ed. *Agriculture, women and land: The African experience,* Boulder, CO: Westview Press. 79–89.

Barker G and Ricardo C. 2005. *Young men and the construction of masculinity in sub-Saharan Africa: Implications for HIV/AIDS, conflict, and violence.* Washington, DC: World Bank.

Barry D, Larson AM and Colfer CJP. 2010. Forest tenure reform: An orphan with only uncles. In Larson AM, Barry D, Dahal GR and Colfer CJP, eds. *Forests for people: Community rights and forest tenure reform.* London and Washington, DC: Earthscan. 19–42.

Brown D, Seymour F and Peskett L. 2008. How do we achieve REDD co-benefits and avoid doing harm? In Angelsen A, ed. *Moving ahead with REDD: Issues, options and implications.* Bogor, Indonesia: Center for International Forestry Research. 107–18.

Colfer CJP. 2011. Marginalized forest peoples' perceptions of the legitimacy of governance: an exploration. *World Development* 39(12): 2147–64.

—— and Minarchek RD. 2013. Introducing "the gender box": A framework for analysing gender roles in forest management. *International Forestry Review* 15(4): 411–26.

Coulibaly-Lingani P, Tigabu M, Savadogo P, Oden P-C and Ouadba J-M. 2009. Determinants of access to forest products in southern Burkina Faso. *Forest Policy and Economics* 11(7): 516–24.

Davison J. 1988. Land and women's agricultural production: The context. In Davison J, ed. *Agriculture, women, and land: The African experience.* Boulder, CO: Westview Press.

Doss CR. 2002. Men's crops? Women's crops? The gender patterns of cropping in Ghana. *World Development* 30(11): 1987–2000.

Dunn A, Mamza J, Ananze F and Gawaisa S. 2000. Sticking to the rules: Working with local people to conserve biodiversity at Gashaka Gumti National Park, Nigeria. In Abbot J. et al. 2000. *Promoting partnerships: Managing wildlife resources in Central and West Africa. Evaluating Eden.* London: International Institute for Environment and Development.

FAO (Food and Agriculture Organization). 2007. *Gender mainstreaming in forestry in Africa.* Forest Policy Working Paper No. 18. Rome: Food and Agriculture Organisation.

Flintan F. 2003. *"Engendering" Eden: Volume II, Women, gender and ICDPs in Africa. Lessons learnt and experiences shared.* London: International Institute for Environment and Development.

Hickel J. 2014. The "girl effect": Liberalism, empowerment and the contradictions of development. *Third World Quarterly* 35(8): 1355–73.

Hopkins J, Levin C and Haddad L. 1994. Women's income and household expenditure patterns: Gender or flow? Evidence from Niger. *American Journal of Agricultural Economics* 76(5): 1219–25.

Ingram V, Schure J, Tieguhong JC, Ndoye O, Awono A and Iponga DM. 2014. Gender implications of forest product value chains in the Congo basin. *Forests, Trees and Livelihoods* 23(1–2): 67–86.

John RM. 2008. Crowding out effect of tobacco expenditure and its implications on household resource allocation in India. *Social Science and Medicine* 66(6): 1356–67.

Kabeer N. 1999. Resources, agency, achievements: Reflections on the measurement of women's empowerment. *Development and Change* 30(3): 435–64.

Kandiyoti D. 1988. Bargaining with patriarchy. *Gender & Society* 2(3): 274–290.

Leach M. 1991. Engendered environments: Understanding natural resource management in the West African forest zone. *IDS Bulletin* 22(4): 17–24.

———. 2000. New shapes to shift: War, parks and the hunting person in modern West Africa. *Journal of the Royal Anthropological Institute* 6(4): 577–95.

Lewis J. 2000. *The Batwa Pygmies of the Great Lakes Region.* London: Minority Rights Group International.

Luttrell C, Loft L, Fernanda Gebara M and Kweka D. 2012. Who should benefit and why? Discourses on REDD+ benefit sharing. Angelsen A, Brockhaus M, Sunderlin WD and Verchot LV, eds. *Analysing REDD+: Challenges and choices.* Bogor, Indonesia: Center for International Forestry Research.

Lwambo D. 2013. "Before the war, I was a man": Men and masculinities in the eastern Democratic Republic of Congo. *Gender and Development* 21(1): 47–66.

Mabsout R and Van Staveren I. 2010. Disentangling bargaining power from individual and household level to institutions: Evidence on women's position in Ethiopia. *World Development* 38(5): 783–96.

Mai Y, Mwangi E and Wan M. 2011. Gender analysis in forestry research: Looking back and thinking ahead. *International Forestry Review* 13(2): 245–58.

Malapit HJL and Quisumbing AR. 2015. What dimensions of women's empowerment in agriculture matter for nutrition in Ghana? *Food Policy* 52: 54–63.

———, Sproule K, Kovarik C, Meinzen-Dick R, Quisumbing A, Ramzan F, Hogue E and Alkire S. 2014. *Measuring progress toward empowerment. Women's empowerment in agriculture index: Baseline report.* Washington, DC: International Food Policy Research Institute.

Meagher K. 2010. *Identity economics: Social networks and the informal economy in Nigeria.* Suffolk, UK and Rochester, NY: Boydell and Brewer.

Meinzen-Dick RS, van Koppen B, Behrman J, Karelina Z, Akamandisa V, Hope L and Wielgosz B. 2012. *Putting gender on the map: Methods for mapping gendered farm management systems in sub-Saharan Africa.* Washington, DC: International Food Policy Research Institute.

Mwangi E, Meinzen-Dick R and Sun Y. 2011. Gender and sustainable forest management in East Africa and Latin America. *Ecology and Society* 16(1): 17.

Nartey WD. 2014. A REDD solution to a green problem: Using REDD plus to address deforestation in Ghana through benefit sharing and community self-empowerment. *African Journal of International and Comparative Law* 22(1): 80–102.

Peach Brown C. 2011. Gender, climate change and REDD+ in the Congo Basin forests of Central Africa. *International Forestry Review* 13(2): 163–76.

Pérez MR, Ndoye O, Eyebe A and Ngono DL. 2002. A gender analysis of forest product markets in Cameroon. *Africa Today* 49(3): 97–126.

Quesada-Aguilar A, Aguilar L, Amunau SB, Blomstrom EH, Dogbe T, Kutegeka S, Nyuyinwi M, Ruta D and Siles J. 2012. Gender and REDD+ road maps in Cameroon, Ghana and Uganda. In Broekhoven G, Savenije H and von Scheliha S, eds. *Moving forward with forest governance*. Wageningen, The Netherlands: Tropenbos International. 149–57.

Quisumbing AR, Payongayong E, Aidoo J and Otsuka K. 2001. Women's land rights in the transition to individualized ownership: Implications for tree resource management in Western Ghana. *Economic Development and Cultural Change* 50(1): 157–82.

Richards P. 1998. *Fighting for the rain forest: War, youth and resources in Sierra Leone*. Oxford: James Currey.

Rocheleau D and Edmunds D. 1997. Women, men and trees: Gender, power and property in forest and agrarian landscapes. *World Development* 25(8): 1351–71.

——, Thomas-Slayter B and Wangari E. 1996. *Feminist political ecology: Global issues and local experiences*. London and New York: Routledge.

Russell S. 1996. Ability to pay for health care: Concepts and evidence. *Health Policy and Planning* 11(3): 219–37.

Schoepf BG and Schoepf C. 1988. Land, gender, and food security in Eastern Kivu, Zaire. In Davison J, ed. *Agriculture, women, and land: The African experience*. Boulder, CO: Westview Press.

Schroeder RA. 1995. Contradictions along the commodity road to environmental stabilization: Foresting Gambian gardens. *Antipode* 27(4): 325–42.

——. 1999. *Shady practices: Agroforestry and gender politics in the Gambia*. Berkeley: University of California Press.

Sefa Dei GJ. 1994. The women of a Ghanaian village: A study of social change. *African Studies Review* 121–45.

Sen A. 1985. Well-being, agency and freedom: The Dewey lectures 1984. *The Journal of Philosophy* 169–221.

——. *Gender and cooperative conflicts*. Helsinki: World Institute for Development Economics Research.

Sudarkasa N. 1973. "Where women work: A study of Yoruba women in the marketplace and in the home." Anthropological paper, Museum of Anthropology. University of Michigan.

Sunderland T, Achdiawan R, Angelsen A, Babigumira R, Ickowitz A, Paumgarten F, Reyes-García V and Shively G. 2014. Challenging perceptions about men, women, and forest product use: A global comparative study. *World Development* 64(1): S56–S66.

Temudo MP. 2012. "The white men bought the forests": Conservation and contestation in Guinea-Bissau, Western Africa. *Conservation and Society* 10(4): 354.

Tiayon FF. 2011. Gender in tropical forestry: Realities, challenges and prospects. In Aguilar L, Shaw DDMP and Quesada-Aguilar A, eds. *Forests and gender*, Gland, Switzerland: International Union for Conservation of Nature; New York: Women Environment Development Organization. 14–17.

Tshombe R, Mwinyihali R, Girineza M and de Merode E. 2000. Decentralising wild-life management in the Democratic Republic of Congo: Integrating conservation and development objectives in a country at war. In Abbot J et al. 2000. *Promoting partnerships: Managing wildlife resources in Central and West Africa. Evaluating Eden.* London: International Institute for Environment and Development.

UN (United Nations). 1997. *Report of the economic and social council for 1997, General Assembly: Fifty-second session 18 September 1997*, New York, NY: United Nations.

Vansina J. 1990. *Paths in the rainforests: Toward a history of political tradition in Equatorial Africa.* Madison: University of Wisconsin Press.

Wan M, Colfer CJP and Powell B. 2011. Forests, women and health: Opportunities and challenges for conservation. *International Forestry Review* 13(3): 369–87.

Whitehead A and Tsikata D. 2003. Policy discourses on women's land rights in sub-Saharan Africa: The implications of the return to the customary. *Journal of Agrarian Change* 3(1–2): 67–112.

Yatchou NA. 2011. Mainstreaming gender into community forestry in eastern Cameroon. In Aguilar L, Shaw DDMP and Quesada-Aguilar A, eds. *Forests and gender.* Gland, Switzerland: International Union for Conservation of Nature; New York: Women Environment Development Organization. 46–7.

7

GENDER AND FOREST DECENTRALIZATION IN CAMEROON

What challenges for adaptive capacity to climate change?

Anne-Marie Tiani, Mekou Youssoufa Bele, Richard Sufo Kankeu, Eugene Loh Chia and Alba Saray Perez Teran

Introduction

There are strong indications that climate change will have considerable global effects on all sectors of development (Stern Review 2006; IPCC 2007). There is a widespread consensus that developing countries will bear the brunt of the adverse consequences from climate change due to high levels of poverty and weak capacity to adapt. In fact, climate change is likely to undermine many years of development efforts in developing countries (IPCC 2007; MEA 2005; Stern Review 2006). In Africa, climate-related risks are expected to intensify existing problems and create new combinations of risks, given the existing widespread poverty and dependence on the natural environment (e.g. Bele et al. 2013; Somorin et al. 2014; Sonwa et al. 2010, 2012a, 2012b). Of particular concern are communities with vulnerable livelihoods, food and environmental insecurity, under-resourced health care, gender inequalities, weak security and governance, poor infrastructure and education, and lack of access to appropriate resources and capacities to deal with extreme events. Women shoulder many of the consequences of climate change, due to gender norms that limit their asset base and make them heavily dependent on natural resources and rain-fed activities sensitive to climate change.

According to the IPCC (2007), climate change impacts will differ across regions, generations, age classes, income groups, occupations and gender. How resources and opportunities are shared in society and who decides how these should be used have a strong gender dimension and are critical factors in adaptation. While much policy attention focuses on the impacts on society's poorest, less attention has been paid to gender differences and ensuring that climate change does not further erode gender equality.

Gender refers to socially constructed roles, responsibilities and opportunities associated with men and women, as well as hidden power structures that govern the relationships between them (UNDP 2010). Understanding the linkages between gender relations and the environment requires achieving a better analysis of patterns of representation and decision-making processes among community organizations concerned with the environment. This entails analyzing equality and equity between women and men in these organizations' governance structures to ensure that decisions are responsive to the needs and interests of both.

In the context of sustainable use, management and conservation of natural resources, gender equality and equity are matters of fundamental human rights and social justice and a precondition for sustainable development. However, gender discrepancies still exist in rights and access to natural resources, including land, trees, water and animals. Within communities and households, women and men have differing levels of adaptive capacity. According to Adger et al. (2004), all decisions on adaptation privilege one set of interests over another and create winners and losers. Women continue to be disadvantaged by insecure access and property rights to forest and tree resources (Meinzen-Dick et al. 1997) and disproportionately experience the impacts of poor forest management (Tiani et al. 2005; Agrawal and Chhatre 2006). Though among the most vulnerable to increasing global challenges (e.g. trade, climate change, urbanization, and energy and food insecurity; IUFRO 2010), women are often excluded from decision making at household, community and national levels (Agarwal 2001) and are poorly placed to influence resource allocation or research priorities (Crewe and Harrison 1998). They are particularly absent from climate change and natural resource-related decision-making processes at all levels (e.g. Brown 2011). Access to new technology, information and training related to natural resource management also remains highly gendered, with most of the related initiatives targeted towards men (Mwangi et al. 2011). The inclusion of women in resource management offers a potential pathway for empowering women both within their private and public lives (Torri 2010). In addition, climate change is projected to magnify existing patterns of gender inequality (WEDO 2007; UNDP 2008) by exacerbating the existing vulnerabilities of individuals and households who already have limited or insecure access to physical, natural, financial, human, social, political and cultural assets (Flora and Flora 2008). Availability of, and access to, assets is socially differentiated, because it is shaped by formal and informal inequalities in many aspects of life (Otzelberger 2011). Therefore, it is crucial that adaptation responses are pro-poor and gender-aware, informed both by gender-based vulnerabilities as well as the unique contributions of women and men to these processes (Devisscher et al. 2013).

Decentralization conceptual framework

Decentralization is a polysemous and evolving concept, whose interpretations have led to different conceptual frameworks, program implementation and implications. Decentralization is usually referred to as the transfer of powers from central government to lower levels (Mawhood 1993), in a political-administrative and territorial hierarchy (Crook and Manor 1998; Agrawal and Ribot 1999; Ribot 2007). This official power transfer can assume several forms (Manor 1999; Larson 2005): (1) bureaucratic or administrative decentralization (authority fragmentation, delegation and devolution; Cheema and Rondinelli 1983), which concerns the transfer of authority, responsibility and financial resources for providing public services among different levels of government; (2) fiscal decentralization (Ribot 2002), which is the transfer of financial powers from the central government to decentralized institutions; (3) economic decentralization (Cheema and Rondinelli 2007), which includes privatization of publicly owned functions and businesses, and (4) democratic decentralization known as a process through which powers and resources are transferred to actors who represent local populations and are, in return, accountable to those populations (Manor 1999; Ribot 2007).

In the context of forest decentralization, this last form is more appropriate as it can generate a greater sense of ecological responsibility among non-state actors, greater environmental justice and better governance (Wellstead et al. 2003; Steel and Weber 2003). The representation and accountability, along with the transfer of decision making to local populations, can increase public participation and efficiency of public service provisions, and empower local citizens (Agrawal and Ribot 1999; Francis and James 2003). In fact, democratic decentralization requires public participation and democracy in the management of the commons (Pitkin 1967). Such participation entails representation that presupposes that a person or a group of persons have powers and rights allocated to them by a larger group of persons at a given time (Oyono 2004). As such, representation-building should be free and transparent and the values and actions of representatives should correspond to those of the general public (Pitkin 1967). In that sense, "representatives mirror who they represent" (Wellstead et al. 2003). In this scenario, women have often lacked representation in community structures in general, and community forest management structures in particular because community representatives have mainly been men. Agarwal (2001) identified key elements that mediate women's capacity to access community forest governance structures. These include rules, social norms, social perceptions, men's claims over community structures and more.

Decentralization in the context of Cameroon

In Cameroon, decentralization was part of a wider process of externally initiated reforms designed to reduce the role of the state and as a precondition for continuing to receive international aid (Karsenty et al. 1997; Karsenty 1999; Oyono 2004). It was promoted as an appropriate means to involve local people in the management of public affairs so as to strengthen equity and democracy (Ribot 2006; Larson 2005), and to implement policies and programs that reflect people's real needs and preferences, as central state authorities usually lack relevant "time and place knowledge" (Hayek, cited in Ostrom et al. 1993). Decentralization has been seen as highly relevant for poverty reduction through increased possibilities for participation, improved access to services and a more efficient way of providing public goods. It was likely that decentralization would allow women to improve their position by taking advantage of a window of opportunity to join the wider circle of decision makers (Bandiaky and Tiani 2010). In this context and under pressure from the World Bank, Cameroon's government initiated forest reforms in the 1990s, after more than a century of colonial and postcolonial forest policies (Kouna Eloundou 2012). These reforms constituted a major policy shift, resulting from the promulgation of the 1994 Forest Law (RoC 1994), its decree of application (RoC 1995), and subsequent legal and administrative instruments. The major innovations in the new policy were the transfer to local councils and communities of statutory rights, authority and responsibilities for acquiring and managing a share of forests and forest revenues.

Mechanisms of forest decentralization in Cameroon

Three basic mechanisms constitute the foundations of decentralized forest management in Cameroon:

1. "*Council forests*"—defined as "any forest that has been classified and assigned to a council concerned or that was planted by the council on council land" (RoC 1994: 6). According to the 1996 constitutional reforms a "council" is the basic decentralized territorial unit.
2. "*Community forests*"—defined as "a forest of the nonpermanent forest estate, subject to a management agreement between a village community and the Administration in charge of forests. The management of this forest is entrusted to the village community concerned, with the technical support of the Administration" (RoC 1995). A local community has the right to manage up to 5,000 ha of forests on a 25-year rotation, according to a simple forest management plan.
3. "*Annual forestry fees redistribution scheme*"—includes "fees paid annually by timber companies from the logging of forest concessions or Forest Management Units, distributed as follows: 50% to the central State, 40% to rural councils in whose domains the exploited forests are located, and 10% to neighboring villages" (RoC 1996).[1]

Each of these mechanisms is overseen by a management committee composed of local residents.

Here, we contribute to the assessment of gender equity in the policy and practices of decentralized forest management in Cameroon. We assess modes of representation in decentralized forest management committees, gender representation and positions in the management committees, and the causes and effects

---

1 In addition, there is a village *eco-tax,* another forestry fee. It is a payment of US$2 per m^3 of wood harvested in smaller concessions (*ventes de coupe*), which are given to the populations residing in the area as compensation for cutting rights or a kind of "royalties" (Oyono 2004).

of women's marginalization with subsequent impacts on their adaptive capacity. According to Smit and Pilifosova (2003: 287) "the forces that influence the ability of a system to adapt are the drivers or the determinants of adaptive capacity." Adaptive capacity in this case is the potential or ability of a system, region, or community to adapt to the effects or impacts of climate change (Adger et al. 2004). It is measured using determinants such as equality and equity, information and skills, institutions, participation and empowerment in decisions.

Methodology

Description of the study site

Field work was carried out in the Boumba and Ngoko Division, in the Eastern Region of Cameroon (Figure 7.1). The Division covers an area of 30,389 square kilometers and in 2001, had a total population of 116,702. It is divided administratively into four councils—Gari-Gombo, Moloundou, Salapoumbé, and Yokadouma—each made up of a number of villages.

The populations, mostly Bantu and Baka Pygmies, are organized following a "nonhierarchical model" (Oyono 2004). Despite the existence of recognized leaders, authority and power are highly dispersed so that Mamdani (1996) called these "stateless communities." However, in the 1990s, the advent of democracy brought

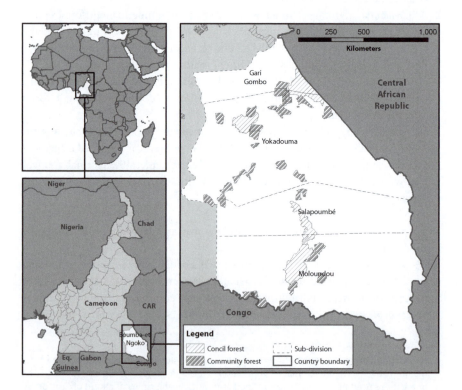

FIGURE 7.1 Map of Cameroon showing the study area

about a number of legal innovations including Law No. 90/53 of December 19, 1990, on freedom of association, and Law No. 92/006 of August 14, 1992, legalizing cooperatives and Common Initiative Groups (CIGs). This gave rise to the proliferation of rural micro-organizations, with their operations often supported by NGOs and bilateral and multilateral cooperation projects.

Key threats to forests in the study area include poaching and trade of protected species, large-scale logging and illegal tree felling by individuals, bush fires and prolonged dry seasons (Betti 2004). Agriculture is the principal activity of most households. Other activities include collection of NTFPs, hunting, livestock rearing, and fishing. The threats caused by climate variability and change are strongly perceived by communities, in particular, dry spells in the beginning of the wet season, a prolonged dry season and strong winds (Devisscher et al. 2013; Bele et al. 2013; Chia et al. 2013).

Policy document content analysis

The 1994 Forest Law (RoC 1994) and its decree of application (RoC 1995), along with subsequent legal and administrative instruments, are major policy documents that comprehensively address the key issues of forest decentralization, biodiversity conservation and sustainable forest management in Cameroon. Analysis focused on their content and explored how gender is integrated in those documents. This information was supplemented by a review of other relevant published texts and unpublished articles related to decentralized forest management in Cameroon.

Fieldwork

Content analysis (Bele et al. 2011) was supplemented with fieldwork examining how gender issues manifest themselves on the ground. We examined women's presence in the management committees (community forest management committee, council forest management committee, and forestry fee management committee) by counting the number of women in those committees, their positions, the causes and effects of women's marginalization, if any, and potential impacts on their capacity to adapt to climate change.

Data collection and analysis

Data were collected in 2013 in all the 24 legalized community forests and four council forests in the Boumba and Ngoko Division, Eastern Cameroon. Data were drawn mainly from focus-group discussions composed of members of community forest, council forest and forestry fee management committees. The groups were first subdivided by sex and subsequently came together in plenary sessions to share results. The number of participants varied by committee. A total of 454 members were involved from community forest committees, 33 members from council forest committees and 22 members from forestry fee management committees.

Focus-group discussions provided information on women's presence and positions in the different management committees and their power in decision making. Furthermore, the discussions helped identify the causes and effects of women's marginalization in the area and their potential impacts on women's capacity to adapt to climate change.

Group discussions were facilitated by a research team comprised of three women and one man (two note takers, one facilitator and one translator when possible). Facilitation involved applying a set of processes and "soft skills" to help groups to attain their objectives (German et al. 2010) by jointly identifying problems related to women's involvement in different management committees and in decision-making processes in general. The facilitator had the responsibility of managing group dynamics, including power imbalances that threaten the "voice" of certain members, mostly women and some elders. The facilitator had the analytical capacity to integrate and synthesize diverse views to distill an emerging consensus or key points of differences.

To deepen the understanding of focus-group discussions, interviews were also carried out with 18 individuals who showed substantial knowledge on forest decentralization and gender issues during focus-group discussions. The interviews followed a predesigned semi-structured questionnaire containing questions regarding gender representation in local forest management decision-making instances. Interviews focused on barriers that prevent women from fully participating in decision making. Interviews were conducted in French and/or local languages, mainly Mpiemo, Pupong and Baka where necessary with the aid of interpreters. Field observations consisted of perceptions of how men and women intervened during group discussions and identification of any barriers relative to women's interventions. Analysis of the data mostly utilized simple descriptive statistics. Quantitative data analysis was limited to the number of men and women in different management committees. Interviews served as convenient supplements to the discussions from focus groups.

Results and discussion

Gender dimensions in the policy and practices of decentralization in Cameroon

Law No. 94/01 of January 20, 1994 on the Forestry, Wildlife and Fisheries Regime and its Decree of Application No. 95-531-PM of August 23, 1995 were key political milestones in the protection of user rights and the responsibilities of local communities in forest management in Cameroon. Though greater involvement of non-state actors was advocated in the new reforms, no guidance was given to local communities as to what they must actually do. Based on our extensive review of the laws, the reforms failed to define modes of representation in the management entities or to encourage gender equality in those entities. As such, institutional arrangements for decentralized forest management were gender blind. Even though indigenous user rights were mentioned specifically in the management of council forests (article 30(2) of the 1994 Forestry Law), nowhere were women's user rights mentioned as a distinct social group in the legislative and regulatory texts.

According to Law No. 92/002 of August 14, 1992, fixing the election conditions in councils in Cameroon, all councils are to be led by a mayor elected from within the council committee. The councils have a responsibility in the management of local affairs under the supervision of the state. Article 3, sub. 2 of the same law requires political parties to take into consideration the various sociological components of their localities, including the representation of ethnic minorities. However, a much tightened interpretation is given to this prescription all over the country as "sociological components" are interpreted as ethnic diversity with no attention to gender balance in the process of constituting electoral lists.

Modes of representation in decentralized forest management committees

The Decree of Application of the Forestry Law No. 94/01 of January 20, 1994 requires a village community to be officially recognized as a legal entity to acquire a community forest and manage forestry fees. Such a legal entity could be an association, a common initiative group, a business group or a cooperative.[2] Legal entities are to be run by the management committees for community forests, council forests and forestry fees. They serve as the interface between the community and the local administration. The main objective of management committees is to fulfill the needs and aspirations of the local communities they represent. This fulfillment depends on the ways such committees are constructed and operated.

In principle, the construction of forest management committees should be democratic. This is contrary to the classical mode of structuring derived from customary traditional power where the main criteria of access to power are birth and masculinity (Diaw 2010). As such, women's representation in management committees would help ensure that their needs and aspirations are taken into account. However, how members of those committees are to be chosen is not explicit in the Decree of Application. For instance, results from our fieldwork show that modes of representation in the forestry fee management committee and council forest management committee are either by statutory appointment, self-appointment, or co-optation.[3] In the case of community forests, only 15 percent of the presidents

2 These are defined as a group of individuals designated by their community to act on their behalf. Nevertheless, each of them bears its own specificity: Association is easy to manage, exonerated from tax, and the benefits are entirely dedicated to social infrastructures. A CIG, however, is entitled to gifts and legacy and the benefits derived from its activities can be distributed to its members. The last two entities are more complex structures.

3 Co-optation occurs when a village chief, already self-appointed as a member, co-opts his dependents, supporters, or individuals who have obligations to him as members. This strategy is common in the formation of forestry fee management committees.

of the 24 community forest management committees investigated were selected through democratic and competitive elections. Statutory, self-appointment, or co-optation could be attributed to the joint Ministry of Economy and Finance/Ministry of Territorial Administration (MINEFI/MINAT) Order of 1998, which encouraged the involvement of administrative and municipal authorities in the functioning of forestry fee management committees. According to Articles 4 and 5 of the Order:

> The management of income destined for local communities is assured by a Management Committee, hereafter designated the "Committee," and instituted in each beneficiary community. The Committee is placed under the guardianship of the nearest administrative authority (the Divisional Officer). The Committee provided for in article 4 is composed as follows: the president, being the mayor of the council or his representative having the status of town councilor; six representatives of the village community; and the local representative of the Ministry of Forests.

Gender representation in management committees

Table 7.1 gives an overview of the number of women in the community forest management committees in 2013.[4] It shows that out of a total of 454 members of

TABLE 7.1 Number of women in community forest management committees

Subdivision	Number of legalized community forests	Number of members in the management committees	Number of women (n)	Proportion of women (%)
Gari–Gombo	1	11	0	0
Yokadouma	20	399	52	13
Moloundou	3	44	14	32
Total	24	454	66	14.5

TABLE 7.2 Number of women in council forest management committees

Council forest	Total number of members in the committees	Number of women in the committees	Positions women occupied in the committees
Gari–Gombo	12	2	Simple members
Moloundou	12	0	0
Yokadouma	9	0	0
Total	33	2 (6%)	

4 The time of the study.

TABLE 7.3 Trends in numbers of women among town councilors

Council	2007			2013		
	Town council	No. of women	Women's positions	Town council	No. of women	Women's positions
Gari–Gombo	25	2	Councilors	25	3	1st Deputy mayor (1) Councilors (2)
Moloundou	25	0	None	25	2	2nd deputy mayor (1) Councilors (1)
Salapoumbé	25	0	None	25	4	Councilors (4)
Yokadouma	41	0	None	41	3	3rd deputy mayor (1) Councilors (2)
Total	116	2		116	12	

all the community forest management committees in the study area, only 66 (14.5 percent) were women.

Results are more striking for the council forest management committees (Table 7.2). In 2013, Gari-Gombo counted only two women among its twelve members. Moloundou and Yokadouma council forests had no women in their management committees. In Salapoumbé, the council forest was still in the legalization process.

With regards to the forestry fee management committee, there was no woman among the 22 members of the three forestry fees management committees investigated in Yokadouma.

Table 7.3 gives an overview of the composition of the town councils in the study area.

Table 7.3 also shows that men have largely dominated the councils and thus the council management committees; but the number of women has slightly increased from 2007 to 2013. This slight increase could be explained by more advocacy by NGOs for gender mainstreaming in all sectors in Cameroon. Figure 7.2 summarizes the number of women in the various management committees in the study area.

Positions occupied by women in the management committees

In the study area, women's representation in the committees was minimal and the positions they occupied were generally of lesser importance with limited influence over decision making. Only one woman was the president and another one the vice president of the community forest committees in Yokadouma (Table 7.4). All the other women participating in the management committees were limited to advisory, financial, or communication roles.

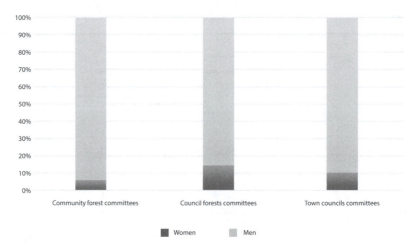

FIGURE 7.2 Percentage of men and women in different management committees

Table 7.4 shows that most women act as advisors, with the main role providing guidance to the group. In fact, this position is very important for women to lobby for their needs and aspirations in the decision-making process. Unfortunately, according to the study participants, the position is mostly *de jure*. In practice, women are not generally consulted. As a matter of fact, lack of such consultation during decision making was identified as one of the barriers that limit women's voice in forest management processes. Furthermore, most women interviewed seemed not to know what this position confers in terms of roles and responsibilities. For more important positions like president or treasurer, women are generally required to prove their capabilities before their selection. This might be by an

TABLE 7.4 Positions occupied by women in the forest management committees in the study area

Positions	Key role	Men	women
President	Run the day-to-day operations of the association; make effective decisions for managing and operating the association by fulfilling the following duties: open the meetings, determine whether there is a quorum, establish the schedule of the meeting, and coordinate the subject matter of the agenda.	23	1
Vice president	Assist the president in his/her role	47	1
Treasurer	Responsible for all the income and savings of the association	22	2
Deputy treasurer	Assist the treasurer in his/her role	21	3
Controller/auditor	Watch over the association income and expenses	40	8
Communication	Inform other members of events (e.g. meetings)	20	4
Advisors	Advise the management committee	40	28

explicit and proven act of bravery or exceptional services rendered to the community. In addition, traditional perceptions of women as inferior to men prevail as many people uphold cultural practices and perceptions that enhance the subordination of women (Agarwal 2001). Consequently, men continue generally to dominate women in all spheres of life (e.g. in political, economic, social and religious realms; Kasomo 2012).

However, women were also believed to have special qualities that can be capitalized on. For instance, a man from a community forest committee in Yokadouma said, "Generally, women care a great deal about social welfare, practice more justice and are more trustworthy than most men." Clearly, equal participation of women and men in decision making also provides a balance that more accurately reflects the composition of society. In addition, women's equal participation in decision making is important for their interests to be taken into account. If a certain level of representation can be achieved, women are able to achieve solidarity of purpose to represent their interests (Dube 2014). However, the current system of representation in the study area does not create opportunities to defend the interests of the communities, especially the interests of women. Most committee members are motivated by individual interest rather than by local communities' substantive interests.

Marginalization of women: causes and implications

Causes of women's marginalization: the weight of traditional systems

Focus group participants identified an array of intrinsic or exogenous factors responsible for women's marginalization in forest management and decision making. Such factors include socio-cultural beliefs, attitudes, biases, stereotypes, the institutional framework guiding the gender division of labor with greater family responsibilities to women, feelings of inferiority, lower levels of education, and less access to information. Additional formidable barriers include the deprivation of rights that has left women with fewer opportunities to acquire political and decision-making experience, and insufficient participation and empowerment in other decisions that affect their lives in political and social processes. Since men dominate public decision-making processes, male values are predominantly reflected in decision-making bodies. Traditional systems set a gendered division of responsibilities: the public domain for men and the domestic sphere for women. In Cameroon, forest management is a public issue, thus falling into the men's portfolio. In addition, in this study site, as in other parts of Cameroon, most communities are built on a patriarchal structure which values men more than women. In Moloundou, a female participant in her fifties said:

> In our communities, women are treated as property. They are to be bought by their male counterparts through the payment of what is known as the "bride-price." No matter how small the amount is, it is a significant determinant of who owns whom and who has leadership and makes decisions.

Our study also reveals that although women are present during the launching of the forest committee establishment process, they are gradually sidelined and barely represented. Either they are not invited or they are no longer informed of the meetings. Tobith and Cony (2006) and Agarwal (2001) found similar results, affirming that even when women attend forest committee meetings, their husbands do not generally allow them to speak out, as they think that is primarily not a woman's job.

As a matter of fact, in the study site, as in other parts of Cameroon (Tobith and Cony 2006), the marginalization of women in political affairs and decisions pre-dates the advent of colonialism in the country. Indeed, pre- and postcolonial traditional cultures and European culture were both deeply rooted in patriarchy. The normative systems they independently produced were male-biased and dominated. The marginalization of women was also evident in all other spheres of life such as the family, economic, social, labor and other relationships. It is widely believed that the marginalization of women in political participation and decision-making processes has been responsible for the exclusion of their interests in governance and development paradigms (Lapelle et al. 2004; Plumwood 1992).

Some effects of marginalization

Women's marginalization in Cameroon in general has led to multiform effects of which some are described below.

Socio-economic effects

Weak representation of women in municipal councils and in forest management committees leads to lack of consideration of women's needs, interests and constraints in the orientation of activities to be carried out in the community. For many researchers, local income from forestry revenues is lower than expected. In this regard, Bigombe Logo (2008) notes that the members of forestry income management committees invest most money in prestigious rather than social activities. Most of the activities carried out are not the priorities of the populations. One dramatic example is buying and distributing televisions to households in an area with no electricity. In fact, the same author showed that only 20 percent of forestry revenues allocated to beneficiary communities are invested in local development activities. Mbotto-Ndomi et al. (2007) confirmed this tendency when they listed and commented on the investments of forestry incomes in three villages of Mbang council in the Southern region of Cameroon: They used the income for

> . . . a shed, the rehabilitation of chapels, the purchase of television sets, generators, even if this decision did not always meet with unanimity. Women gave priority to the purchase of soap and salt; only one village invested a little bit in education.

Men themselves were unconscious of the impacts of such exclusion as they believed that they legitimately represented women in public. It was common to hear speeches such as: "Why ask women about their needs? We are their husbands and so know better what their needs are."

Ecological impacts

In the study area, the management of community forests is completely oriented towards the exploitation and commercialization of wood, ignoring other opportunities offered by forests. Decentralization has not brought tangible positive ecological effects on forests. The main reasons are the absence of internal arrangements rooted in the search for ecological sustainability, and the attitudes of the forest populations whose prime concern is to obtain financial benefit from the forests (Oyono 2004) and for whom the concept of sustainable management arouses less enthusiasm (Porro et al. 2001). Many people, especially the youth, want immediately to earn money from the forest and favor rapid exploitation. For these people, logging should generate enough money to change their lives quickly. However, since women heavily rely on forest products other than timber, their more equal representation in management committees would promote useful negotiation of management objectives and interests. As such, women and men could more easily find a balance among management for timber, collection of NTFPs and promotion of other services offered by the forest.

Some impacts of marginalization on women's adaptive capacity to climate change

In the study area, marginalization drastically affects women's capacity to adapt to climate change in the following ways:

1. Limited roles in decision making—limitations are placed on women's voice and participation in public and household decision making, which in turn creates constraints on their adaptive capacity. This limits the ability of families and communities to realize the potential contribution of women's specific knowledge and skills to adaptation efforts (Djoudi and Brockhaus 2011).
2. Women's needs and priorities are neglected (Brown 2011; Shanley et al. 2011)—without participation by women, programs to replace traditional crops with those better suited to the changing environment might focus only on the needs of men's fields and not address the problems women face with their food crop fields and home gardens. For instance, in 2011, the Strategic Committee at the Senior Divisional Office in Boumba and Ngoko decided after consultation with mostly men to support the expansion of cocoa farms in order to contribute to the fight against climate change and poverty reduction. However, cocoa farms are principally men's activity. If given a chance, women may have asked for a climate-resilient crop such as fast-growing cassava that is resistant to African cassava mosaic virus.
3. Limited access to resources (Bandiaky-Badji 2011)—in the study area, access to land and security of tenure were often cited as important causes of women's

vulnerability, as most often women do not own land and therefore cannot make decisions regarding its use. They are also in danger of losing what access they have if they are abandoned, divorced, or widowed.

4. Dependence on rain-fed agriculture and natural resources—in the study area, women depend on the resources most at risk from climate such as rain-fed agriculture and the exploitation of natural resources for their subsistence. Projected climate changes such as increases in temperature and reductions in precipitation will change the availability of natural resources such as forests and fisheries and potentially affect the growth of staple crops. These changes will negatively affect women's adaptive capacity.

5. Limited access to education and to information (Lewark et al. 2011). In the study area, priority is still placed on boys' education. Girls suffer the consequences of resource shortages experienced by households as they are likely to be the first ones pulled out of school. As a result, girls receive fewer years of education than boys. Without education, women are in a disadvantaged position, as they have less access to crucial information and limited ability to interpret available information. This can affect their ability to understand and to act on information concerning climate risks and adaptation measures. As such, women's adaptive capacity is jeopardized. In addition, limited education further reinforces women's subordination relative to men (Appleton 1996).

6. Conflicting interests in natural resources management are often solved at the expense of women (Shackleton et al. 2011; Sun et al. 2011)—for instance, *moabi* (*Baillonnella toxisperma*) and caterpillar trees (*Triplochiton scheroxylon*) are highly valuable trees for men (timber) and for women (NTFPs). According to Ndzomo Abanda (2006), an adult *moabi* tree produces 327 kilos of fruit per year, from which women can extract 104.6 liters of oil for US$313. Yet, although sustainably harvested by women, these trees are the first to be cut in community forests.

Conclusion

The new Forestry Law promulgated in Cameroon in 1994 transferred to local councils and communities statutory rights, authority and responsibilities for acquiring and managing a share of forests and forest revenues. However, this policy is gender blind and provides only vague rules regarding the management of community and council forests. The law is concerned only with bureaucratic technicalities and the administrative relations between public powers (notably the Ministry of Forests) and the formal managers of the local forests. All the resulting management committees created are male dominated and are not organically rooted in the collective organization of the forest societies onto which they were grafted. This situation can encourage women's disinterest in public affairs. In addition, the lack of consideration of women's needs, priorities and aspirations in decision making in the 1994 Forestry Law has the potential to affect women's ability to adapt to climate change. This situation makes it imperative for policy makers to ensure that sectoral policies are gender sensitive. When policies are blind to gender realities, they are likely to reinforce existing inequality and inequity.

In Cameroon, efforts are being made at various levels to push gender equity in the decentralization agenda. Recent electoral processes have increased the inclusion of women in different political parties aiming to manage decentralized entities. Decision makers are more and more aware that (1) gender equity could help bring about gains in sustainability and can help ensure greater returns on investments in Sustainable Development Goals, and (2) gender equity in forest management could strengthen women's adaptive capacity and sustain households and communities. It is therefore possible to develop a true gender-integration policy that can both draw upon and strengthen the capacities women possess. For that, there is need for a political will at the national level, a strict application of ratified international conventions and a deliberate policy on quotas in decision-making contexts. In addition, women's access to information, public participation and justice should be promoted.

Acknowledgments

This study was embedded within an ongoing project by the Center for International Forestry Research (CIFOR). Started in July 2010, COBAM (Climate Change and Forests in the Congo Basin: Synergies between Adaptation and Mitigation) is a 5-year project funded by the African Development Bank and the Economic Community of Central African States under PACEBCo (*Programme d'Appui à la Conservation des Ecosystèmes du Bassin du Congo*).

The International Model Forest Network Secretariat (IMFNS) also kindly made their materials available to us. We express our special thanks to Gagoe Julie, Chimere Diaw and to two anonymous reviewers. Special thanks also go to the men and women of all the community forests investigated in our research.

References

Adger WN, Brooks N, Bentham G, Agnew M and Eriksen S. 2004. New indicators of vulnerability and adaptive capacity. Technical report 7. Final project report. Tyndall Centre for Climate Change Research. School of Environmental Sciences, University of East Anglia, Norwich, UK.

Agarwal B. 2001. Participatory exclusions, community forestry, and gender: An analysis for South Asia and conceptual framework. *World Development* 29(10): 1623–48.

Agrawal A and Chhatre A. 2006. Explaining success on the commons: Community forest governance in the Indian Himalaya. *World Development* 34(1): 149–66.

—— and Ribot J. 1999. Accountability in decentralization: A framework with South Asian and West African environmental cases. *The Journal of Developing Areas* 33: 473–502.

Appleton S. 1996. Women-headed households and household welfare: An empirical deconstruction for Uganda. *World Development* 24: 1811–27.

Bandiaky S and Tiani AM. 2010. Gendered representation and participation in rural decision-making in decentralized forest management: Case studies from Cameroon and Senegal. In German L, Karsenty A and Tiani AM, eds. *Governing Africa's forests in a globalized world*. London: Earthscan. 144–59.

Bandiaky-Badji S. 2011. Gender equity in Senegal's forest governance history: Why policy and representation matter. *International Forestry Review* 13(2): 177–94. doi:10.1505/146554811797406624.

Bele MY, Somorin OA, Sonwa DJ, Nkem JN, Locatelli B. 2011. Forests and climate change adaptation policies in Cameroon. *Mitigation and Adaptation Strategies for Global Change* 16: 369–85.

——, Tiani AM, Somorin OA and Sonwa DJ. 2013. Exploring vulnerability and adaptation to climate change of communities in the forest zone of Cameroon. *Climatic Change* 119: 875–89.

Betti JL. 2004. *Contribution à l'élaboration des règles de gestion de la zone communautaire du Parc National de la Lobeke.* Rapport consultation WWF-Jengi, Sud-Est Cameroun.

Bigombe Logo P. 2008. Foresterie communautaire et réduction de la pauvreté rurale au Cameroun: Bilan et tendances de la première décennie. *World RainForest Movement* 126.

Brown PHC. 2011. Gender, climate change and REDD+ in the Congo Basin forests of Central Africa. *International Forestry Review* 13(2): 163–76. doi:10.1505/146554811797406651.

Cheema GS and Rondinelli DA, eds. 1983. *Decentralization and development: Policy implementation in developing countries.* Beverly Hills, CA, London and New Delhi: Sage Publications.

—— and Rondinelli DA, eds. 2007. *Decentralizing governance: Emerging concepts and practices.* Washington, DC: Brookings Institution Press and Ash Institute for Democratic Governance and Innovation.

Chia EL, Somorin AO, Sonwa DJ and Tiani AM. 2013. Local vulnerability, forest communities and forest-carbon conservation: Case of southern Cameroon. *International Journal of Biodiversity and Conservation* 5(8): 498–507.

Crewe E and Harrison E. 1998. *Whose development? An ethnography of aid.* London: Zed Books.

Crook RC and Manor J. 1998. *Democracy and decentralization in South East Asia and West Africa: Participation, accountability and performance.* Cambridge: Cambridge University Press.

Devisscher T, Bharwani S, Tiani AM, Pavageau C, Kwack NE and Taylor R. 2013. Current vulnerability in the Parc Tri-National de la Sangha landscape, Cameroon. Working Paper 107. Bogor, Indonesia: Center for International Forestry Research.

Diaw MC. 2010. Elusive meanings: Decentralization, conservation and local democracy. In German L, Karsenty A and Tiani AM, eds. *Governing Africa's forests in a globalized world.* London: Earthscan. 56–78.

Djoudi H and Brockhaus M. 2011. Is adaptation to climate change gender neutral? Lessons from communities dependent on livestock and forests in northern Mali. *International Forestry Review* 13(2): 123–35. doi: 10.1505/146554811797406606.

Dube S. 2014. *Support women to build their adaptive capacity and resilience. Conflict-sensitive adaptation: Use human rights to build social and environmental resilience.* Brief 9. Gland, Switzerland: Indigenous Peoples of Africa Co-ordinating Committee and IUCN Commission on Environmental, Economic and Social Policy.

Flora CB and Flora JL. 2008. *Rural communities: Legacy and change*, 3rd edition. Boulder, CO: Westview Press.

Francis P and James R. 2003. Balancing rural poverty reduction and citizen participation: The contradictions of Uganda's decentralization program. *World Development* 31(2): 325–37.

German LA, Tiani AM, Daoudi A, Maravanyika TM, Chuma E, Jum C and Yitamben G. 2010. *The application of participatory action research to climate change adaptation in Africa.* Reference guide. Bogor, Indonesia: International Development Research Centre and Center for International Forestry Research.

IPCC (Intergovernmental Panel for Climate Change). 2007. Contribution of Working Group II to the Third Assessment Report of the Intergovernmental Panel on Climate

Change. In Parry ML, Canziani OF, Palutikof JP, van der Linden PJ, Hanson CE, eds. *Climate change 2007: Impacts, adaptations and vulnerability.* Cambridge: Cambridge University Press.

IUFRO (International Union of Forestry Research Organization). 2010. Making African forests fit for climate change: A regional view of climate-change impacts on forests and people, and options for adaptation. Vienna, Austria: IUFRO.

Karsenty A. 1999. Vers la fin de l'état forestier? Appropriation des éspaces et partage de la rente forestière au Cameroun. *Politique Africaine* 75: 5–106.

——, Mebenga LM and Pénélon A. 1997. Spécialisation des espaces ou gestion intégrée des massifs forestiers? *Bois et Forêts des Tropiques* 251(1): 43–54.

Kasomo D. 2012. Factors affecting women's participation in electoral politics in Africa. *International Journal of Psychology and Behavioral Sciences* 2(3): 57–63.

Kouna Eloundou CG. 2012. Décentralisation forestière et gouvernance locale des forêts au Cameroun: Le cas des forêts communales et communautaires dans la Région Est. Thèse pour obtenir le grade de Docteur de l'Université du Maine, Maine, France.

Lapelle PR, Smith PD and McCool SF. 2004. Access to power and genuine empowerment? An analysis of three community forest groups in Nepal. *Human Ecology Review* 11(1): 1–12.

Larson AM. 2005. Democratic decentralization in the forestry sector. Lessons learned from Africa, Asia and Latin America. In Colfer CJP, Capistrano D, eds. *The politics of decentralization: Forests, power and people.* London: Earthscan, Center for International Forestry Research. 32-62.

Lewark S, George L and Karmann M. 2011. Study of gender equality in community based forest certification programmes in Nepal. *International Forestry Review* 13(2): 195–204. doi: 10.1505/146554811797406633.

Mamdani M. 1996. *Citizen and subject: Contemporary Africa and the legacy of late colonialism.* Princeton, NJ: Princeton University Press.

Manor J. 1999. *The political economy of democratic decentralization.* Washington, DC: The World Bank.

Mawhood P. 1993. Decentralization: The concept and the practice. In Mawhood P, ed. *Local government in the Third World: Experience of decentralization in tropical Africa.* 2nd edn. Johannesburg: Africa Institute of South Africa.

Mbotto-Ndomi AA, Mandigou E and Ngonde B. 2007. La contribution de la redevance forestière au développement local: Les communes peuvent-elles faire mieux lorsqu'elles sont responsabilisées? In Dongmo Tsobzé A, Hiehorst T, Mfou'ou JM, eds. *Entre désespoir et espoir: Les défis de la gouvernance et de la décentralisation dans les réalisations des investissements publics locaux au Cameroun.* Yaoundé, Cameroun.

MEA (Millennium Ecosystem Assessment). 2005. *Ecosystem and human well-being: Our human planet. Summary for policy makers.* Washington, DC: Island Press.

Meinzen-Dick R, Brown L, Feldstein H and Quisumbing A. 1997. Gender, property rights, and natural resources. *World Development* 25(8): 1303–15.

Mwangi E, Meinzen-Dick R and Sun Y. 2011. Gender and sustainable forest management in East Africa and Latin America. *Ecology and Society* 16(1): 17.

Ndzomo Abanda G. 2006. Utilisation des produits forestiers non ligneux dans le Dja. In *Home-Grown Plus. Connaissances locales, bien-être des populations, politiques et sciences: Les actes de l'atelier régional sur l'agroforesterie dans le landscape du Dja, Cameroun.* Lomié, 21–22 Juin 2006. World Agroforestry Centre, West and Central Africa, Humid Tropics, Yaoundé, Cameroon. 28–9.

Ostrom E, Schroeder LD and Wynne SG. 1993. *Institutional incentives and sustainable development: Infrastructure policies in perspective.* Boulder, CO: Westview Press.

Otzelberger A. 2011. *Gender-responsive strategies on climate change: Recent progress and ways forward for donors.* Brighton: BRIDGE/Institute of Development Studies.

Oyono PR. 2004. *Institutional deficit, representation and decentralized forest management in Cameroon: Elements of natural resource sociology for social theory and public policy.* Washington, DC: World Resources Institute and Center for International Forestry Research

Pitkin HF. 1967. *The Concept of Representation.* Berkeley: University of California Press.

Plumwood V. 1992. Feminists and ecofeminists: Beyond the dualistic assumptions of women, men and nature. *The Ecologist* 22: 8–13.

Porro R, Tiani AM, Tchikangwa B, Sardjono MA, Salim A, Colfer CJP and Brocklesby MA. 2001. Access to resources in forest-rich and forest-poor contexts. In Colfer CJP and Byron Y. eds. *People managing forests: The links between human well-being and sustainability.* Washington, DC: Resources for the Future and CIFOR. 250–73.

Ribot JC. 2002. *La décentralisation démocratique des ressources naturelles, institutionnaliser la participation populaire.* Washington, DC: World Resources Institute.

——. 2006. Décentralisation démocratique des ressources naturelles: Choix institutionnels et transferts de pouvoirs discrétionnaires en Afrique sub-saharienne. In Bertrand A, Montagne P and Karsenty A. eds. *L'état et la gestion locale durable des forêts en Afrique Francophone et à Madagascar.* Paris: L'Harmattan.

—. 2007. *Dans l'attente de la démocratie: La politique des choix dans la décentralisation de la gestion des ressources naturelles.* Washington, DC: World Resources Institute.

RoC (Republic of Cameroon). 1994. Law No 94/01 of 20 January 1994 establishing forestry, wildlife and fisheries regulations. Presidency of the Republic, Yaoundé, Cameroon.

——. 1995. Decree No. 95/531/PM of 23 August 1995 establishing the modalities for the implementation of forestry regulations. Prime Ministry, Yaoundé, Cameroon.

——. 1996. Law No. 96/8 of 1 July 1996 establishing the Finance Law of Cameroon for 1996/1997. National Assembly, Yaoundé, Cameroon.

Shackleton S, Paumgarten F, Kassa H, Husselman M and Zida M. 2011. Opportunities for enhancing poor women's socioeconomic empowerment in the value chains of three African non-timber forest products (NTFPs). *International Forestry Review* 13(2): 136–51. doi: 10.1505/146554811797406642.

Shanley P, Da Silva FC and Macdonald T. 2011. Brazil's social movement, women and forests: A case study from the National Council of Rubber Tappers. *International Forestry Review* 13(2): 233–44.

Smit B and Pilifosova O. 2003. From adaptation to adaptive capacity and vulnerability reduction. In Smith JB, Klein RJT and Huq S, eds. *Climate change, adaptive capacity and development.* London: Imperial College Press.

Somorin OA, Visseren-Hamakers IJ, Arts B, Sonwa DJ and Tiani A-M 2014. REDD+ policy strategy in Cameroon: Actors, institutions and governance. *Environmental Science & Policy* 35(0): 87–97.

Sonwa DJ, Bele MY, Somorin OA, Nkem J. 2010. Central Africa is not only carbon stock: Preliminary efforts to promote adaptation to climate change for forest and communities in Congo Basin. In Bojang F, ed. *Nature and Fauna: Enhancing Natural Resources Management for Food Security in Africa* 25(1).

——, Nkem JN, Idinoba ME, Bele MY and Cyprain J. 2012a. Building regional priorities in forests for development and adaptation to climate change in the Congo Basin. *Mitigation and Adaptation Strategies for Global Change* 17(4): 441–450. doi:10.1007/s11027-011-9335-5.

——, Somorin OA, Jum C, Bele MY and Nkem JN. 2012b. Vulnerability, forest-related sectors and climate change adaptation: The case of Cameroon. *Forest Policy and Economics* 23: 1–9.

Steel BS and Weber E. 2003. Ecosystem management, decentralization, and public opinion. *Global Environmental Change* 11(2): 119–31.

Stern Review. 2006. *Stern Review final report on the economics of climate change.* Cambridge and New York: Cambridge University Press.

Sun Y, Mwangi E and Meinzen-Dick R. 2011. Is gender an important factor influencing user groups' property rights and forestry governance? Empirical analysis from East Africa and Latin America. *International Forestry Review* 13(2): 205–19.

Tiani AM, Akwah G and Nguiebouri J (2005). Women in Campo-Ma'an National Park: Uncertainties and adaptations in Cameroon. In Colfer CJP, ed. *The equitable forest.* Washington, DC: Resources for the Future and CIFOR. 131–49.

Tobith C and Cony P. 2006. Gender issues and community forests in Cameroon: Perspectives for women. *Bois et Forêts des Tropiques* 289(3): 17–26.

Torri MC. 2010. Power, structure, gender relations and community-based conservation: The case study of the Sariska region, Rajasthan, India. *Journal of International Women's Studies* 11(4): 1–18.

UNDP (United Nations Development Programme). 2008. *Resource guide on gender and climate change.* New York: UNDP. http://www.un.org/womenwatch/downloads/Resource_Guide_English_FINAL.pdf.

———. 2010. *Gender, climate change and community-based adaptation.* New York: UNDP.

WEDO (Women's Environment and Development Organization). 2007. Changing the climate: Why women's perspectives matter. http://www.wedo.org/wp-content/uploads/changing-the-climate-why-womens-perspectives-matter-2008.pdf.

Wellstead AM, Steadman RC and Parkins JR. 2003. Understanding the concept of representation within the context of local forest management decision making. *Forest Policy and Economics* 5(1): 1–11.

8

GENDER AND VULNERABILITY TO MULTIPLE STRESSORS, INCLUDING CLIMATE CHANGE, IN RURAL SOUTH AFRICA

Sheona Shackleton and Leigh Cobban

A woman is brave enough to hold a knife even on the side where it cuts.

(Pedi proverb)

Introduction

The world is facing an era of accelerated change and greater uncertainty brought about by multiple powerful and interacting drivers and shocks. Levels of insecurity and vulnerability are rising, driven by factors such as the global economic downturn, rising corruption and weak governance, escalating poverty and food insecurity, failing health systems, and more extreme weather events amongst others (IPCC 2013). At the same time, the longevity of the goods and services provided by the planet's ecosystems is questioned (Rockström et al. 2009), with climate change merely one facet of global environmental change. These global risks and changes interact with, and may be compounded by, country-level structural, economic and political processes (e.g. poorly conceived national policies and corruption) that create and sustain inequities (Ribot 2014), and by localised contextual dynamics including institutional breakdown, declining human health through diseases such as HIV/AIDS, forest degradation and biodiversity loss (Fraser et al. 2011). Such situations result in differentiated vulnerability and adaptive capacity amongst households and individuals at the local level. This heterogeneity is further aggravated by unequal access to livelihood options and assets (Goh 2012), with gender differences often being particularly stark. Consequently, many commentators argue that there is an urgent need, especially in the growing field of climate change adaptation research, for more studies that capture the distinct, localised and interactive effects of multiple stressors on livelihoods, vulnerability and adaptive capacity across gender and other intersecting social categories such as age, ethnicity, income and class (Carr et al. 2013; Drimie and Gillespie 2010; Goh 2012).

In this chapter, we investigate differentiated vulnerability and responses, amongst men and women and households with different gender structures in two different local contexts, paying attention to the many stressors that influence local livelihoods, especially at the nexus of HIV/AIDS and climate change. Specifically, we consider gender-differentiated perceptions, perspectives and 'lived experiences' of vulnerability, including various influences on both generic and specific adaptive capacity (Eakin et al. 2014), as well as the common types of responses to shocks and stressors adopted by men and women. We reflect on whether particular types of female-headed households and women are more vulnerable and less able to cope and adapt (Arora-Jonsson 2011).

We begin below by defining vulnerability and its links to gender as used in this chapter. We then provide background on the South African context in relation to multiple stressors focussing on HIV/AIDS and climate change, and briefly describe the study sites. This is followed by a section that provides information on our multiple data sources and methods. We then present results and insights from the study through a gendered analysis of: (a) household livelihoods and asset holdings, including land (as indicators of sensitivity and adaptive capacity), (b) perceptions and experiences of vulnerability and food security in relation to multiple stressors (exposure to risk), and (c) types of responses employed when faced with shocks and stresses, including those based on ecosystem services. We present discussion alongside our results, highlighting the complexities associated with determining vulnerability amongst heterogeneous households in different contexts, before offering conclusions on the policy implications of the findings.

Conceptualising vulnerability

Drawing on both risk-hazard perspectives (that locate vulnerability within external risk) and entitlements-livelihoods and political ecology perspectives (that trace vulnerability to multiple social, political and economic factors at different scales; Ribot 2014), we consider vulnerability as consisting of exposure to multiple shocks and stressors, the susceptibility to harm from these (sensitivity), and the capacity to respond and recuperate from such adverse impacts (adaptive capacity; O'Brien et al. 2009). This combination of underlying cause and susceptibility in the conceptualisation of vulnerability is important when analysing how socially constructed factors such as gender influence people's or households' ability to respond to stressors. From a multiple-stressor perspective, vulnerability is increasingly being analysed through the lens of integrated social-ecological systems, as the availability of ecosystem services affects the vulnerability of society while at the same time society can positively or negatively influence the vulnerability of ecosystems (Fraser et al. 2011; Shackleton and Shackleton 2012). Likewise, there is growing appreciation that some groups may be more vulnerable than others to particular combinations of stressors; influenced by *inter alia* the livelihood activities they engage in, their assets, the sociocultural norms and institutions that structure their rights, roles and responsibilities and that influence power relations, and their access to knowledge and information (Goh 2012). These areas need more attention if vulnerability studies are to capture the complexities and nuances of local reality.

Setting the context: interactions between multiple stressors, ecosystem services and gender

South Africa has deeply ingrained inequality, largely due to its history of segregation (Özler 2007), with the former 'homelands' in rural areas remaining under-developed in comparison with the previous 'white' South African areas (Bank and Minkley 2005). Rural areas in South Africa are home to 36 per cent of the total population, yet 59 per cent of all poor individuals in the country live in these areas (Armstrong et al. 2008). As well as being highly racialised, this inequality takes on a gendered dimension. In South Africa, as elsewhere, women are more vulnerable due to various social norms and institutions (Goh 2012). Forty-five per cent of all female-headed households lived below the 'lower-bound' poverty line in 2007, compared to 25 per cent of male-headed households (Armstrong et al. 2008). Women make up a higher proportion of the unemployed and generally receive lower wages as a result of lower skills and education (CGE 2010; Oxfam 2014). Despite earning less, women take on responsibilities such as caregiving within the home, maintaining the household, and securing food, whether through trade, cultivation, or collection (CGE 2010; Goh 2012). Linked to this, rural women are closely reliant on ecosystem services. Indeed, 61 per cent of agriculturalists in South Africa are women (Ruiters and Wildschutt 2010). They are also the primary harvesters of natural resources for subsistence purposes, especially fuelwood and wild foods (Shackleton and Shackleton 2004), as well as the collection of water (Aggarwal et al. 2001).

Rural women's significant reliance on ecosystem services for food security, through agricultural production and/or natural resource harvesting, could place them at greater risk to the negative effects of climate change, although men as livestock farmers are also threatened. South Africa is currently warming at about twice the average global rate. This trend is projected to continue leading to most of the inland landscape becoming drier, while heavier rainfall is anticipated along the eastern coastline over the next fifty years (Taylor 2009; Scholes 2011). These trends are predicted to result in a decrease of 38–55 per cent in the country's biomes, reducing plant species and potential for natural resource harvesting (Turpie et al. 2002). These changes will also impact agricultural production; for instance, the production of maize, the staple for the majority of the population, could decrease by 30 per cent by 2030 (Lobell et al. 2008). Livestock are an important source of income, food and security for rural households (Shackleton et al. 2005). As air temperatures approach the body temperature of livestock, the production of milk and the animal's reproduction rate decrease (Scholes 2011). Bush encroachment into rangelands can also decrease the amount of grazing land available (Turpie et al. 2006).

The strain on ecosystem services has ramifications for food security and women's human capital in terms of time, health and labour, particularly in rural households, which rely more on crop production, livestock, fuelwood for energy, and other natural resource harvesting to sustain livelihoods or as safety nets in the face of stress (Shackleton and Shackleton 2004; Shackleton et al. 2009; Goh 2012). In South Africa, some 37 per cent of the rural poor experience hunger, while 32 per cent are at risk of hunger. In general, this is highest amongst female-headed households (21 per cent) and in the Eastern Cape (Oxfam 2014). Food security is vital to support a functioning immune system to decrease the risk of disease and infection, and so is key when considering HIV/AIDS (Gillespie and Drimie 2009).

South Africa is home to the largest population of people living with HIV/AIDS in the world (UNAIDS 2009), with prevalence higher among women (AVERT 2011). Women are biologically more susceptible to infection, as well as more vulnerable through unequal power and chronic gender-based violence (Gillespie and Drimie 2009). South Africa has the highest rape rate in the world, with 40 per cent of reported cases being victims under the age of 18 (CGE 2010). Male migrancy has also increased the risk of contracting HIV/AIDS amongst married women (Babuguru 2010). Furthermore, women are vulnerable to the secondary effects of HIV/AIDS, often caring for the sick, and with older women often caring for orphans (Makiwane and Chimere-Dan 2010). HIV/AIDS can also affect household food entitlements both because a sick household is less productive and less capable of producing its own food, or of earning an income, and because income must be diverted towards health-related expenses (Gillespie and Drimie 2009).

Description of study sites

We selected two sites for this study to allow us to unpack the contextual factors that may influence vulnerability in different settings. These sites included Gatyana in Mbashe Local Municipality and Lesseyton in Lukanji Local Municipality in the Eastern Cape Province, South Africa (Figure 8.1). Both sites are typical of the province, which has one of the highest unemployment rates in the country, low income levels, poor education, inadequate infrastructure and service delivery (Makiwane and Chimere-Dan 2010) and a high prevalence of HIV/AIDS (AVERT 2011). It is, thus, considered one of the most vulnerable regions to climate change (DST 2010). The study sites were chosen to correspond to a gradient of increasing rainfall and decreasing accessibility to urban amenities and markets; all of which can influence livelihoods and vulnerability variously. The inland, peri-urban site, Lesseyton, is 20 kilometres from the city of Queenstown, and is semi-arid with a mean annual precipitation of between 350 and 500 millimetres. Homesteads are built in a typical village layout and generally have only a small garden space for vegetable cultivation. There is access to communal land, used primarily for grazing by livestock owners. The coastal site, Gatyana, is about 30 kilometres from the town of Willowvale. It still lags behind in infrastructure such as tarred roads, and has marginal local markets and poor transport systems. Most homesteads are widely dispersed with large plots of land for subsistence cultivation. The use of more distant fields has declined dramatically in recent years (Shackleton et al. 2013), although livestock grazing takes place in these and surrounding communal rangelands. The site is wetter, receiving a mean annual rainfall of between 950 and 1100 millimetres.

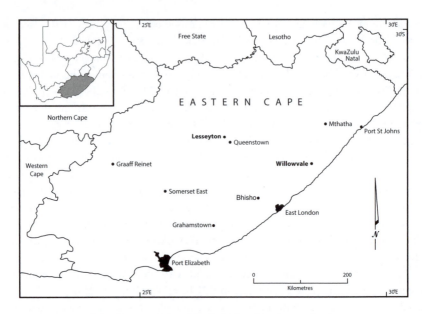

FIGURE 8.1 Map of the Eastern Cape, South Africa showing study sites

Households in both sites derive cash and other direct household contributions ('in kind income') from a variety of sources such as grants, formal and casual labour, self-employment, natural resource harvesting, crops and livestock, with the latter being more prominent in rural Gatyana, while employment is a key source of income in Lesseyton. Commercial farming is practically non-existent in both sites, while state welfare grants form the most widespread source of income. The proportion of male- and female-headed households is similar in both sites, at around 50 per cent in each category, although Gatyana has double the number of households with only adult females relative to Lesseyton (28.4 per cent versus 14.7 per cent).

Approach and methods

Our analysis draws on and synthesises a suite of gender related data from sub-studies undertaken by postgraduates who employed both qualitative and quantitative methods in their research (see Abu-Basutu 2013; Clarke 2012; Ndlovu 2012; Stadler 2012; Trefry 2013). Specifically, we use data from a baseline survey (Stadler 2012), a survey targeted specifically at poorer (more vulnerable) households[1]

---

1 Households were purposively selected by comparing their physical attributes. The presence and size of livestock pens, the physical appearance/condition of the house and the presence of costly assets (fences, vehicles, satellite dishes) were used as wealth indicators; these households were avoided. Fifty households per site were interviewed. Life histories were targeted at the most vulnerable of these households.

(Clarke 2012), focus-group discussions linked to a social learning process (Cundill et al. 2014), participatory workshops with different gender groups (Stadler 2012; Clarke 2012), andtwenty life histories narrated by vulnerable men and women (Clarke 2012). We briefly describe these methods below, with further details in the tables.

In early 2011, we administered a household survey to 170 randomly selected households in each site, and collected information on household demographics, health and HIV/AIDS proxy indicators, livelihood assets (social, natural, financial, human and physical capital), welfare perceptions, shocks and responses, livelihood activities and household income (both cash and in kind) from farming and gathering. Heads of households or the next most senior adult were interviewed. For this chapter, we compare results for households with different gender structures going beyond simple male- and female-headed household dichotomies, as these are often insufficient for capturing heterogeneity, especially among female-headed households (Carr et al. 2013; Fuwa 2000; Klasen et al. 2015). Indeed, the empirical relationship between female headship and vulnerability remains variable and contentious (Klasen et al. 2014). Furthermore, headship is only one aspect of household structure (Fuwa 2000); households also vary in terms of the gender and ages of other members, with implications for production and adaptive capacity. For instance, some of the external gendered constraints faced by female-headed households, e.g. limited property rights, may be alleviated by adult male presence in the household. Conversely, adult males in female-headed households may impose internal constraints on innovation through reinforcement of restrictive gender norms. Male-only households similarly may face a unique set of constraints related to, for example, food production and, in South Africa, access to grants (Dubbeld 2013). In recognition of this, we considered four household gender categories based on the sex of the adults in the household, with an adult being those over 18. Adult members were considered those people who mostly ate at home or regularly supplied remittances. The categories were: households with only male adults, male-headed households with adult females (typical family structure), female-headed households with adult males and households with only female adults.

In addition to the survey, participatory mental maps were used in both sites to record the stressors believed by local men and women to be creating vulnerability, and their perceptions of their internal linkages (Clarke 2012). We held four small focus groups (five participants each) with men and women separately. Mental maps are 'qualitative representations of a system consisting of variables and the causal relationships between them' (Bunce et al. 2010, 414). For our study, these took the form of a spider-gram drawn on large sheets of paper, with key stressors being linked via directional lines. HIV/AIDS was suggested to participants as an initial item, to which they added the various causes and effects of this disease. They then added further stressors and began connecting these, until they felt they had exhausted all the key factors contributing to their vulnerability. All links and decisions were discussed.

Parallel to the research, an ongoing participatory process grounded in social learning theory was underway in both sites with mixed groups of ten to fifteen people. This process aimed to build problem-solving capacity and strengthen local practices that had evolved in response to multiple stressors (see Cundill et al. 2014). In some of the foundational meetings, we asked participants what it meant to be vulnerable in the local context, and to share stories of vulnerability based on their experiences. These discussions, as well as twenty life histories with vulnerable households undertaken by Clarke (2012), added layers of nuance to our understanding of the quantitative findings.

The South African Rand-US Dollar exchange rate at the time of the study averaged about R7 to US$1 (now double that).

Results and discussion

Household livelihoods and asset holdings

Livelihood and income sources

Households with only female adults had the lowest mean quarterly income (Table 8.1) in Lesseyton compared to other headship types. In Gatyana, a decline in income was evident as household structure shifted from male to female dominated. These results suggest that female-headed households have less income and lower access to employment (Table 8.2), often a result of women's role as domestic carers, inequality in job access and lack of opportunities locally (CGE 2010).

Most households across all gender categories in both sites relied on government grants as their primary source of income (Table 8.1). The exception was male-headed households with adult females in peri-urban Lesseyton, who derived, on average, 45.3 per cent of household income from formal employment. By contrast, households with only female adults in Lesseyton had the lowest proportion of income from formal employment (11.7 per cent), but the highest in the informal sector (10.3 per cent). In Gatyana, on the other hand, households with only male adults earned a larger proportion of household income from formal employment (28.7 per cent), whilst male-headed households with adult females derived the lowest proportion of income from this source (6.5 per cent) and the largest from grants (63.8 per cent). In a subset of poorer households, female-headed households showed lower levels of employment than male-headed households in the settlement close to Queenstown, whereas this difference was less distinct for the rural setting probably due to fewer employment opportunities (Table 8.2). Overall, while means-tested social grants can help to protect against the impacts of climate and other shocks by providing a safety net, they are generally insufficient (current value of a pension is R1,410, and child grant, R330) to offer more than just a means to cope (Clarke 2012; Stadler 2012), particularly amongst female-headed households, with low access to other cash sources:

I was living with my mother-in-law who was blind. She was getting a social grant from the government which assisted us to get food and other necessities. She then passed away. Without her social grant we struggled even more. Still my husband couldn't get permanent employment. He got a job as a casual in the construction industry in Saldana. He would send us whatever he could afford R300 or R400 per month. Time went by and my husband returned home as he was ill. He passed away three years back. Things got a little bit better when I received a social grant for the younger children. Children attend school, they need uniforms, shoes and books; all must come from the child social grant.

(Gatyana, female, 52 years old)

Bank and Minkley (2005) and Dubbeld (2013) similarly recognised that the grant system alone is unable to change rural people's lives, arguing that it is not designed to lift a significant proportion of individuals out of poverty. Moreover, old-age pension and child support grants are not secure, being lost when the pensioner dies or the child matures.

In Lesseyton, the proportion of household income derived from livestock was higher in male-headed than female-headed households (from 9.2 to 0.3 per cent; Table 8.1). Interestingly, in Gatyana, households with only female adults earned a similar proportion of income from livestock (10.1 per cent) relative to other gender categories, possibly due to wider ownership and widows inheriting animals. In both sites, female-headed households obtained double the amount of income from remittances (8.1 and 7 per cent) compared to male-headed households (3.2 and 4.3 per cent).

We argued earlier that ecosystem services are of crucial importance for women and female-headed households in South Africa. Our findings, however, show that in these Eastern Cape sites, similar marginal income is derived from crops and wild natural resources across the different household gender categories. But natural resources form a slightly higher share of total income for female-headed households (Table 8.1), and more especially female-only households (despite these consuming minimal bushmeat). Specifically selecting poorer households, a higher proportion of female-headed households were found to be more actively engaged in growing home gardens and using forest products (Clarke 2012) than male-headed households. This suggests that any climate change impacts on ecosystem services could impact women's vulnerability, threatening one of their primary means of coping, especially for female-headed households and those without adult males. That said, both men and male and female-headed households dependent on livestock are also likely at risk from climate change.

This year – yes we did experience drought. Livestock died. Those who have money buy feed for their stock.

(Lesseyton, female, 77 years old)

TABLE 8.1 Mean quarterly household income (ZAR) for different income sources and mean percentage contribution (in parentheses) of this income to total income (cash and in kind), disaggregated by household gender category in Lesseyton and Gatyana[#]

Lesseyton

	Only male (N = 47) (mean hh size = 5.5)	Male-headed with female (N = 45) (mean hh size = 4.3)	Female-headed with male (N = 53) (mean hh size = 5.9)	Only female (N = 25) (mean hh size = 4.2)	★ = significant at p <0.1 NS = not significant
Total income					
Household	6952	7869	7926	4905	★
per capital	1268	1830	1343	1167	★
Grants	3445	2272	3903	2920	★
	(49.6)	(28.8)	(49.2)	(60.0)	
Formal employment	1433	3566	1928	573	★
	(20.6)	(45.3)	(24.3)	(11.7)	
Casual employment	734	616	278	288	★
	(10.6)	(7.8)	(3.5)	(5.9)	
Self-employment	214	350	168	504	NS
	(3.1)	(4.5)	(2.1)	(10.3)	
Remittances	221	342	641	342	★
	(3.2)	(4.3)	(8.1)	(7.0)	
Crops	28	5	3	36	NS
	(0.4)	(0.1)	0	(0.7)	
Livestock	640	473	410	13	NS
	(9.2)	(6.0)	(5.2)	(0.3)	
Natural resource use	225	231	556	228	NS
	(3.2)	(2.3)	(7.0)	(4.6)	

Gatyana

	Only male (N = 36) (mean hh size = 5.5)	Male-headed with female (N = 41) (mean hh size = 5.3)	Female-headed with male (N = 43) (mean hh size = 4.9)	Only female (N = 48) (mean hh size = 4.1)	★ = significant at p <0.1 NS = not significant
Total income					
Household	9157	8162	7005	5505	★
per capita	1665	1540	1429	1343	
Grants	4011	5207	3784	2804	★
	(43.8)	(63.8)	(54)	(50.9)	
Formal employment	2630	526	837	1000	NS
	(28.7)	(6.5)	(12)	(18.2)	
Casual employment	172	134	252	31	★
	(1.9)	(1.7)	(3.6)	(0.6)	
Self-employment	135	357	286	140	NS
	(1.5)	(4.4)	(4.1)	(2.5)	
Remittances	305	115	576	354	NS
	(3.3)	(1.4)	(8.2)	(6.4)	
Crops	54	169	123	44	★
	(0.6)	(2.1)	(1.7)	(0.8)	
Livestock	993	1115	578	556	NS
	(10.9)	(13.7)	(8.2)	(10.1)	
Natural resource use	853	54	546	553	NS
	(9.3)	(6.6)	(7.8)	(10.0)	

Natural resources included all harvested local products such as fuelwood, palm leaves, rocky shore molluscs, fish, thatch and wild foods. Most were used for direct household consumption. Self-employment relates to informal sector production and trading activities. Depending on the normality of the data either an ANOVA or Kruskal Wallis test was used to determine significant differences between household gender categories.

TABLE 8.2 Employment amongst the most vulnerable households in the study sites

Employment amongst hh members (most vulnerable hh)	Lesseyton % (n = 50)		Willowvale % (n = 50)	
	Male-headed (n = 27)	Female-headed (n = 23)	Male-headed (n = 23)	Female-headed (n = 27)
Unemployed	42	74	87	88
Employed	58	26	13	12
Employment disaggregated further				
Full-time	29	4	13	4
Part-time	29	22	0	4
Self-employed	0	0	0	4

Source: Clarke 2012.

Asset access and stocks

Measures of household stocks of human, social, financial and physical capital revealed little difference between household types (Stadler 2012). For human capital, no significant differences were found between various measures (adults per household, education, language skills, health) in either site, with the exception that female-only households had the fewest adults, higher child dependency, lower health scores and, in Lesseyton specifically, fewer additional skills (e.g. English proficiency). Regarding physical capital, homes were of similar size and quality (based on construction materials) across all household types. For widows, this likely reflects investment in their homes prior to their husbands' deaths. The value of large household assets (measured as a set of 14 items and priced according to current sales value), however, was lower for female-only households in both sites. The results for financial capital, measured as household savings, mirror this finding, being lower in female-headed households, although balanced by fewer debts. In terms of social capital, female-only households in Lesseyton had a slightly higher score for cognitive social capital (trust, social cohesion, reciprocity) than other household categories (Table 8.3), whereas there were no differences in structural social capital (group membership, leadership, access to advice) between household categories or sites. For the former, the difference in score can be attributed mainly to a larger proportion of female-only households agreeing or agreeing strongly with items related to mutual trust and 'counting on their neighbours'. This was supported by qualitative results, which demonstrated that women often undertake collective coping activities, especially related to caring for the ill, gardening, and other small income-generating and food security projects such as chicken farming (Trefry 2013). Social capital has been associated with the exchange of resources, collective action and innovative thinking and so is seen as critical for enhancing adaptive capacity (Pelling and High 2005).

Male-headed households, consistent with the literature, have larger plots than female-headed households on both a household and per capita basis (Table 8.4).

TABLE 8.3 Mean cognitive social capital scores,[#] disaggregated by household gender category in Lesseyton and Gatyana

		Lesseyton	Gatyana
Only male adults	Mean	20.9	22.2
	N	47	37
Male-headed with female adults	Mean	20.7	22.0
	N	45	41
Female-headed with male adults	Mean	20.9	22.7
	N	53	43
Only female adults	Mean	21.2	22
	N	25	48
★ = significant at p < 0.1		★	★

Scores were calculated from seven Likert-scale items related to community trust, reciprocity and cohesion (e.g. 'People in this neighbourhood can be trusted'; 'If I had to borrow R50 in an emergency my neighbours would help'; 'People in this neighbourhood generally get along'). Responses were weighted thusly: strongly disagree = 1, disagree = 2, agree = 3; agree strongly = 4. A total score for all statements was obtained: maximum score, 28; minimum, 7; mode, 21.

In Lesseyton, households with only female adults had a considerably smaller mean garden size than households with only male adults (47 versus 259 square metres). In Gatyana, the two female-headed household categories had roughly 2000 square metres less land than male-headed households (around 4,000 versus 6,000 square metres). This inequity in respect to land access in South Africa is an ongoing concern, as widows and divorcees frequently lose rights to land, while single and unmarried women are often never granted these rights or receive less land than men (Cousins 2010). This suggests that the smaller land areas owned by female-headed households, even if not fully used presently, may limit their ability to increase food security and diversify their farming in the future, especially if food prices continue to escalate with a changing climate (Vermeulen 2014).

TABLE 8.4 Mean garden area[#] (in square metres) of households and per capita (in parentheses), disaggregated by household (hh) gender category in Lesseyton and Gatyana

		Lesseyton	Gatyana
Only male adults	Mean area per	259.3 (47.1)	6018.5 (1094.2)
	hh (per capita)	46	35
	N		
Male-headed with female adults	Mean	123.9 (28.8)	6303.5 (1189.2)
	N	45	40
Female-headed with male adults	Mean	119.0 (20.2)	3885.1 (792.9))
	N	50	40
Only female adults	Mean	47.1 (11.2)	4418.5 (1077.6)
	N	25	47
★ = significance at p < 0.1, NS = not significant		★	NS

Garden sizes were measured by enumerators and converted to area.

Overall, the findings suggest that the most definitive gendered livelihood differences relate to flows of income rather than capital stocks (except land which is limited for female-headed households). This results in female-headed households being potentially more vulnerable as they have lower access to cash resources and fewer options to boost these or their food security through, for instance, intensifying and expanding arable production. Consequently, their coping responses may be limited to those requiring few inputs, and are often climate sensitive, such as natural resource harvesting and gardening. However, due to more evidence of cooperation and higher levels of self-employment, our results also suggest that women and female-headed households may be more ready than men to modify their behaviour and adopt new strategies in relation to future uncertainty. Indeed, in focus-group discussions, men tended to indicate the lack of jobs and job insecurity as the main causes of vulnerability and saw few alternatives to these jobs.

Perceptions and experiences of shocks, stressors, vulnerability and food security

Multiple stressors and vulnerabilities

The gender-mixed social learning groups in both sites identified rape and violence against women as defining features of vulnerability in their communities. Women were also depicted (by both men and women) as being susceptible to sudden loss of income through the household breadwinner's death. These stories focused on the difficulty new female-headed households had in maintaining authority over children, who often turned to prostitution or stealing. The same concern was emphasised in a separate mental map activity where men's and women's groups worked independently (Figure 8.2). Transactional sex is no longer unusual as a coping strategy amongst young women in marginalised settlements in South Africa (Hunter 2010). In addition to gender-related crimes and prostitution, both male and female participants identified other social drivers of vulnerability in their communities: a lack of education, low levels of knowledge and skills, corruption, theft, poor policing, poor health care, dependence on others, and heavy drug and alcohol consumption (Figure 8.2). Both groups also saw climate-related uncertainty and shocks as an important characteristic of local vulnerability. For example, one narrative flowing through the discussions was how women used to rely on their crops for food and on remittances for other purchased goods. However, they can no longer do this because of perceived increased frequency of drought, greater variability in the weather and increasing unemployment; so they have become dependent on social grants. Moreover, food prices are becoming unaffordable, so people expressed concerned about future food security (Stadler 2012).

> We are starving because we cannot get a good harvest. I use my pension to buy basics from the shop.
>
> (Gatyana, female, 65 years old)

The participatory mental map activity also revealed that, while there was much agreement between men and women, the identification and expression of some stressors was defined by gender and site (Figure 8.2). Both groups of women identified poverty as the main driver of vulnerability (B, D), while the men, being those most often employed, spoke about unemployment rather than poverty (A, C). Both groups of women saw food insecurity as an important aspect of vulnerability and linked it to ill health and HIV/AIDS. By contrast, men in Gatyana linked hunger to crime (A), whilst men in Lesseyton recognised the negative impacts of climate variability and water shortages on agriculture and food security (C). Both women's groups and men in Gatyana also emphasised factors that make it difficult to farm, such as drought and livestock illness. These were less mentioned by the men's group in Lesseyton, possibly reflecting rural households' and women's more prominent role in farming. The lack of electricity as a stressor was identified only by the women's group in Gatyana (B), their argument being that it added physical and health burdens to their lives. Overall, however, it was factors such as poverty, violence, theft and corruption that were underscored most by both men and women in this mental mapping activity.

Food security

Food security was identified as one key dimension of vulnerability in the qualitative work and is important to consider in terms of sensitivity to climate change, especially when linked to HIV/AIDS. Specifically, we asked households their perceptions of their own food security. While it might be expected that households with higher mean monthly income and more land would perceive their food security to be adequate, scores for the different gender categories indicated that, in each site, households with only female adults (lowest income and land) reported on average the highest perceived food security (Table 8.5). In contrast, households with only male adults reported the lowest perceived food security. As reported elsewhere, community members in a meeting attributed this to women's tendency to prioritise their families, whereas men sometimes use their income for purposes like entertainment and alcohol. This observation was supported by results that showed higher alcohol expenditure in male-headed households (Stadler 2012). Detailed dietary studies undertaken in the same sites demonstrated that only young men were not obtaining their full calorific needs (Abu-Basutu 2013); possibly explaining the low perception of food security in male-only households. Furthermore, these households often have low access to grants. Social grants certainly play an important role locally in ameliorating food insecurity (Ndlovu 2012). However, people did comment that food costs are high, which places a strain on household resources (Clarke 2012).

HIV/AIDS as a stressor

HIV/AIDS was mentioned as a major stressor in both sites, linked to many others (Figure 8.2). The results show distinctive patterns in HIV/AIDS experiences across

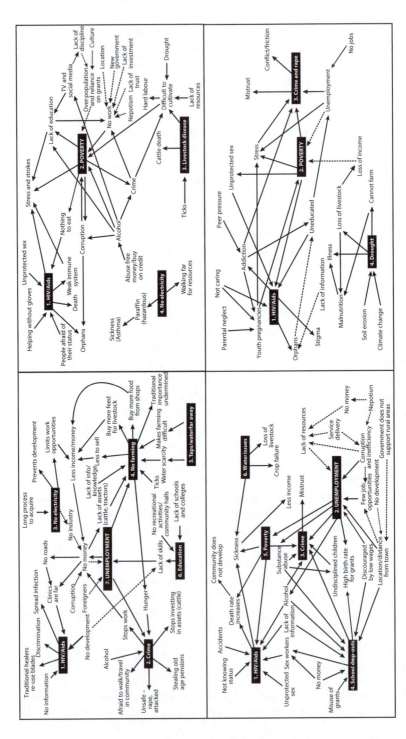

FIGURE 8.2 Participatory mental modelling spider-diagrams of local stressors done by men in Gatyana (A), women in Gatyana (B), men in Lesseyton (C) and women in Lesseyton (D)

gender categories, although not statistically significant in all cases (Table 8.6). Overall, female-headed households from both categories had a lower proportion of non-affected households and higher HIV/AIDS proxy scores than male-headed households. The higher proportion of female-headed households with de facto orphans reflects a trend observed by Makiwane and Chimere-Dan (2010) who found that the Eastern Cape has a high rate of older women taking responsibility for children. In Gatyana, the considerably higher proportion of households with only female adults experiencing an illness-related death could reflect mortality associated with migrant labour from this rural site (also found in other parts of South Africa; O'laughlin 1998). Interestingly, while a lower proportion of male-headed households with adult females was affected by HIV/AIDS in each site, a relatively high proportion of households with only male adults were.

Responding to shocks and stressors

A differentiated response to various stressors and shocks is common and, as emphasised above, is related to assets, vulnerability and poverty levels, access to cash income, and gendered roles. In both sites, a lower proportion of households with only female adults reported doing nothing in response to a short-term shock[2] (40 and 44.4 per cent) compared to other categories (Table 8.7). The more common self-employment in female-only households in Lesseyton suggests a greater willingness for experimentation, perhaps an outcome of more profound need. Women's collective responses to some of their stresses also suggest a gender bias regarding

TABLE 8.5 Mean weighted perceptions of food security,[#] disaggregated by household gender category in Lesseyton and Gatyana

		Lesseyton	*Gatyana*
Only male adults	Mean	0.91	0.65
	N	47	37
Male-headed with female adults	Mean	0.91	0.70
	N	45	40
Female-headed with male adults	Mean	1.00	0.69
	N	53	42
Only female adults	Mean	1.08	0.75
	N	25	48
NS = not significant at p < 0.1		NS	NS

Households were asked if their income and production could meet their food needs. Responses were rated from 0–2, with 2 indicating households saw themselves as food secure.

2 Shocks here are considered temporary disruptions as opposed to enduring shifts (stressors) and included both co-variate (weather, pests, diseases, bushfires) and idiosyncratic shocks (illness and death in the household, job loss, theft of assets, homestead damage) within two years of the survey.

TABLE 8.6 Percentage of households experiencing different HIV/AIDS impacts[#] and mean number of HIV/AIDS proxy indicators per household, disaggregated by household gender category in Lesseyton and Gatyana

Lesseyton

Type of HIV/AIDS impact (%)	Only male adults (N = 47)	Male-headed with female adults (N = 45)	Female-headed with male adults (N = 53)	Only female adults (N = 25)	Chi-Square test; ★ = significant at p < 0.1; NS = not significant
Non-affected	40.4	66.7	28.3	32.0	★
Chronic illness and receiving free care	46.8	28.9	58.5	48.0	★
Illness-related death in previous 10 years	17.0	8.9	26.4	20.0	NS
Presence of de facto orphans	25.5	13.3	30.2	28.0	NS
Mean number of indicators	0.89	0.51	1.15	0.96	★

Gatyana

Type of HIV/AIDS impact (%)	Only male adults (N = 36)	Male-headed with female adults (N = 41)	Female-headed with male adults (N = 43)	Only female adults (N = 48)	Chi-Square test; ★ = significant at p < 0.1
Non-affected	37.8	41.5	37.2	25.0	NS
Chronic illness and receiving free care	45.9	48.8	41.9	52.1	NS
Illness-related death in previous 10 years	21.6	14.6	18.6	35.4	NS
Presence of de facto orphans	13.5	17.1	18.6	20.8	NS
Mean number of indicators	0.81	0.80	0.79	1.06	NS

Proxy indicators of HIV/AIDS impacts were used (see http://www.sadc.int/english/fanr/food_security/Documents/HIV_AIDS_Report.pdf) and households with positive responses for each of these indicators counted. The number of positive indicators were then summed and averaged across household categories. A figure of 0 indicates no affliction, while 3 indicates that the household has chronic illness, had experienced a recent illness-related death and looks after orphans.

active engagement in tackling problems. Interestingly, in Lesseyton, despite lower levels of savings, a large proportion of female-only households listed using their savings as a response option, while in Gatyana more male-only households listed this as an option. The reliance on social safety nets differs between household categories. Male-headed households with adult females in Lesseyton (14.8 per cent) utilised the support of their friends and family in response to a shock far less than the other three gendered categories (Table 8.7). In both sites, more households with only female adults turned to friends and family for assistance (30 and 40.7 per cent) compared to the other gender categories, suggesting that social capital is an important safety net for this group:

> Neighbours and relatives assist with whatever they can afford, from financial assistance to perishables. They give when they are in a position to do so.
>
> (Lesseyton, female, 70 years)

Coping and adaptive strategies were also discussed within the social learning group. With regards to HIV/AIDS, it emerged that women often took in orphans, with some starting home-based care groups to assist child-headed households and the elderly by providing food. Several were also involved in gardening and other support groups (Trefry 2013). Women's high reliance on bonding social capital as a means of coping with stress could be under threat as vulnerability increases with new, co-variate risks to livelihoods (e.g. increasing extreme weather events where all households may be impacted simultaneously).

Relatively few respondents stated harvesting of natural resources as a coping strategy, except for female-only households in Lesseyton (Table 8.7). When specifically targeting poorer households, Clarke (2012) found that women ranked

TABLE 8.7 Percentage of households using various coping strategies following a shock, disaggregated by household gender category in Lesseyton and Gatyana (org. = organisation)

Lesseyton

	Only male adults (N = 31)	*Male-headed with female adults (N = 27)*	*Female-headed with male adults (N = 32)*	*Only female adults (N = 10)*
Spent savings	29.0	18.5	18.8	40.0
Sold assets	9.7	3.7	6.2	10.0
Extra work	0	14.8	3.1	10.0
Kin/friend assistance	29.0	14.8	28.1	30.0
Org. assistance	3.2	3.7	9.4	0
Loan	16.1	3.7	15.6	10.0
Nothing	61.3	55.6	65.6	40.0
Other (natural resource and self-employment)	19.4	25.9	21.9	50.0

Gatyana

	Only male adults (N = 19)	Male-headed with female adults (N = 23)	Female-headed with male adults (N = 27)	Only female adults (N = 27)
Spent savings	47.4	21.7	14.8	25.9
Sold assets	5.3	4.3	3.7	0
Extra work	5.3	0	7.4	3.7
Kin/friend assistance	21.1	17.4	25.9	40.7
Org. assistance	5.3	0	3.7	3.7
Loan	21.1	26.1	14.8	33.3
Nothing	63.2	65.2	63	44.4
Other (natural resource and self-employment)	5.3	21.7	25.9	11.1

natural resource gathering highly as a coping response. Similarly, pair-wise ranking of coping strategies by focus groups showed that women chose natural resource harvesting in preference to other options more often than men (Table 8.8).

The comparatively higher proportional contribution of ecosystem service-related income amongst female-headed households (mentioned earlier) also suggests higher dependence on this option, perhaps in response to the lower incomes and higher health burden in these households. Continued access to forest and other wild resources is therefore critical for reducing poor women's vulnerability.

Conclusions

This study has highlighted gender differences in vulnerability and responses to multiple stressors, including climate change. Rural women are often conceived to

TABLE 8.8 Pair-wise ranking* of coping strategies by men and women in Gatyana

	Loan	Assistance	Change role	Harvest NRs	Sell assets
Loan					
Women	X	Assistance	Change role	Harvest NRs	Sell assets
Men		Loan	Loan	Loan	Loan
Assistance					
Women		X	Assistance	Harvest NRs	Sell assets
Men			Assistance	Assistance	Sell assets
Change role			X		
Women				Harvest NRs	Change role
Men				Change role	Change role
Harvest NRs				X	
Women					Harvest NRs
Men					Harvest NRs
Sell assets					X

*Items in columns are compared against items in rows and the better of the two indicated in the table.

be highly vulnerable because of their heavier reliance on ecosystem services, lower income levels, labour constraints and poorer health, which combine to make them more susceptible to shocks (Arora-Jonsson 2011; Goh 2012). This is generally the case in our two study sites, but it is not as simple as that.

Female-headed households, especially those with no adult males, tended to have fewer adult household members, less land, savings and possessions, less income, and higher HIV/AIDS impacts, while women, more broadly, faced a larger number of stressors, especially those related to gender violence. However, many women are also actively managing resources to address their vulnerabilities. Women demonstrate agency in their responses to some of the stresses in their communities by developing support organisations and by turning to local social networks (cf. Vincent et al. 2010, in Limpopo Province), a key component of adaptive capacity (Pelling and High 2005; Agrawal 2008). Further, female-headed households often diversify their food and income sources in difficult times by turning to ecosystem services and to local, small-scale sources of self-employment (in the face of few alternative opportunities), whereas men tend to place an emphasis on the need for, but lack of, formal employment. In fact, men in Lesseyton saw themselves as being particularly vulnerable to job loss. Additionally, female-headed households did not perceive their households to be more food insecure than male-headed households. In South Africa, this could be the result of women using their government grants responsibly to support dependants and the absence of such grants in male-only households. Only young men were found to be food insecure in our study sites (Abu-Basutu 2013).

These results thus suggest that, while women, with lower capital stocks and income to draw on, appear to be exposed to more stressors relative to men, men are also vulnerable in these communities and need to be included in adaptation planning and action. Further examples of men's vulnerability are their greater reliance on income from livestock, making them particularly sensitive to climate change effects on livestock, and the high incidence of several types of HIV/AIDS impacts in male-only households. Regarding the latter, it has been suggested that men be included in training for caregiving, which not only gets them involved in caring activities, but also eases women's burdens (Peacock et al. 2010; UNAIDS 2008). Moreover, engaging men in combating gender-based violence could contribute to reducing the vulnerability of women and girls, ultimately affecting social relations and adaptive capacity in communities as a whole, as illustrated in the mental maps.

The mental maps and the social learning groups' descriptions of vulnerability in the two sites highlight the complexity involved in identifying appropriate community interventions. Many of the key drivers local people identified were not climate or weather related, although these concerns were raised. Thus, given that vulnerability and poor adaptive capacity in our sites seems to arise principally from socio-economic conditions, interventions in the social and political domains may be more important than many climate change studies and policies would imply. Different approaches are needed that shift focus onto the structures and institutions that govern access to resources. This requires reversing the continued neglect of the previous homelands regarding health, education, policing, extension support, access to credit, and facilitating better access to grants (McDowell and Hess 2012). Furthermore, development and adaptation policies and planning need to address the multiplicity of interactions across scales. In doing so, vulnerability must be recognised, as a variable phenomenon, with subtly different types of vulnerability in different contexts, and that neither men nor women form homogeneous groups. This is seldom the case. National policies on climate change in southern Africa tend to focus on the technical and biophysical aspects of climate change, often neglecting socio-economic dimensions and embedded structural issues such as poverty, inequality and gender discrimination (Madzwamuse 2010). The role of healthy ecosystems for the delivery of regulating and provisioning services is also often neglected. South Africa's new National Climate Change Response White Paper (DEA 2011), for instance, still separates climate impacts by sector, ignoring interactions between social and environmental systems and various types of stressors, and the heterogeneity of impacts across space, context and different segments of society. Furthermore, few climate change policies specifically consider the needs of women, children and other vulnerable groups, particularly climate impacts on nutrition, food security and health (PHI et al. n.d.). Strong political will to change the gendered status quo will be required if the persistent social and structural dimensions of vulnerability are to be adequately addressed (Eriksen and O'Brien 2007; Prowse and Scott, 2008).

Acknowledgements

We are grateful to the International Development Research Centre (IDRC) Ecohealth Programme and the National Research Foundation (NRF), South Africa for funding. A first version of this chapter was published in *AGENDA: Empowering women for gender equity* 28(3): 73–89 (Taylor and Francis). We also acknowledge Monde Nshudu for invaluable assistance in the field and the communities of Lesseyton and Gatyana without whose cooperation we could not have completed this work.

References

Abu–Basutu K. 2013. *Relative contribution of wild foods to individual and household food security in the context of increasing vulnerability due to HIV/AIDS and climate variability* [Masters thesis]. Rhodes University, South Africa.

Aggarwal R, Netanyahu S and Romano C. 2001. Access to natural resources and the fertility decision of women: The case of South Africa. *Environment and Development Economics* 6(2): 209–36.

Agrawal A. 2008. *The role of local institutions in adaptation to climate change.* Washington, DC: Social Development Department, World Bank.

Armstrong P, Lekezwa B and Siebrits K. 2008. Poverty in South Africa: A profile based on recent household surveys. Economic Working Paper 04/08. Stellenbosch University, South Africa.

Arora-Jonsson S. 2011. Virtue and vulnerability: Discourses on women, gender and climate change. *Global Environmental Change* 21: 744–51.

Avert. 2011. *South Africa HIV/AIDS statistics.* Accessed 19 January 2014. http://www.avert.org/safricastats.htm.

Babuguru A. 2010. *Gender and climate change: South Africa case study.* Cape Town: Heinrich Boll Foundation Southern Africa. Accessed 28 February 2013. http://www.boell.de/downloads/ecology/south_africa.pdf.

Bank L and Minkley G. 2005. Going nowhere slowly? Land, livelihoods and rural development in the Eastern Cape. *Social Dynamics* 31(1): 1–38.

Bunce M, Rosendo S and Brown K. 2010. Perceptions of climate change, multiple stressors and livelihoods on marginal African coasts. *Environment, Development and Sustainability* 12: 407–40.

Carr ER, Thompson MC and Biodiversity International. 2013. *Gender and climate change adaptation in agrarian settings.* Washington, DC: USAID.

CGE (Commission on Gender Equality). 2010. A gendered review of South Africa's implementation of the Millennium Development Goals. South Africa: CGE. Accessed 5 November 2011. http://www.cge.org.za.

Clarke C. 2012. *Responses to the linked stressors of climate change and HIV/AIDS amongst vulnerable rural households in the Eastern Cape, South Africa* [Masters thesis]. Rhodes University, South Africa.

Cousins B. 2010. The politics of communal tenure reform: A South African case study. In Anseeuw W and Alden C, eds. *The struggle over land in Africa: Conflicts, politics and change.* Cape Town, South Africa: HSRC Press.

Cundill G, Shackleton S, Sisitka L, Ntshudu M, Lotz-Sisitka H, Kulundu I and Hamer N. 2014. *Social learning for adaptation: A descriptive handbook for practitioners and action researchers.* South Africa: Rhodes University.

DEA (Department of Environmental Affairs, South Africa). 2011. National Climate Change White Paper, Department of Environmental Affairs, Government of South Africa.

Drimie S and Gillespie S. 2010. Adaptation to climate change in southern Africa: Factoring in AIDS. *Environmental Science and Policy* 13(8): 778–84.

DST (Department of Science and Technology). 2010. *South African risk and vulnerability atlas.* SARVA, Pretoria: DST.

Dubbeld B. 2013. How social security becomes social insecurity: Unsettled households, crisis talk and the value of grants in a KwaZulu-Natal village. *Acta Juridica*: 197–217.

Eakin HC, Lemos MC and Nelson DR. 2014. Differentiating capacities as a means to sustainable adaptation. *Global Environmental Change* 27: 1–8.

Eriksen S and O'Brien K. 2007. Vulnerability, poverty and the need for sustainable adaptation measures: Integrating climate change actions into local development. *Climate Policy* 7(4): 337–352.

Fuwa N. 2000. Poverty and heterogeneity among female-headed households revisited: the case of Panama. *World Development* 28(8): 1515–42.

Fraser EDG, Dougill AJ, Hubacek K, Quinn CH, Sendzimir J and Termansen M. 2011. Assessing vulnerability to climate change in dryland livelihood systems: Conceptual challenges and interdisciplinary solutions. *Ecology and Society* 16(3): 3. Accessed 17 January 2016. http://dx.doi.org/10.5751/ES-03402-160303.

Gillespie S and Drimie S. 2009. 'Hyperendemic AIDS, food insecurity and vulnerability in southern Africa: A conceptual evolution'. Working paper. Washington, DC: RENEWAL and IFPRI.

Goh AHX. 2012. 'A literature review of gender-differentiated impacts of climate change on women and men's assets and well-being in developing countries'. CAPRI Working Paper No. 106. Washington DC: IFPRI.

Hunter M. 2010. *Love in the time of AIDS: Inequality, gender and rights in South Africa*. Bloomington: Indiana University Press.

IPCC (Intergovernmental Panel on Climate Change). 2013. *Climate change 2013: The physical science basis*. Contribution of Working Group I to the Fifth Assessment Report of the Intergovernmental Panel on Climate Change. Cambridge University Press. Cambridge, United Kingdom and New York: IPCC.

Klasen S, Lechtenfeld T and Povel F. 2015. A feminization of vulnerability? Female headship, poverty and vulnerability in Thailand and Vietnam. *World Development* 71: 36–53.

Lobell DB, Burke MB, Tebaldi C, Mastrandrea MD, Falcon WP and Naylor RL. 2008. Prioritizing climate change needs for food security in 2030. *Science* 319(5863): 607–10.

Madzwamuse M. 2010. *Drowning voices: The climate change discourse in South Africa*. Cape Town, South Africa: Heinrich Boell Stiftung.

Makiwane MB and Chimere-Dan DOD. 2010. *The people matter: The state of the population in the Eastern Cape*. Eastern Cape Department of Social Development. East London, South Africa.

McDowell JZ and Hess JJ. 2012. Accessing adaptation: Multiple stressors on livelihoods in the Bolivian highlands under a changing climate. *Global Environmental Change* 22: 342–52.

Ndlovu P. 2012. Effects of social grants on labor supply and food security of South African households: Is there a disincentive effect? [Master's thesis.] University of Alberta, Edmonton, Canada.

O'Brien K, Quinlan T and Ziervogel G. 2009. Vulnerability interventions in the context of multiple stressors: Lessons from the South Africa Vulnerability Initiative (SAVI). *Environmental Science and Policy* 12: 23–32.

O'laughlin B. 1998. Missing men? The debate over rural poverty and women-headed households in southern Africa. *Journal of Peasant Studies* 25(2): 1–48.

Oxfam. 2014. Hidden hunger in South Africa: The faces of hunger and malnutrition in a food-secure nation. Oxfam. London. www.oxfam.org.

Özler B. 2007. Not separate, not equal: Poverty and inequality in post-apartheid South Africa. *Economic Development and Cultural Change* 55(3): 487–529.

Peacock D, Stemple L, Sawires S and Coates TJ. 2010. Men, HIV, and human rights. *Journal of Acquired Immune Deficiency* 51(3): 119–25.

Pelling M and High C. 2005. Understanding adaptation: What can social capital offer assessments of adaptive capacity? *Global Environmental Change* 15: 308–19.

PHI (Public Health Institute), World Food Programme, United Nations Standing Committee on Nutrition and Action Against Hunger. n.d. *Enhancing women's leadership to address the challenges of climate change on nutrition security and health*. Rome: PHI, WFP, UN.

Prowse M and Scott L. 2008. Assets and adaptation: An emerging debate. *IDS Bulletin* 39(4): 42–52.

Ribot J. 2014. Cause and response: Vulnerability and climate in the Anthropocene. *Journal of Peasant Studies* 41(5): 667–705.

Rockström J, Steffen W, Noone K, Persson Å, Chapin F, Lambin E, Lenton T, Scheffer M, Folke C, Schellnhuber H, Nykvist B, de Wit C, Hughes T, van der Leeuw S, Rodhe H, Sörlin S, Snyder P, Costanza P, Svedin U, Falkenmark M, Karlberg L, Corell R, Fabry V, Hansen J, Walker B, Liverman D, Richardson K, Crutzen P and Foley J. 2009. A safe operating space for humanity. *Nature* 461(24): 472–5.

Ruiters M and Wildschutt A. 2010. Food insecurity in South Africa: Where does gender matter? *Agenda* 86: 8–24.

Scholes B. 2011. *Adaptation to climate change: Eastern Cape*. Presentation, the Eastern Cape Climate Action Conference, East London, Eastern Cape, South Africa.

Shackleton CM and Shackleton SE. 2004. The importance of non-timber forest products in rural livelihood security and as safety nets: A review of evidence from South Africa. *South African Journal of Science* 100: 658–64.

——, Shackleton SE, Netshiluvhi TR and Mathabela FR. 2005. The contribution and direct-use value of livestock to rural livelihoods in the Sand River catchment, South Africa. *African Journal of Range and Forage Science* 22(2): 127–40.

Shackleton R, Shackleton C, Shackleton S and Gambiza J. 2013. Deagrarianisation and forest succession in abandoned fields in a biodiversity hotspot on the Wild Coast, South Africa. *PLoS ONE* 8(10): e76939.

Shackleton SE and Shackleton CM. 2012. Linking poverty, HIV/AIDS and climate change to human and ecosystem vulnerability in southern Africa: Consequences for livelihoods and sustainable ecosystem management. *International Journal of Sustainable Development and World Ecology* 19(3): 275–86.

Stadler LT. 2012. *Assessing household assets to understand vulnerability to HIV/AIDS and climate change in the Eastern Cape, South Africa* [Master's thesis]. Rhodes University: South Africa.

Taylor T. 2009. *Climate Change, Development and Energy Problems in South Africa: Another World is Possible*. Durban: Earthlife Africa.

Trefry A. 2013. *Local institutional structures and food security in South Africa* [Master's thesis]. University of Alberta: Canada.

Turpie J, Winkler H, Spalding-Fletcher R and Midgley G. 2002. *Economic impacts of climate change in South Africa: A preliminary analysis of unmitigated damage costs*. Research Report. Cape Town, South Africa: University of Cape Town.

UNAIDS (Joint United Nations Programme on HIV/AIDS). 2008. *Caregiving in the context of HIV/AIDS*. United Nations, Geneva: UNAIDS.

——. 2009. Annual report 2009. United Nations, Geneva: UNAIDS.

Vermeulen SJ. 2014. *Climate change, food security and small-scale producers*. CCAFS Info Brief, CGIAR Research Programme on Climate Change, Agriculture and Food Security. Copenhagen: CCAFS.

Vincent K, Cull T and Archer E. 2010. Gendered vulnerability to climate change in Limpopo Province, South Africa. In Dankelman I, ed. *Gender and Climate Change: An Introduction*. London: Earthscan. 160–67.

9

UNVEILING THE COMPLEXITY OF GENDER AND ADAPTATION

The "feminization" of forests as a response to drought-induced men's migration in Mali[1]

Houria Djoudi and Maria Brockhaus

Introduction

Observed and projected climate change will have significant and varying impacts on regions, ecosystems, societies and individuals. "Multiple stresses" occurring at various levels, which include endemic poverty, complex and overlapping governance, limited access to capital, as well as ecosystem degradation, increase vulnerability of several countries and regions (IPCC 2014). Various local adaptation strategies to cope with current climate variability do exist but may not be sufficient for future changes of climate (IPCC 2007, 2014).

The gender dimension of climate change is recognized by the fifth and seventh assessment reports of the Intergovernmental Panel on Climate Change (IPCC). The contribution by Working Group II notes that those marginalized in society are especially vulnerable to climate change as a result of intersecting social processes that result in inequalities due, for example, to discrimination on the basis of gender, class, race/ethnicity, age and (dis)ability. It explicitly mentions that women "often experience additional duties as laborers and caregivers as a result of extreme weather events and climate change, as well as responses (e.g., male outmigration)" (IPCC 2014, 50).

Literature on gender and climate change is scarce. The existing studies link women's vulnerability mostly to poverty, lack of assets, lack of access to and control over resources (Tompkins and Adger 2004; Aguilar 2009; Shackleton, this

1 An extended version of this chapter was first published in the *International Forestry Review* 13(2): 123–35. 2011, under the title "Is Adaptation to Climate Change Gender Neutral? Lessons from Communities Dependent on Livestock and Forests in Northern Mali," http://dx.doi.org/10.1505/146554811797406606.We are grateful to the Commonwealth Forestry Association for permission to reprint this revised and updated version of the chapter.

volume). The disaster reduction literature provides further evidence and local case studies show gendered impacts (Dankelman 2002). However, evidence about the complex and intertwined linkages on gendered climate change impact and adaptation, especially at the local level, is at an early stage of understanding. Particularly missing is how different types of adaptation responses will affect the gender determinants and relationships, the gendered division of labor and access to and control over resources.

This chapter provides further evidence on adaptation-related gender dynamics using a case study from Northern Mali. We first show women's context-specific vulnerability and adaptation in the socio-ecological system around Lake Faguibine. Women's preferences and contributions for adaptation and adaptive strategies are explored, as well as the impact of men's specific strategy, migration, on women's lives and activities as well as their involvement in forest-based livelihoods activities. We aim to highlight the complexity of interactions within the socio-ecological system. Focusing on local realities and evidence, we aim to improve understanding of complex gendered interactions and linkages in the area of climate change adaptation and natural resources.

Gender, vulnerability and adaptation

Vulnerability—the susceptibility to harm associated with environmental and social change—is a result of high-level interaction between ecological and human factors (Turner et al. 2003; Adger 2006). Vulnerability drivers may be economic (lack of entitlement), political (disempowerment) and structural (exploitation of one group by another; Davies 1993; Sen 1987; Watts and Bohle 1993). Those social and political aspects balance the initially common "eco-centric" approaches to environmental change (Ribot 1996) and reframe vulnerability from a household perspective, highlighting the differential vulnerabilities of social groups and individuals.

Linkages among gender, poverty and vulnerability are very common in the climate change adaptation discourse (Nelson et al. 2002; Arora-Jonsson 2011). However, we argue that gender subordination and the resulting vulnerability occur globally in different ways depending on social, economic and political contexts, rather than as a result of poverty per se. Several publications (e.g. Cannon 2002; Skutsch 2002) ask therefore *how much of women's vulnerability is due to poverty and how much is apportioned to gendered roles and restrictions?* Poverty affects women and men disproportionately, due to the interaction of context-specific social and political conditions with global factors, and hence the assumption that "poverty leads automatically to vulnerability" is challenged by some scholars (Arora-Jonsson 2011). Furthermore this author points out that the common assumption that vulnerability is equal to poverty hinders deeper analysis of power relations and context-specific gender inequalities.

There is a general view that marginalized groups of men and women are likely to be most vulnerable to climate change because of the socially and politically driven lack of participation in decision making and access to power. The link between women's vulnerability and their limited access to resources is another perspective for understanding gender-specific climate change impacts. In the absence of diversification strategies and because of their role in food and energy household supply, women are dependent on natural resources. However, their decision-making involvement and control over those resources are socially restricted across societies and cultures. Besides the existing gender complexity, the contextual differences in access to land, and how such rights are acquired and transferred, there is general agreement that women's participation in decision making over resource use and management is limited (Rocheleau and Edmunds 1997; Fortmann and Rocheleau 1997; Colfer 2005; Porro and Stone 2005). These links between access to resources and decision making and vulnerability seem evident but are not considered in most national climate change policy frameworks. This is reflected, for instance, in the absence of gender-specific issues in national adaptation and mitigation plans, such as the National Adaptation Program of Action (NAPA). Few NAPAs link climate change impacts to women's economic, political and social status. Even fewer incorporate women as key stakeholders in NAPA activities (UNFPA and WEDO 2009). Even when women are asked for their perspectives, these may be overlooked when the project is implemented (e.g. Porro 2010). Gender equality as an emerging mandatory item in adaptation projects follows the same structures and schemes as in most development projects and is de facto, insufficient.

Ecosystem-based adaptation denotes "policies and measures that take into account the role of ecosystem services in reducing the vulnerability to climate change, in a multi-sector and multi-scale approach" (Vignola et al. 2009: 2). Evidence is growing that sustainably managed ecosystems are very effective for coping with climate change (Pramova et al. 2012). Additionally, ecosystem-based adaptation is readily available to the rural poor who could address many of the most vulnerable peoples' priorities and concerns, if integrated into community-based adaptation (Colls et al. 2009).

In several rural areas of Africa, Asia and Latin America, women's activities are strongly interlinked with the services provided by local ecological systems. While women have a daily need of forest products for nutrition, health and income, they can hardly influence processes at play or decisions related to natural resources dynamics (Agarwal 1985; Gautier 2011; Rocheleau and Edmunds 1997; Fortmann and Rocheleau 1997; Colfer 2005; Porro and Stone 2005). This reliance on natural resources increases women's ability to acquire and disseminate knowledge and information about ecosystems, sustained practices and conservation techniques (Sydie 1994). Beyond romanticized discourses around women and nature, it is evident that women often have the greatest stake in the long-term development of forest and tree resources (Fortmann and Rocheleau 1985). Hence, women could be key actors and agents of change in ecosystem adaptation success. However, several authors have highlighted the tendency to present women more as victims than agents of change in the climate change adaptation debate and call for more differentiated analysis beyond generalizations and assumptions of women as a homogenous group (Emetriades and Esplen 2008; Terry 2009; Arora-Jonsson 2011).

Migration, climate change and gender

Several studies confirm the role of seasonal and circular migration as a well-established adaptive strategy (De Haan 1999; De Haan and Rogaly 2002; Mertz et al. 2009). Structural resource scarcity, however, seems to be more important than environmental degradation (Carr 2005; van der Geest 2011). Several authors point out the alarmist debate on migration and environmental changes and argue that cumulative causation (environmental, social and political) as well as the inter-temporal dimension of migration have to be considered more (Jónsson 2010; Doevenspeck 2011).

Migration is known to bring notable impacts and changes in the gender composition of communities (Elmhirst 2000); It is obvious that this strategy affects gendered roles and responsibilities. However, these impacts and especially their links to natural resources and forests have received little attention. In the context of the global South, migration may be formal or informal, permanent or circular, internal or external, rural to urban, or vice versa. Each type may have various impacts and outcomes for landscapes and livelihoods. Women are the ones who are mostly "left behind," and shifts in social and institutional structures can have different gender outcomes. Migration can contribute to women's empowerment through remittances and increased involvement in decision making. But it can also engender new economic and social inequalities or exacerbate existing ones (Chant and Radcliffe 1992; Chant 1998). In other cases, women may gain new spheres of influence to fill the institutional gap created by men's absence.

Little evidence exists on the combined impacts of climate change and migration on natural resource management. Turner (1999) showed that migration of men combined with changes in access to land led to a "feminization" of livestock keeping in the Sahel. This "feminization" is reflected in changes in herd composition, which have shifted away from cattle towards small ruminants (goat and sheep) because culturally, cattle are associated with male herders. These changes have a direct impact on land-use patterns as cattle and small ruminants have different grazing behavior and affect vegetation differently. Cattle are also well known to be more sensitive to drought. The "feminization" of livestock is likely to produce shifts in land-use patterns as well as in the sensitivity of the livestock systems, two decisive factors in planning ecosystem-based adaptation.

The impact of remittances on gender relations and on the management of natural resources is also not well established. In the specific context of Mali, the international financial inflow of remittances had a share of 5.1 percent of GDP in 2010, and the flows increased (Figure 9.1) from US$154 to 405 million between 2003 and 2009 (World Bank 2011; Scheffran et al. 2012).

These numbers capture only the formal flow of remittances; the total volume is surely larger.

Migration statistics are scarce in Mali, particularly related to uses of remittances, since most of such transfers flow through informal channels. The existing data on remittances shows a fundamental shift in migrants' livelihood activities. While

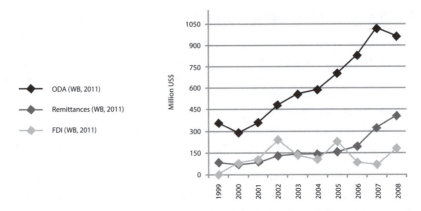

FIGURE 9.1 Time series for Official Development Assistance (ODA), Foreign Direct Investment (FDI) and remittances Mali (WIR, 2009, in Scheffran et al. 2012)

the majority of people were gaining their income from agriculture, hunting and forestry before they migrated, trade and other non-natural resources-based services were the most important after their return (Figure 9.2). This is an important shift linked to how migration impacts natural resources and consequently ecosystem-based adaptation.

In Northern Mali, migration has been driven by a multitude of social and environmental changes. Political insecurity drives a "collective" migration, whereas climate-driven migration is generally individual, except for extreme large-scale droughts (e.g. in the 1970s and 1980s). Men migrate individually in search of employment to the urban areas of Mali (Bamako) or neighboring countries (Mauritania, Ivory Coast, Algeria). Timbuktu has the second highest emigration rate in Mali, with more than 50 percent of migrants going to Bamako, 25 percent to other African countries and the rest outside of Africa (IOM 2009). Meanwhile, the International Organization for Migration refers to a "migration crisis" in Northern Mali, including large-scale and unpredictable flows and mobility patterns caused by armed rebellion and the January 2012 military coup (IOM 2013).

Knowledge of the linkages between adaptation and migration is still at an early stage. However, many studies argue that migration can support climate adaptation and build social resilience by providing new opportunities for diversification and by taking advantage of networks of migrants in the host regions. Migrants in some regions in Mali play an important role in co-financing the majority of development projects, such as water systems, dikes, schools, cooperatives, grain mills, transportation and health centers (Scheffran and Gioli 2013; Galatowitsch 2009; Daum 2007; Scheffran et al. 2012). Such health-care facilities quadrupled between 1980 and 1995, using co-financing from migrants (Daum 2007).

The linkages between gender and migration in Mali show different outcomes. De Haan (1999) argues that it may not be possible to generalize about the effects

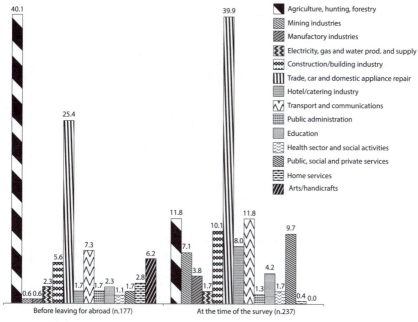

FIGURE 9.2   Main job sector of employed returnees in Mali, % (source: Cross-Regional Information System (CRIS) on the Reintegration of Migrants in their Countries of Origin, 2012). See more at: http://rsc.eui.eu/RDP/research-projects/cris/#sthash.UEOKtGwe.dpuf (accessed January 18, 2016)

of migration on broader development, inequality, or poverty. Some authors found that generally male out-migration did not significantly change the patriarchal patterns of decision making and the gender division of labor or women's ability to make decisions about agriculture and natural resource activities (David 1995; Ahlin and Dahlberg 2010). Furthermore, male out-migration has diverse impacts, in each case depending on the context-specific gender division of labor, land tenure, women's decision-making power and women's workloads. Tacoli (2002) shows that high levels of migration, especially among younger generations and, increasingly, among young women, have both positive and negative aspects.

A Case Study from Northern Mali

Research area

Figure 9.3 shows the location of our study site. The average population density around Lake Faguibine is low, with 1.1 persons per square kilometer in the Timbuktu region and 1.7 persons per square kilometer in Goundam district

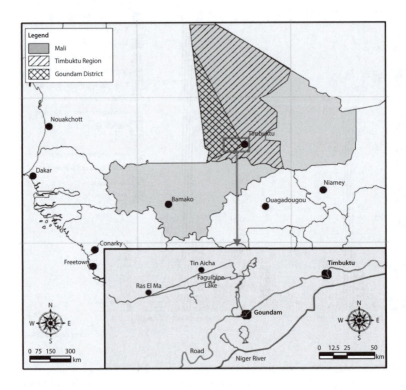

FIGURE 9.3 Location of the study area in Northern Mali

(DRPSIAP-T 2006), with wide spatial variation. Settlements are concentrated around Lake Faguibine and along the Niger River. Different ethnic groups with different livelihoods make up the population around the lake. Mainly Arabo-Berber livestock keepers live in the western and northern parts, where research was conducted. In Tin Aicha in the north, most residents are from Kel Tamacheq, a Berber sub-group. Most households belong to the *Iklan* class, with the lowest social status, descended from slaves. In Ras El Ma to the west, most people belong to the Arab (Moorish) group, Tormoz, and the Berber ethnic group Kel Tamacheq. Members of the Bozo ethnic group are traditionally fishers.

Livelihoods depend mostly on mobile and sedentary livestock breeding. In the Timbuktu region, around 72 percent of the land is used as pasture and the rest is reported as forested (DRPSIAP-T 2006). Two kinds of mobile livestock breeding systems are practiced: transhumant and nomadic. Livestock breeding is also associated with sedentary agropastoralism.

Lake Faguibine is part of a Niger River-fed lake system. It was once a productive area for agriculture and fisheries, but experienced wet and dry phases in the twentieth century. The lake has been almost completely dry since the mid-1970s (UNEP 2009). Lake Faguibine has drastically transformed from a water-based to a forest ecosystem. More than a third of the lake area has naturally reforested with *Acacia spp.* and *Prosopis spp. Prosopis* was introduced by an NGO-led development project, the Association Sahel, in the 1980s to protect the lake against siltation. After the lake dried out, the highly invasive *Prosopis* occupied the former bed more quickly than did local species such as *Acacia* (Brockhaus and Djoudi 2008; Djoudi et al. 2013). *Acacia* is prevalent in the lake's western part (Ras El Ma community) and *Prosopis* in the northern part (Tin Aicha community). *Prosopis* is controversial and perceived either as a "curse" or "blessing" (Laxen 2007).

Since the lake dried out, several programs have sought to "reflood the lake" and restore water-based economic activities. UNEP (2009) describes such projects as aiming to restore the ecosystem functions' sustainable services delivery of Lake Faguibine.

Approaches and methods

From July to October 2008, research was conducted in two Lake Faguibine communities. Six participatory workshops were organized in each community—Tin Aicha (a sedentary farmer community) and Ras El Ma (a pastoral community). Each workshop had 25–35 participants. Various perspectives were captured through workshops with three groups in each community: adult men, adult women and youth. Vulnerability, adaptation strategies and measures were assessed using tools from Participatory Rural Appraisal (PRA). Tools included fodder calendars, resource maps, historical axes and ranking exercises.[2]

Vulnerability is a theoretical concept and difficult to translate locally. We framed vulnerability in a palpable way adapted to local realities, by taking a historical perspective. We started with past extreme climatic events, moving on to the socio-economic, institutional and political factors influencing past and present coping strategies. Historical timelines were used to identify strategies used in the past to cope with droughts. The strategies were bundled and ranked according to preferences of different groups and the link of the strategies to forest and trees.

Results and discussion

Impact of climate change and variability on women's livelihood strategies

During the six community workshops, participants identified the main events of recent history on a historical axis. Droughts in the 1970s and 1980s were always mentioned first, by both men and women. Participants clearly argued that they were still dealing with the social and environmental consequences of those events. People adapted spontaneously by diversifying their livelihoods, rebuilding their

---

2 These methods were supplemented qualitatively by both of our long-term experience in and familiarity with this region.

herds and migrating. However, a series of major climatic and political events, such as Lake Faguibine drying out and the rebellion (1995–2003), put pressure on the socio-ecological system. These events reduced adaptive capacity by reducing resource availability, and altered the social structure of the community vis-à-vis migration. Layers of vulnerability have resulted from multiple successive stressors, aggravated by their cumulative effects. Additionally, as identified in the historical axes, most state or aid organization interventions focused on emergency relief rather than building the community's adaptive capacity in the medium or long term.

The historical axes show that the major environmental change—the lake drying out—shifted livelihoods from water-based to forest and livestock-based systems (Figure 9.4). Results indicated increasing daily vulnerabilities due to restricted food availability and greater health risks. The quality and quantity of food, as well as its availability during the year, have deteriorated drastically. Especially during the dry season when Niger River levels are low, the supply of vegetables and fish from the south (Mopti) is interrupted. Nutritional rations in households are very poor for eleven months of the year, with August as the exception when more milk is available. Due to the strongly gendered social rules in distributing household meals,

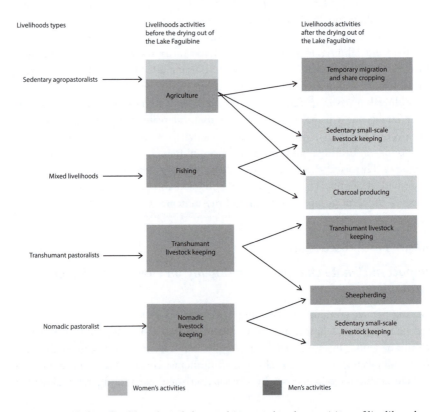

FIGURE 9.4 Shifting livelihoods and the resulting gendered repartition of livelihood activities before and after the drying out of Lake Faguibine

men are served first, followed by the elderly and children. Men also have more opportunities to access food outside the household, e.g. during market days. Food scarcity during the drought, therefore, is likely to burden women more.

As livelihoods shifted from water- to forest-based, divisions of labor along gender lines also shifted, increasing women's workload. Several activities explicitly associated with men, such as livestock herding and charcoal production, have been added to women's responsibilities (Figure 9.4). Sedentarization as a coping strategy—a consequence of drought and migration of men—increased women's burdens as they take on traditionally male activities such as tending livestock. The transformation from a lake to a forest also brings new income-generating activities, such as charcoal production. In the absence of male labor, women have also undertaken those new activities. With more work and fewer men to do it, participants in the women's workshops explained that they were increasingly vulnerable to environmental and economic shocks.

Women's views of adaptive strategies

Our results show that men and women have different preferences in coping with climate change and variability. The identified past and present strategies are based to a different degree on the use of the available forest resources in the two communities (see Figure 9.5). Migration, though practiced outside the communities and thus outside the forest, is still linked to the forest resource. This is because revenues are reinvested in livestock (to different degrees in the two communities and groups) with forest and trees as a main source of fodder.

Women's preferences are more focused on enhanced education for girls and boys to allow them a secure income from activities that are less directly dependent on the ecosystem. Men's response to climate change seems highly influenced by the political discourse at the meso- and national level promoting the large project to "reflood the lake." This strategy was highlighted and ranked as the most important adaptive strategy in the men's workshops. Men see this as an opportunity to go back to the former production system based on agriculture and fishing. Women in both communities did not consider these as potential strategies for the future. The women's perceptions appeared to be based on a more realistic assessment of what is possible. This can be related to women being isolated from the political discourse and influences. Therefore their strategic choices are based instead on their experienced realities.

Participants highlighted the time dimension of vulnerability (long and short term). They also highlighted the interdependency between current strategies and future vulnerabilities. Achieving long-term strategies was highly dependent on the success or failure of short- and medium-term strategies.

Women's short-term adaptation practices serve as a starting point to address the urgent vulnerability to longer-term climate change. Those strategies involve, for example, technical measures to reduce work time and improve health and nutrition. One example is technological improvement in daily activities such as mills for

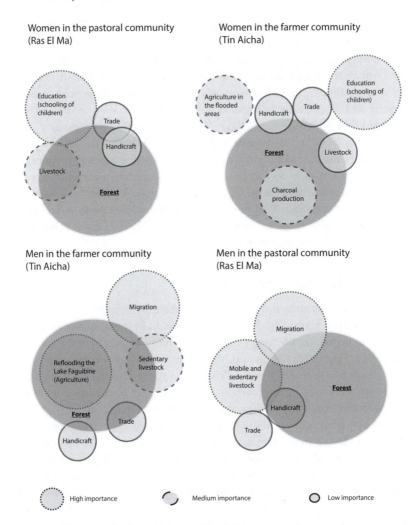

FIGURE 9.5 Adaptive strategies and their dependency on forest, developed by women and men in pastoral and farmer communities during the PRA workshops

millet or local transport for water (donkeys and camels). The focus-group discussions showed that women's workload is very high, but they suffer poor nutrition and therefore low energy, especially during droughts. Technologies and improved nutrition and health could improve women's conditions. They would be able to achieve their goals in livelihood diversification and reduce structural vulnerability.

Medium-term strategies are based on the diversification of livelihoods. Wood, charcoal, handicrafts and agricultural activities in still-flooded neighboring regions contribute to diversification. However, our results show that those activities must simultaneously fulfill two separate objectives. The first is to respond to immediate food and subsistence needs. The second is to cover the cost of future adaptation strategies. In the current situation, households struggle with trade-offs between investing in future strategies and securing their basic needs.

The long-term strategies of women, focused on the education of both girls and boys, were based on the rationale that natural resources like forests are highly sensitive to climate change. Therefore, future perspectives should reduce dependency on natural resources.

Workshop participants pointed to a vicious circle between doing the activities needed to meet urgent livelihood needs and achieving long-term adaptive strategies. For instance, the investment in future human capital by schooling children constantly conflicts with the need for person power and the financial means to secure daily needs. One example that emerged in the workshops is the NGO flock rebuilding programs started after droughts. These provide farmers with animals of good stock. By selling the offspring, farmers could contribute to paying children's school fees. But cereal prices increased and seed stocks were lost because of drought. This forced the benefiting farmer to sell the assets (female animals) and therefore fail to invest in future strategies. Male migration also meant most households stopped schooling their children in part or completely to replace some of the lost person power. Other studies have identified a positive correlation between the number of available adults in a household and the schooling rate of children (Konate et al. 2004).

Defeminization of agriculture and the "feminization" of forests

A defeminization of agricultural activities took place following the drastic decrease in agricultural areas in the formerly flooded northern and southern parts of the lake. Sedentary agriculture households, mainly from the *Iklan* social class, lost access to water and arable land. The workshop resource maps show that only the still-flooded third of the area (the lake's eastern part) can be irrigated for agriculture, with some irregularities spanning years. Communities are coping by migrating temporarily to the former lake's southeast, close to the Niger River, to cultivate land under a sharecropping system (Figure 9.4). In this new system, land access is no longer regulated by traditional mutual arrangements. Instead, it is regulated by monetarily capitalized and annually negotiated contracts. As the demand for land is greater than the supply, financial speculation is common. Thus, changes in resource availability, or the loss of former assets because of the loss of their arable land, increase migrant households' vulnerability.

Power struggles related to land access and acquisition, monetary speculation and the distance from the community area mean women no longer have access to agricultural land. In this new institutional system of access, women have no networks or decision-making opportunities and no access to irrigated agricultural land. However, harvesting and transportation are still women's responsibilities. Women's workload is therefore further burdened by the long distance between the village and the fields during harvest. The defeminization of agriculture did not result automatically in decreasing women's workloads, because the loss of agricultural tasks was replaced by new responsibilities in activities previously affiliated with men. Women had to go further and spend more time accessing water supplies. The higher workload and decreased access to assets and decision making on the newly acquired agricultural land increased women's vulnerability.

The historical axis shows on the other hand a "feminization" of forest use due to the shift in forest-related activities from men to women. With the lake drying out, short- and long-term migration by men became a part of local coping strategies. In the sedentary communities, charcoal production, now managed by women, was strictly men's work in the past. This activity was taken up exclusively by the *Iklan* women, which represent a "lower class" in the hierarchical structures of northern Mali. The *Illelan* (higher-class) women were culturally restricted from adopting this strategy, because of their higher social status and also their lack of needed skills. Women have assumed responsibility for producing and selling charcoal; however, challenges related to land tenure and market access hinder them in optimizing their income as charcoal producers (see also Ingram et al., this volume).

The workshop results clearly indicated the institutional dimensions of inadequate control over the production and sale of charcoal. Due to the transition from lake to forest, rights and tenure over the forest remain unclear. Forest resources are subject to speculation and profit seeking from some individuals. Charcoal production is theoretically regulated by state agreements, but the lack of transparency and equity in regulating access has led to ambiguous, opaque access and use rights. Some local political leaders use this ambiguity to profit personally at the expense of the local producer. One example is the distribution of charcoal production permits, which is determined by power and personal networks. Local government institutions lack transparency and accountability in regulating access. Generally, women are not included in personal and political networks and are therefore sidelined to produce charcoal in an "illegal way." This makes them more dependent on the arbitrariness of some local authorities and individuals, and it increases their vulnerability and production costs. This leads us to question local government's role in creating or hindering incentives for sustainable charcoal production as part of an ecosystem based adaptation. Adaptive capacity is undermined by institutional failures at various scales.

Women's socially restricted mobility and access to markets hinder their optimal charcoal commercialization. Women only have access to small local markets, or they sell the charcoal in the village to traders. Our observations at the big regional markets (Diré and Goundam) reveal prices that are five times higher than in the villages (Tin Aicha and Ras El Ma). As in other studies (Little et al. 2001), men sell regardless of the distance to market while women find it more difficult the further away they are.

The second activity reflecting the "feminization" of forest use is sedentary livestock herding. In the Timbuktu region, livestock, especially nomadic flocks, were men's responsibility. With the migration of men, women and children had to take over this task. Livestock in the arid area is based on trees and shrubs from the savannah woodlands. However, women's mobility restrictions hinder them in moving long distances with animals and most herds directly managed by women are grazing close to the village. This restriction implies an increase of vulnerabilities because of the higher pressure of animals on the closer areas of village territories. An even more restricting factor is water. With the drying-out of the lake, the few remaining water sources have been reallocated according to power structures, social status and wealth. Women were not a part of this new redistribution and women herders struggle to gain access to water. Women must walk long distances, often at night when pressure on water use is less, for such access.

Gender, ethnicity and identities shaping vulnerability to climate change

The comparison between the two communities shows differences in women's adaptive capacity. Pastoral communities are mostly represented by the *Illelan* social group, the highest socio-cultural group in the hierarchical Tamacheq society. Despite their higher societal roles, *Illelan* women seem to face more barriers as they diversify their livelihoods than *Iklan* women. Diversification using charcoal production is not practiced by *Illelan* women. Charcoal production in *Illelan* communities is perceived to be "beneath them," a result of cultural and societal barriers related to identities and hierarchal roles. In addition, *Illelan* women experience stronger mobility restrictions and seclusion than *Iklan* women, and are therefore more constrained in taking on new diversification opportunities. Power relationships and their interlinkages with gender and class are evolving in very complex ways. Class structure was once strongly correlated with social status and wealth, though this is changing, due to the migration induced by drought and rebellion (see e.g. Randall 2005). Women from the former *Iklan* social class are more able to diversify their livelihoods than *Illelan* women.

Adaptive capacity is not only determined by wealth (assets) but also by the ability to seize livelihood diversification opportunities. This ability is inhibited by gender restrictions. It is also inhibited by rules and norms for labor division and self-perceptions in class and ethnic groups. Vulnerability to climate change is dynamic and can shift following social, ecological, economic and/or political changes. This means linear assumptions and conclusions, such as socio-economically higher classes have more assets and therefore higher adaptive capacity, must be reconsidered.

Identities, expressed in class, gender and ethnicity, affect the options that are available and socially feasible for a group or individuals. Similar evidence has also been found in other ethnic and cultural groups in West Africa (Nielsen and Reenberg 2010). Recognition of those contexts is crucial for climate change adaptation. Sound, differentiated and gender-sensitive vulnerability assessments are useful, and need to be successfully incorporated into adaptation and development.

Conclusion

This case study provides evidence of vulnerability of women in northern Mali in an ecosystem shifting due to climate change. Higher risks are related to increasing workloads without increasing incomes and the defeminization of agricultural activities due to the loss of former arable land around the lake. This is accentuated by the loss of access to land in the new institutional land allocation and access systems. However, the "feminization" of the use of forest products and services has the potential to minimize the risk of vulnerability by providing new income-generating activities like charcoal production and fodder provision. Women's workload has clearly been increased by the climate event itself (drought) and by the responses to it. But some climate change-induced effects on women remain unclear.

Charcoal production, especially based on the invasive species *Prosopis*, offers an opportunity for women to improve their income. However, women are impeded by limitations related to insecure land tenure and social restrictions on access to markets. Current gender roles in decision making are emerging in local communities. Those new roles are enhanced by the forced migration of men and the emergence of new roles and responsibilities for women. Gender relations are therefore in flux. Whether this translates into women contributing more to decision making depends on how local and national governments facilitate empowerment and policy enhancement. To support women in turning short-term negative impacts into long-term positive developments, gender-sensitive analysis of adaptive strategies is required. However, broader societal and political changes are needed to realize the potential. In addition, investment in women's capacity and knowledge will avoid gender roles changing with negative impacts on the forest ecosystem, for example non-sustainable forest management for charcoal production.

Male and female traditional roles and activities are evolving faster under recurrent drought and migration. Male migration is increasing women's workload in the short term. But in the long term, it could give women the social space to assume more leadership in household decision making. It could also allow women to get more involved in activities that were once strictly a male domain. Women are increasingly undertaking "male" activities, but without acquiring automatically the same rights. Successful adaptation means reconsidering the usual theoretical dichotomies, in classifying "male" and "female" activities. We argue that emerging new societal roles could empower women to negotiate new institutional arrangements to access and control resources.

Additional social and societal changes are necessary to make use of the new opportunities inside and outside the forest ecosystem, and to remove barriers to the shift in vulnerability. The link between migration, the "feminization" of forests and adaptation need to be further explored. It is crucial that those local-level dynamics are considered in climate change and development and migration policies. Those policies and plans need to enhance women's abilities and adaptive capacities. Education seems one component outside the field of natural resource management, which is revealed (again) to be crucial to decrease women's and households' vulnerabilities. Enhancing a context adapted with equitable access to and control over land and trees is one of the most determinant factors in enhancing women's efforts to overcome environmental and other global changes. This cannot happen without active participation of women and other marginalized groups in local, regional and national political processes and decision making.

Acknowledgments

The authors thank all participants and interviewees in the local communities (Tin Aicha, Ras El Ma), Goundam, Timbuktu and Bamako. We also thank Barun Gurung, Esther Mwangi, Carol Colfer and two anonymous reviewers for their valuable comments and suggestions for the extended version of this chapter (published in 2011 in the *International Forest Review* 13(2): 123–35).

This document has been produced within the framework of the "Tropical Forests and Climate Change Adaptation" (TroFCCA) project executed by CATIE and CIFOR and funded by the European Commission under contract EuropeAid/ENV/2004-81719. We thank the ACFAO project funded by the French Fund for World Environment ("Fond Français pour l'environnement Mondial—FFEM") for funding contribution to this research. The contents of this document are the sole responsibility of the authors and can under no circumstances be regarded as reflecting the position of the European Union.

References

Adger N. 2006. Vulnerability. *Global Environmental Change* 16: 268–81.

Agarwal B. 1985. *Women and technological change in agriculture: The Asian and African experience* (pp. 67–114). London: George Allen and Unwin.

Agrawal A. 2008. The role of local institutions in adaptation to climate change. Paper prepared for the Social Dimensions of Climate Change, Social Development Department, The World Bank, Washington, DC, March 5–6, 2008.

Aguilar L. 2009. Women and climate change: Vulnerabilities and adaptive capacities. In *World Watch Institute 2009 State of the World*. Washington, DC: World Watch Institute.

Ahlin S and Dahlberg A. 2010. Migration, remittances and the women left behind: A study on how women in Mali are affected by migration and remittances from their migrated husbands. G3-paper in Political Science, Institution of Social Sciences Linnaeus University, Sweden. Accessed on August 1, 2015. http://www.diva-portal.se/smash/get/diva2:389996/FULLTEXT01.pdf.

Arora-Jonsson S. 2011. Virtue and vulnerability: Discourses on women, gender and climate change. *Global Environmental Change* 21(2): 744–51.

Brockhaus M and Djoudi H. 2008. Adaptation at the interface of forest ecosystem goods and services and livestock production systems in Northern Mali. *CIFOR Info brief* No. 19. Accessed November 9, 2009. http:/www.cifor.cgiar.org/publications/pdf_files/Infobrief/019-infobrief.pdf.

Cannon T. 2002. Gender and climate hazards in Bangladesh. *Gender and Development* 10(2): 45–50.

Carr ER. 2005. Placing the environment in migration: Environment, economy, and power in Ghana's central Region. *Environment and Planning* 37(5): 925–46.

Chant S. 1998. Households, gender and rural–urban migration: Reflections on linkages and considerations for policy. *Environment and Urbanization* 10(1): 5–22.

———. 2001. Migration and development: The importance of gender. In S Chant, ed. *Gender and migration in developing countries*. 1–29.

Colls A, Ash N, and Ikkala N. 2009. *Ecosystem-based adaptation: A natural response to climate change*. Gland, Switzerland: IUCN.

Colfer CJP. 2005. *The equitable forest: Diversity, community and resource management.* Washington, DC: Resources for the Future and CIFOR.

Dankelman I. 2002. Climate change: Learning from gender analysis and women's experiences of organizing for sustainable development. *Gender and Development* 10(2): 21–9.

Daum C. 2007. Migration, retour, non-retour et changement social dans le pays d'origine. In Petit V, ed. *Migrations Internationales de Retour et Pays d'Origine.* Nogent-sur-Marne: CEPED. 157–69.

David R. 1995. *Changing places? Women, resource management and migration in the Sahel: Case studies from Senegal, Burkina Faso, Mali and Sudan.* London: SOS Sahel International.

Davies S. 1993. Are coping strategies a cop out? *IDS Bulletin* 24: 60–72.

De Haan A. 1999. Livelihoods and poverty: The role of migration—a critical review of the migration literature. *Journal of Development Studies* 36(2): 1–47.

—— and Rogaly B. 2002. *Labour mobility and rural society.* New York: Psychology Press.

Djoudi H, Brockhaus M, and Locatelli B. 2013. Once there was a lake: Vulnerability to environmental changes in northern Mali. *Regional Environmental Change* 13(3): 493–508.

Doevenspeck M. 2011. The thin line between choice and flight: Environment and migration in rural Benin. *International Migration* 49(s1): 50–68.

DRPSIAP-T. 2006. Annuaire statistique Année 2006. Timbuktu, Mali: Direction Régionale de la Planification, de la Statistique, de l'Informatique, de l'Aménagement du Territoire et de la Population.

Elmhirst R. 2000. A Javanese diaspora? Gender and identity politics in Indonesia's transmigration resettlement program. *Women's Studies International Forum* 23(4): 487–500.

Fortmann L and Rocheleau D. 1997. Women and agroforestry: Four myths and three case studies. In Sachs CE, ed. *Women working in the environment.* Washington, DC: Taylor & Francis. 193–211.

Galatowitsch D. 2009. Co-development in Mali. *ISP collection* 737. Accessed January 7, 2016. http://digitalcollections.sit.edu/isp_collection/737.

Gautier D, Hautdidier B and Gazull L. 2011. Woodcutting and territorial claims in Mali. *Geoforum* 42(1): 28–39.

IOM (International Organization for Migration). 2009. Migration au Mali, Profil National 2009. Accessed January 7, 2016. http://publications.iom.int/bookstore/free/Mali_Profile_2009.pdf.

——. 2013. The Mali migration crisis at a glance. Accessed August 5, 2015. https://www.iom.int/files/live/sites/iom/files/Country/docs/Mali_Migration_Crisis_2013.pdf.

IPCC (Intergovernmental Panel on Climate Change). 2007. *Climate change 2007 – The IPCC Fourth Assessment Report (AR4).* Accessed August 2010. http://www.ipcc.ch/pdf/assessment-report/ar4/syr/ar4_syr_spm.pdf 11.

IPCC (Intergovernmental Panel on Climate Change). 2014. *Climate change 2014: Synthesis Report.* Contribution of Working Groups I, II and III to the Fifth Assessment Report of the Intergovernmental Panel on Climate Change (Core Writing Team, RK Pachauri and LA Meyer eds.) Geneva, Switzerland: IPCC.

Jónsson G. 2010. The environmental factor in migration dynamics: A review of African case studies. *International Migration Institute Working Paper* 21: 1–34.

Konate MK, Guéye M, and Nseka Vita T. 2004. Scolarisation des enfants au Mali selon le profil des ménages et étude de leur maintien à l'école. Background paper prepared for the Education for All Global Monitoring Report 2003/4. Gender and Education for All: The Leap to Equality. Accessed August 2010. http://unesdoc.unesco.org/images/0014/001467/146798f.pdf.

Laxen J. 2007. Is Prosopis a curse or a blessing? An ecological-economic analysis of an invasive alien tree species in Sudan. University of Helsinki, Viikki Tropical Resources Institute. *Tropical Forestry Report* 32: 203.

Little PD, Smith K, Cellarius BA, Coppock DL, and Barrett C. 2001. Avoiding disaster: Diversification and risk management amongst East African herders. *Development and Change* 32(3): 401–33.

Mertz O, Mbow C, Reenberg A, and Diouf A. 2009. Farmers' perceptions of climate change and agricultural adaptation strategies in rural Sahel. *Environmental Management* 43(5): 804–16.

Nelson V, Meadows K, Cannon T, Morton J, and Martin A. 2002. Uncertain predictions, invisible impacts and the need to mainstream gender in climate change adaptation. *Gender and Development* 10: 51–9.

Nielsen JØ and Reenberg A. 2010. Cultural barriers to climate change adaptation: A case study from Northern Burkina Faso. *Global Environmental Change* 20(1): 142–52. Accessed January 7, 2015. doi:10.1016/j.gloenvcha.2009.10.002.

Porro NM. 2010. For a politics of difference. In Tsikata D and Golah P, eds. *Land tenure, gender and globalisation: Research and analysis from Africa, Asia and Latin America.* New Delhi: Zubaan. 271–94.

—— and Stone S. 2005. Diversity in living gender: Two cases from the Brazilian Amazon. In Colfer CJP, ed. *The equitable forest: Diversity, community and resource management.* Washington, DC: Resources for the Future and CIFOR. 242–55.

Pramova E, Locatelli B, Brockhaus M, and Fohlmeister S. 2012. Ecosystem services in the national adaptation programmes of action. *Climate Policy* 12(4): 393–409.

Reyes R. 2002. Gendering responses to El Niño in rural Peru. *Gender and Development* 10(2): 60–69.

Randall SC. 2005. Demographic consequences of conflict, forced migration and repatriation: A case study of Malian Kel Tamasheq. *European Journal of Population* 21(2): 291–320.

Rankin K. 2001. Governing development: Neoliberalism, microcredit, and rational economic woman. *Economy and Society* 30(1): 18–37.

Ribot JC. 1996. Climate variability, climate change and vulnerability: Moving forward by looking back. In Ribot JC, Magalhães AR and Panagides SS, eds. *Climate variability, Climate change and social vulnerability in the semi-arid tropics.* Cambridge: Cambridge University Press.

Rocheleau D and Edmunds D. 1997. Women, men and trees: Gender, power and property in forest and agrarian landscapes. *World Development* 25(8): 1351–71.

Scheffran J and Gioli G. 2013. The role of remittances in building climate-resilient communities: Migration-for-adaptation in Western Sahel. Accessed January 7, 2016. ssrn.com/abstract=2250741.

——, Marmer E, and Sow P. 2012. Migration as a contribution to resilience and innovation in climate adaptation: Social networks and co-development in Northwest Africa. *Applied Geography* 33: 119–27.

Sen A. 1987. *Hunger and entitlements: Research for action.* Helsinki: World Institute for Development Economics Research.

Skutsch MM. 2002. Protocols, treaties and action: The climate change process viewed through gender spectacles. *Gender and Development* 10(2): 30–39.

Sun Y, Mwangi E, and Meinzen-Dick R. 2010. Gender, institutions and sustainability in the context of forest decentralisation reforms in Latin America and East Africa. *CIFOR Infobrief.* Bogor, Indonesia: Center for International Forestry Research.

Sydie RA. 1994. *Natural women, cultured men: A feminist perspective on sociological theory.* Vancouver, Canada: UBC Press.

Tacoli C. 2002. *Changing rural-urban interactions in sub-Saharan Africa and their impact on livelihoods: A summary* (Vol. 4). Working Paper Series on Rural-Urban Interactions and Livelihood Strategies, *IIED Working Paper 7*.

Terry G. 2009. No climate justice without gender justice: An overview of the issues. *Gender and Development* 17(1): 5–18.

Tompkins EL and Adger WN. 2004. Does adaptive management of natural resources enhance resilience to climate change? *Ecology and Society* 9(2): 10. Accessed November 29, 2010. http://www.ecologyandsociety.org/vol9/iss2/art10.

Turner BL, Kasperson RE, Matson PA, McCarthy JJ, Corell RW, Christensen L, Eckley N, Kasperson JX, Luers A, Martello ML, Polsky C, Pulsipher A, and Schiller A. 2003. A framework for vulnerability analysis in sustainability science. *Proceedings of the National Academy of Sciences of the United States of America* 100(14): 8074–9.

Turner MD. 1999. Merging local and regional analysis of land-use change: The case of livestock in the Sahel. *Annals of the Association of American Geographers* 89(2): 191–219.

UNDP (United Nations Development Programme). 1995. Human Development Report. Accessed November 12, 2010. http://hdr.undp.org/en/media/hdr_1995_en_contents.pdf.

——. 2009. Gender and climate change: Impact and adaptation. UNDP Asia-Pacific gender community of practice annual learning workshop. Accessed November 8, 2010. http://asia-pacific.undp.org/practices/gender/publications/Gender_and_Climate_Change-Impact_and_Adaptation-5.pdf.

UNEP (United Nations Environment Programme). 2009. Ecosystem management for improved human well-being in the Lake Faguibine system: Conflict mitigation and adaptation to climate change (draft). Accessed November 10, 2010. http://www.unep.org/pdf/Lake-Faguibine.pdf.

UNFPA and WEDO (United Nations Family Planning Association and Women's Environment and Development Organization). 2009. Making NAPAs work for women, climate change connections. Accessed October 25, 2010. http://www.unfpa.org/webdav/site/global/shared/documents/publications/2009/climateconnections_4_napas.pdf.

Van Der Geest K. 2011. North-south migration in Ghana: What role for the environment? *International Migration* 49(sl): 69–94.

Vignola R, Locatelli B, Martinez C, and Imbach P. 2009. Ecosystem-based adaptation to climate change: What role for policy-makers, society and scientists? *Mitigation and Adaptation of Strategies for Global Change* 14: 691–6. Accessed January 7, 2016. doi:10.1007/s11027-009-9193-6.

Watts MJ and Bohle HG. 1993. The space of vulnerability: The causal structure of hunger and famine. *Progress in Human Geography* 17: 43–67.

WIR. 2009. *World investment report 2009*. New York and Geneva: UNCTAD.

World Bank. 2011. Migration and remittances handbook, 2nd edition. Accessed January 18, 2016. http://siteresources.worldbank.org/INTLAC/Resources/Factbook2011-Ebook.pdf.

PART III

Gender and tenure

10

WOMEN AND TENURE IN LIBERIA AND CAMEROON

Solange Bandiaky-Badji, Cécile Ngo Ntamag-Ndjebet, Julie T. B. Weah and Jonah Meyers

Introduction

Over the last two decades, African countries (primarily in eastern and southern Africa) that have undergone land tenure reform have recognized the importance of women's land rights and increased tenure security, but also the formidable challenge of reconciling statutory and customary tenure regimes (Chigbu et al. 2015). In West and Central African countries, which have lagged behind due to political crisis and civil conflicts, some countries (Cameroon, Gabon, the Democratic Republic of Congo, Liberia, Ivory Coast and Senegal) have recently started new waves of land reforms and other sectoral reforms to move away from the colonial and postcolonial legacy.

Women in statutory and customary systems face particular challenges in securing their tenure rights to land and forest in national reform processes. While continued gender inequality in African local customs and institutions affect women's ownership and control of land and natural resources, some gains have been made in specific countries. The aim of this chapter is to draw examples from Liberia and Cameroon to highlight the advances and challenges to women's tenure recognition. We focus on these countries because both Cameroon and Liberia have undergone major reforms in the forest and land sectors over the last decade. Specifically, we discuss a promising collaborative initiative promoting gender justice and tenure by the Rights and Resources Initiative (RRI)[1] and the African

---

1 RRI is a global coalition of 13 core partners and more than 150 collaborator organizations engaged in forest and land policy reform in Africa, Asia, and Latin America. RRI's mission is to support local communities' and Indigenous Peoples' struggles against poverty and marginalization by promoting greater global commitment and action towards policy, market and legal reforms that secure their rights to own, control and benefit from natural resources, especially land and forests.

Women's Network for Community Management of Forests (REFACOF[2]—in French *Réseau des Femmes Africaines pour la Gestion Communautaire des Fôrets*) through its members in Cameroon and Liberia (such as the Foundation for Community Initiatives, FCI).[3] This initiative is an example of community initiatives advancing new vehicles for raising awareness, preventing reversals and advising policy in matters of tenure reform.

Over the past five years, RRI, its partners and collaborators have been actively involved in advancing Cameroonian and Liberian women's tenure and rights and their voices in reform processes. RRI raises awareness about the importance and challenges of promoting gender justice, and provides tools for addressing gender in forest tenure and governance reforms. In turn, the aims of REFACOF in Cameroon and FCI in Liberia include clear safeguards and language on women's rights to forest lands, and their protection. They insist that whether rights refer to ownership, access, control, or merely use of forests, they must be clearly stated in new laws and policies. They also focus on the gender-accountability mechanisms needed to ensure that women's rights to land and forest are protected.

Our research was carried out in the Republic of Cameroon and Liberia, two countries in which RRI has formed strategic and collaborative partnerships. While RRI partners and collaborators engage in myriad issues on land tenure, we focus here on gender and women's tenure rights. Cameroon and Liberia offer insights into the challenges that statutory, customary and community natural resources management systems pose for realizing women's land and forest rights and the security of their livelihoods in Africa.

Our findings are based on personal observations, reports from RRI partners and collaborators, consultations with local communities and women's groups, advocacy toward government officials, parliamentarians and traditional chiefs, and gender analysis of forest and land policy documents. These sources highlight the practices by which women's tenure rights are articulated, foregrounded and positioned in land and forest reform.

RRI adopts a gender-justice approach, which we define as a process seeking security for women's rights (access and control) to land and resources that seeks broad, representative and equitable participation of women and men in institutions that affect the governance, use and tenure of land and forest resources. We understand gender justice, at the national level, to be essential to any successful legislative or policy development process affecting the local expression of community rights. Women's forest and land tenure rights will not be secured unless the reform processes are able to address gender justice—embedding this as a goal of reforms. The reforms themselves should contain a requirement for monitoring and

2 REFACOF was founded in 2009 and is composed of members from West and Central Africa. Its mission is to carry out advocacy at the level of governments and the international development community to include women's land and forest tenure rights in reform processes and in their agendas.
3 FCI, founded in Liberia in 2004, works with women's groups in rural forest communities to help them gain a voice in how natural resources in their communities are managed.

evaluation of gender-disaggregated data (e.g. requiring the collection of statistics on the implementation of land reform) and that subsequent projects should measure whether the well-being, improved livelihoods, and other benefits of property ownership are increased among men and women as a result of the reforms. Our gender-justice approach fills a gap in current discourse by advancing analysis of the status of women's rights in legal and statutory frameworks. The goal is to better support advocacy related to women's tenure rights and gender-responsive policy making and implementation at the national and local levels.

In what follows, the first section situates gender/women as central to the understanding of land and forest rights in statutory and customary regimes across Africa where women's rights are at the nexus of the two systems. The second section focuses on women's tenure rights in examining the context of statutory and customary regimes as they pertain to women in Cameroon and Liberia. The third section highlights the strategic partnerships that have ensued in the process of advancing gender. The cases we present can help guide similar endeavors in other African countries, mainly in West and Central Africa, that are going through reform processes.

Women's forest and land tenure rights in sub-Saharan Africa

The constitutions and new legal frameworks in most African countries recognize equal rights of men and women. In Mozambique, the 1996 Land Law explicitly recognizes women's rights to inherit land, stating "the right of land use and benefit may be transferred by inheritance, without distinction by gender" (Ntsimi 2015: 16). In Rwanda, the 1999 inheritance law (incorporated into the Constitution in 2003) aimed to eliminate bias against female land ownership. The Kenyan 2010 Constitution espouses the "elimination of gender discrimination in law, customs and practices related to land and property in land . . . and regulates the recognition and protection of matrimonial property and in particular the matrimonial home during and on the termination of marriage." Emeka-Mayaka (2009) states that these provisions are crucial in ending widespread discrimination against unmarried, widowed and divorced women, often justified by customary law, though technically outlawed even by the old Kenyan Constitution.

However, the implementation of progressive laws has been a major challenge. Uganda has made limited progress in implementing its ambitious constitutional and legal reforms for land and forest rights, with similar lags experienced in Kenya (Almeida 2015; Kaarhus and Dondeyne 2015). While constitutions establish broad protections, there are still gaps in the statutes governing land rights and inheritance that leave many women vulnerable, and some countries' statutes still actively discriminate against women. What is more, many of the statutory tenure laws in East Africa have favored those women who are *legally* married, but have not taken into account the situation of *traditionally* married women and widows—common practices in those countries (Daley et al. 2010; Brown and Gallant 2014). Often, only legally married women are entitled to inherit land as a result of a death or divorce under the statutory system. When a marriage is not legally registered, there is a reversion to customary practices, which may be implemented differently from region to region within the same country (Brown and Gallant 2014). For instance, in Rwanda, the 1999 inheritance/succession law gives sons and daughters equal inheritance rights, provides women who are in a legally registered marriage property rights and requires spousal consent for the transaction of matrimonial property; it does not protect the rights of women in non-registered marriages, customary or religious marriages, polygamous marriages and other conjugal arrangements (Daley et al. 2010; Ali et al. 2011). Moreover, the Rwanda 1999 succession law was not retroactive, therefore women who were widowed prior to 1999 do not gain ownership of their deceased husbands' land, nor do their sons or daughters.

In other countries, statutory laws that appear gender-neutral can be harmful to women, as when based on patrimonial customary laws (Landesa 2012). For instance as stated by Almeida (2015), in Tanzania, brothers and sisters do not have equal inheritance rights and statutory law does not guarantee women the right to keep the matrimonial house in case of the death of their spouse. Women may access land through inheritance, but this is not guaranteed. Likewise, in Mozambique, there is no legal stipulation for women to inherit (Almeida 2015).

In customary systems, too, women face particular challenges in advancing their tenure and rights. Their traditional land rights come from ties to kin and husbands, and are contingent upon their status within the household and community. Older wives and women with more children may have higher status than younger women. Women also can lose rights upon divorce, especially if they have no sons (Gray and Kevane 1999).

Some scholars (Ali et al. 2011) assert that customary land practices formerly provided strong security in land tenure, but that threats such as public officials using eminent domain for private benefit have occurred, reducing this security. Customary regimes that are patriarchal and limit women's rights to own land have not adapted well to economic, social and political changes. Statutes that are over-complicated and that do not offset discrimination under customary regimes do not allow for women's realization of their legal rights (Gray and Kevane 1999).

There are conflicts and contradictions between women's rights to land in African statutory and customary laws. However, in some cases, like Cameroon, customary law has been incorporated into some statutes, as the Civil Status Ordinance requires marriages to be legally registered following customary procedure (Fonjong et al. 2012). In Uganda, despite the Land Amendment Act (Section 41) requiring spousal and/or adult children's (depending on the circumstances) approval before the transfer or sale of land, these rights are often ignored (Akullo 2015). The legal pluralism across statutory, customary and religious dictates create multiple and competing levels of authority and often contradictory rules that do not generally contribute to securing women's tenure rights. This is the case, for instance, in Cameroon and Liberia (Fonjong et al. 2012; Ntismi 2015). As Meinzen-Dick and Pradhan (2002, 27) argue, "Legal pluralism is likely to be especially significant where the resource base fluctuates (associated with ecological uncertainties) and the population exploiting it is changing (livelihood or social and political uncertainties)."

However, as stated by Almeida (2015, 6):

> . . . in countries like Mozambique and Tanzania, gender-based discrimination in customary law is superseded by the principle of non-discrimination in the Constitution. This means that, even when laws are not explicit about which rights women have, any customary right that discriminates against women can be deemed unlawful. In Tanzania, the laws are very specific and define exactly what rights women are entitled to within community-based tenure regimes.

Village rules in Tanzania are explicit about the requirement of female membership and participation in the Village Council, in which at least one-third of the 25 members should be women (Kimesera Sikar 2014). The Kenyan Constitution commits to "secure inheritance rights of unmarried daughters in line with practices of respective communities." But this effort to ensure unmarried daughters inherit a portion of their parents' land—though it is already happening in some communities—was among the divisive issues that led to the defeat of the proposed new Constitution in the 2005 referendum. The Kenyan Land Policy adopted in 2009 effectively creates/highlights a confrontation between tradition and modernity (Emeka-Mayaka 2009).

Reform processes have not been effective in achieving gender justice in securing tenure, improving livelihoods and incomes. Insecure ownership and limited control over assets result in missed opportunities for women to exercise decision-making power and thrive economically. In parallel, national legal protections and gender-advocacy programs have failed to advance women's rights in customary regimes and legally pluralistic systems. More secure access would improve women's access to food, water, raw timber and non-timber forest products, or allow them to use land and forests as assets for income generation and enterprises.

Women's forest and land tenure rights in Cameroon and Liberia

Cameroon and Liberia are currently reforming their land and forest sectors. Liberia is very advanced in the reform process, with the adoption of the Community Rights Law (CRL) in 2009 and the new Land Rights Policy, which was approved in 2013. A final draft of the Land Rights Act (LRA) was submitted to the president in July 2014. It is currently being reviewed by government agencies and ministries as well as by civil society groups; a new Land Law was expected to be enacted in 2015. Liberia, a post-conflict country, had undergone a lengthy peacebuilding and reform process.

Cameroon, meanwhile, has just finalized the review of its 1994 forest law. The review started in 2006, and a draft is now being scrutinized in the Prime Minister's Office. Cameroon officially launched its land reform process in 2011. Since then, civil society organizations, traditional chiefs, parliamentarians, indigenous networks and women's groups have submitted their position documents to government officials, describing how the new land law should take into account local communities', indigenous peoples' and women's rights. However, unlike Liberia,

Cameroon's land reform process has not been participatory. Unfortunately, stakeholders have heard about a leaked draft land law that was never discussed with the various interest groups. There is also the possibility that the new land law could be adopted by the president as an ordinance.

Cameroon and Liberia are signatories to the Convention on Biodiversity (Article 10c), the Millennium Development Goals (MDGs), the Convention on the Elimination of All Forms of Discrimination against Women (CEDAW) and other international documents, declarations and conventions that call for governments to take action and achieve women's equality. Both Cameroon's and Liberia's Constitutions recognize equal rights for men and women. In Liberia, the 1986 Constitution states, "all persons (men and women), irrespective of ethnic background, race, sex, place of origin or political opinion, are entitled to the fundamental rights and freedoms of the individual" (UN Women 2014). In Cameroon, the 1996 Constitution states that all persons have a right to own property and mandates equality of the sexes and principles of nondiscrimination. However, in forest policies and laws, women's specific rights are not clearly defined. Despite constitutional protections in place, most of the laws have not provided substantive and procedural requirements for the protection of women's rights.

The case of Cameroon

Cameroon's major regulatory and legal land and forest frameworks are the following: the Land Tenure and State Lands Ordinance (*Le Régime Foncier et Domanial du Cameroun*) of 1974 (currently being reformed); the Forest Law of 1994 and its Decree of Implementation of 1995 (*La Loi Forestière de 1994 et son Décret d'application, 1995*); Framework No. 96/12 of August 5, 1996, on the management of the environment (*La loi cadre n° 96/12 du 05 Août 1996 relative à la gestion de l'environnement*); and the National Forestry Action Plan (*Le Plan d'Action Forestier National*—PAFN) of 2003. A gender analysis of these laws by RRI (2012) reveals that no mention of women is made in these laws. Law no. 94/01 regulates forest, fauna and fisheries and contains a massive 171 articles, none of which mention gender or women's rights. The Forest Law of 1994 (under review) does not make any acknowledgment of women or the non-timber forest product (NTFP) industry on which many rural women rely for their livelihoods and financial independence.

Statutes governing marriage also have significant implications for women's rights to access and own land. The USAID country profile for Cameroon states that official marriage and marital property laws determine that

> . . . ownership of marital property depends on a marriage contract, which can allow for either: (1) separate ownership, in which each person manages separately the property each brings into the marriage; or (2) common ownership. The latter means that the husband holds the rights to manage the property, including the right of transfer without the consent of the wife.
>
> *(USAID 2011, 9)*

However, the vast majority of marriages in Cameroon are unregistered, recognized only in local and customary systems. This unfortunately results in most women's inability to keep their property or gain access to their spouse's property upon his death or divorce. Such gaps in policy and implementation as well as discriminatory practices make it nearly impossible for most women to own land even when they understand it is their basic right. For this reason, in 2011, the Government of Cameroon (GoC) presented a National Gender Policy Document for the period 2011–20 (DPNG—*Document de politique nationale genre 2011–2020*), which emphasized the government's commitment to act on the international conventions it has signed and incorporate gender equality into law.

Despite this National Gender Policy and progress in the GoC's attitude towards gender, little has changed and steps have not been taken to ensure that customary systems do not overpower vague national statutes. Although women are subject to a wide variety of customary land processes in Cameroon, typically their access to land and forests is very limited. Women's land rights included in statutory law are often outweighed by local customs, and may either not exist at all or depend upon the goodwill of local leaders, traditional chiefs, or the head of household. In most cases, women have only usufruct land rights, not a right to own land and rent it, sell it, or make dramatic changes to it.

One reason for this customary position is the view of women as "strangers" in patrilineal societies. Families and clans have an interest to keep land in the family, so allowing a woman who can marry into another family (whether daughter or widow) to own land is often disallowed. If the woman is allowed to own land or even inherits her deceased or divorced husband's land, many customary systems oblige her to forfeit the land upon remarriage. Women may also be denied ownership rights in favor of usufruct rights due to land scarcity, complex legislation allowing only one owner of a parcel or inter-community conflict, among other factors (Fonjong et al. 2012; RRI 2012).

Implementation of women's land rights remains a challenge for technical reasons as well. Rural women are more likely than men to be illiterate or to only speak a language other than French or English, the legal languages of Cameroon. Even if women have a grasp of English or French, they may have no knowledge of laws granting them land rights, and may not feel empowered to learn about their rights. Poorly written statutes combined with oppressive customary law and technical challenges are formidable barriers for Cameroonian women to realize their land rights.

The case of Liberia

Liberian statutes regarding forest management and land rights, like those in Cameroon, are somewhat vague.

The 2006 Forestry Reform Law was aimed at improving a former forestry law; however, it only mentioned gender in passing, stating that the Forestry Management Advisory Committee (FMAC) established by the new statute must "ensure that the interests of women and youth are fairly represented" (FAO 2006a, 14). Other government acts and documents, such as the Forest Development Authority's (FDA—the agency charged with leading forest reform in Liberia) Ten Core Regulations, use vague attempts to include all excluded groups. The Ten Core Regulations state that "The Authority shall use its best efforts to involve women, youths, and other historically excluded groups in each regional public meeting" (FAO 2007, 7). Yet, the FDA and other forest reform actors have been criticized for their lack of gender mainstreaming. Attempts at involving historically marginalized groups in tenure policy formulation and forest management have been superficial, often falling short of identifying specific mechanisms to increase the participation and representation of these groups in public meetings. Although there are several provisions noting the special situations and needs of women, none of these provisions considers a woman's challenges in keeping land a violation of her human rights.

For example, of the 17 Community Forest Development Committees (CFDCs) established by forestry legislation in Liberia by 2012, the maximum number of women members on any committee was two out of ten. Although by law, every committee must have female representation, this representation is limited to a small proportion of the CFDCs and their participation is also nominal; women typically hold positions such as chaplain or treasurer, and in some instances they are stripped of their responsibilities in these positions. As is the case in Cameroon, poor access to information about forest law reforms and illiteracy prevent rural women from effectively participating in decision-making processes (Weah 2012).

It is important to mention the political will of Liberian President Ellen Johnson Sirleaf, the first democratically elected female head of state in Africa, and her support of the Land Commission for the legal recognition of women's land rights. During a Reuters interview in 2013, President Sirleaf made a promise to Liberian women that "Women will have the full right to own their land like anyone else." Nevertheless, Liberian statutes have not done an adequate job of protecting women's rights, instead relegating that protection to customary law. The Land Rights Policy (LRP) "aims to give equal protection to the land rights of men and women," but under the section of Customary Laws (2.1). Nor does the LRP give any specifications about *how* women's tenure rights will be protected. The result is a reinforcement of women's usufruct rights but not their rights to own land. The soon-to-be-enacted New Land Law (2015/2016, NLL) will not only be a milestone along the road to gender equity but also an opportunity to influence both the legislation process and the implementation phase, to ensure effective ownership rights for women, with clear safeguards. According to the FCI and other women's organizations in the country, the NLL will be a "game changer" for the land reform process in Liberia.

The Liberian government has taken steps to address the exclusion and marginalization of women in forest and land policy. The Inheritance Law of 2003 seeks to remove the disparity between statutory and customary laws, giving women in customary marriages the same legal status as those in statutory marriages in terms of property and inheritance rights. The Inheritance Law states that a widow can now own one-third of her deceased husband's property including land, irrespective of the time of their marriage, and daughters have equal rights as sons to inherit land.

Despite this legislation, evidence on the ground shows that implementation remains a challenge because of a lack of knowledge at the grass-roots level and resistance on the part of customary elites. At the local level, decision-making power rests in the hands of older men, who may be opposed to women owning or even co-owning land. Women who have managed to realize the land rights guaranteed to them by law seem to depend on the approval of their menfolk (husbands, fathers, uncles, etc.) and the relationship with their in-laws upon the death of their husband, if applicable, to realize these.

Similar to Cameroon, there are longstanding traditions that put women in a position dependent upon the will of community leaders. Despite the previously mentioned requirement that women be present on CFDCs and participate in meetings regarding community forest management, their representation is poor and of little substance. Women may not be taken seriously if they do attend meetings with recommendations and meetings are typically held in English, which is less widely spoken by women than it is by men. Finally, many men believe that women do not have significant knowledge about forests and the livelihood options they supply to make informed decisions, despite their traditional exploitation of NTFPs.

Women organizing and strategizing for change

In both statutory and customary regimes, in the domestic and international arenas, rural women and women's advocates adopt various approaches and collaborate with other activists to advocate for stronger land tenure rights (Brown and Gallant 2014). For instance, some advocates believe that a multifaceted approach with dialogue among governments, affected women and customary leaders—among other stakeholders—is needed to strengthen women's land rights. They argue that focusing on only one system is a non-starter (Whitehead and Tsikata 2003). Bottom-up, grass-roots approaches are more effective if rural women position themselves as leaders and contributors in the process and if they partner with traditional leaders, headpersons and chiefs, who are generally men, to transform customary practices (Lankhorst 2012; Pritchett 2015). Empowering women to lead community mapping efforts is a prerequisite for improving land governance across a community.

Cameroon

Some progress has been made in Cameroon in recent years as rural women have worked towards obtaining their land, property and inheritance rights. REFACOF is a regional women's organization promoting women's forest and land rights, with country branches throughout West and Central Africa. REFACOF-Cameroon has organized women's groups throughout the country and formed several strategic alliances to help advance women's tenure rights. REFACOF-Cameroon organized meetings with the Network of Parliamentarians for the Management of Forests and Ecosystems in Central Africa (REPAR—*Réseau des Parlementaires pour la Gestion Durable des Écosystèmes Forestiers d'Afrique Centrale*), resulting in a position document for use by parliamentarians who will make recommendations in the forthcoming land law reform. They are advocating for easing the land title registration process to make it more accessible to vulnerable populations, to allow women to register land titles in spite of oppressive customary rules, and to develop implementation guidance more favorable to women and youth. REPAR also recommends the amendment of statutes to enable heirs of customary lands to become legal owners and to incorporate the recent national gender policy in the land law reform.

Additionally, several events incorporating women's land and forest rights have resulted in fruitful collaborations and targeted interventions. An advocacy campaign produced a new alliance between traditional chiefs and parliamentarians, with the inclusion of women members. The RRI coalition in Cameroon also organized the "Week on Tenure" campaign (in 2013 and in 2014), which helped increase public awareness on Cameroon's land reforms currently underway and broaden discussions on improving land registration to include customary practices. Among those participating were government officials, members of civil society, academics, tribal communities and parliamentarians. The 2013 campaign also produced the Common Position document of Traditional Chiefs in Cameroon. The position paper, the first of its kind in Cameroon, was presented before the Ministry of Domains, Territory and Land Affairs. The minister gave a speech recognizing the importance of including customary rights in current land reforms and committed to making the reform process as inclusive and participatory as possible.

REFACOF-Cameroon also organized meetings between women civil society organizations and the National Council of Traditional Chiefs of Cameroon (CNCTC) at both national and local levels to discuss the challenges women face in customary systems. The CNCTC released a list of positions, including the following aims directly relevant to women's land rights: "That custom is not an obstacle for women in land tenure. May the traditional leaders who administer the land customarily sensitize heads of families for an end to the frustration experienced by women holding land," and "That the heads of villages and customary authorities hold meetings with their subjects involving their daughters and women to inculcate in them the usual virtues of customary rights related to land" (Center for Environment and Development 2013, 8). The chiefs considered violations of women's rights to be the result of inaccurate interpretations of some customs and argued that the original customs of the forest zones of Cameroon are protective of women's rights to land ownership. They also mentioned the overlaps in religious practices and customs, which further compound the problem for women. The statements reflect progress in traditional leaders' views of women's rights and the need to incorporate women into statutory and customary law formation and interpretation.

Within the government, there are seven strategic ministries that have each designated REFACOF focal points, namely: Forests and Wildlife (MINFOF), Environment, Land Affairs, Agriculture, Tourism, Social Affairs and Small and

Medium Enterprises. REFACOF has also been engaged with the Ministry for Women's Affairs around the land reform process. Ministry focal points have attended meetings between REFACOF and REPAR, and were audience to REPAR's positions regarding women's forest and land rights in the reform process. REFACOF-Cameroon also organized a meeting with the Working Group of MINFOF to present its advocacy document with propositions and recommendations on how the review of the 1994 Forest Law should take into account women's tenure rights. Most of the propositions were included in the first draft of the new law.

Ongoing REFACOF-Cameroon advocacy and sensitization activities include dialogs with Cameroon's traditional women leaders (Queen Mothers) and wives of male traditional leaders. REFACOF-Cameroon has organized and held meetings with these groups of influential women, local and regional government officials and national ministry representatives to educate leaders on women's concerns vis-à-vis land and forest rights. The organization has noted a strong desire on the part of Queen Mothers and chiefs' wives to influence women's status.

Liberia

The Foundation for Community Initiatives (FCI) from Liberia has gained recognition by policy and decision makers as true advocates for women's tenure rights and forest governance through their multi-stakeholder engagements and participation in key decision-making workshops and events. FCI has engaged with local leaders and regional governments to strengthen the land and forest rights of women and other marginalized groups.

FCI undertook several actions to advance women's role in decision making on the Community Forest Development Committees (CFDCs) and increase women's understanding of current processes on forest use and governance in Liberia, including information-sharing meetings to spread awareness about the 2006 Forest Law Reform. These forums allowed women to actively engage with members of local and national governance structures like the Community Forestry Development Committee (CFDC), County Forest Forums (CFFs), the National Forest Forum (NFF) and other decision-making bodies with authority over natural resource management.

The FCI formed, built capacity and provided support to women's forest management platforms, and connected the platforms to the Ministry of Gender and Development (MoGD), the Forest Development Authority (FDA), and other government bodies. For the first time, women's NGOs came together to share information on their activities and resolved to work together in the future. The participation of MoDG in the activities of the project led to a promise by the ministry to become more engaged with women's groups.

Women's improved participation and representation in CFDCs has begun to have a significant impact on their lives. For instance, they have become increasingly proactive in their communities, taking leadership positions in various natural resource governance structures, while some local leaders advocate for and support them. This type of engagement has helped women's group members gain knowledge and access relevant forest-related documents. Local women can now voice their views on natural resource governance, creating greater accountability and transparency in the management of natural resources.

Conclusion

Scholars and women's rights advocates have made recommendations on how to secure women's tenure rights in statutory laws and in customary practice. Akullo (2015) suggests documenting customary laws in existence that uphold the land rights of women and their children, and having those laws adopted by more customary leaders. Ntsimi (2015) recommends specific legal provisions that take into account women's disadvantaged positions and provide incentives for addressing discriminatory practices, and clear responsibility for enforcement. Many women's rights pertaining to land tenure are dependent on their status and are subject to loss upon divorce or if no sons are born to them (Gray and Kevane 1999). As Scalise and Kamusiime (2015, 4) conclude, "Land rights must be legitimate, not vulnerable to change, granted for an extended period of time, enforceable, and exercisable without an additional layer of approval required."

RRI collaborators in Cameroon and Liberia such as REFACOF-Cameroon and the FCI have made progress in their efforts to secure stronger women's land and forest rights. As Liberia is about to adopt a new Land Law in Parliament and Cameroon will soon reveal its new land and forest laws and their implementation decrees, further lessons will soon be learned about women's comprehension of relevant legislation and their ability to claim their rights. Lessons about the types of necessary safeguards, how the laws will allow different groups (legally married, widows, not legally married, etc.) to benefit, and how to avoid potential rollback will also be forthcoming (Daley et al. 2010; Fombe et al. 2013).

For gender equality, social equity and environmental sustainability to be effective in Africa, women's rights to resources must be recognized in both customary and statutory regimes. Women's groups, networks and their advocates must collaborate with decision makers such as governments, local and traditional leaders, and parliamentarians as part of their advocacy strategy. Issues associated with land and forest tenure are cross-cutting and can only be resolved with the collaborative efforts of all relevant stakeholders. Women's full participation in decision making related to land and natural resources will reduce their marginalization and strengthen their ability to defend those rights. Especially in heavily forested countries like Cameroon and Liberia, where women rely on NTFPs for livelihood security, women's forest tenure rights are as valuable as women's land tenure rights. Specific attention needs to be given to both sets of rights to ensure women's livelihoods and independence and promote sustainable development. With continued buy-in from relevant stakeholders and targeted intervention, women's land and forest rights security should improve on the African continent.

Bibliography

Akullo LG. 2015. *Linking land tenure and use for shared property in Kaabong Karamoja.* Conference, Land and Poverty, World Bank, Washington, DC, United States.

Ali DA, Deininger K and Goldstein M. 2011. *Environmental and gender impacts of land tenure regularization in Africa: Pilot evidence from Rwanda.* The World Bank Development Research Group Agriculture and Rural Development Team and Africa Region Gender Team. Policy Research Working Paper 5765. Washington, DC: World Bank.

Almeida F. 2015. Are women's tenure rights an element of legal recognition of community-based tenure rights? The Rights and Resources Gender Working Paper (unpublished). Washington, DC: RRI.

AWC (African Woman and Child Feature Service). 2010. *Women's gains in the proposed constitution of Kenya.* Nairobi: Noel Creative Media. Accessed January 7, 2016. http://www.awcfs.org/dmdocuments/books/Women_Gains_Doc.pdf.

Belobo Belibi M, van Eijnatten J and Barber N. 2015. Cameroon's community forests program and women's income generation from non-timber forest products: Negative impacts and potential solutions. In Archambault CS and Zoomers A. *Global trends in land tenure reform.* New York: Routledge.

Brown J and Gallant G. 2014. *Engendering women's access to justice: Grassroots women's approaches to securing land rights.* Nairobi, Kenya: Huairou Commission, United Nations Development Program.

CED (Centre pour l'Environnement et le Développement). 2013. *Une proposition des chefs traditionnels pour la reforme du foncier rural au Cameroon.* Rights and Resources Initiative. Validée lors de l'Atelier de réflexion des Chefs Traditionnels et leaders autochtones sur le Foncier Rural au Cameroun Yaoundé, December 11–12, 2013.

Chigbu JE, Chigbu N, Ihesiaba NC and Emenging S. 2015. *Mainstreaming the female-gender participation in economic development of sub-Saharan Africa through non-discriminatory land titling and secured tenure.* Conference, Land and Poverty, World Bank, Washington, DC, United States.

Daley E, Dore-Weeks R and Umuhoza C. 2010. Ahead of the game: Land tenure reform in Rwanda and the process of securing women's land rights. *Journal of Eastern African Studies* 4(1): 131–52.

Emeka-Mayaka G. 2009. Kenyan daughters set to inherit family land. *Daily Nation* (Nairobi). Accessed January 7, 2016. http://allafrica.com/stories/200906261075.html.

FAO. 2006a. *Time for action: Changing the gender situation in forestry.* Report of the UNECE/FAO Team of Specialists on Gender and Forestry. Rome.

———. 2006b. *The state of food and agriculture 2010–2011: Women in agriculture: Closing the gender gap for development.* Rome: FAO. Accessed January 7, 2016. http://www.fao.org/docrep/013/i2050e/i2050e.pdf.

———. 2007. *Gender mainstreaming in forestry in Africa.* Regional Report. Rome.

Fombe LF, Sama-Lang IF, Fonjong L and Mbah-Fongkimeh A. 2013. Securing tenure for sustainable livelihoods: A case of women land ownership in anglophone Cameroon. *Ethics and Economics* 10(2): 73–86.

Fonjong L, Sama-Lang IF and Fombe LF. 2012. Implications of customary practices on gender discrimination in land ownership in Cameroon. *Ethics & Social Welfare* 6(3): 260–74.

———, Sama-Lang IF, Fombe LF and Abonge C. 2015. *Disenchanting voices from within: Interrogating women's resistance to large-scale agro-investments in Cameroon.* Conference, Land and Poverty. Washington, DC: World Bank.

Ghai JC and Ghai YP. 2010. Kenya: Don't waste the new Constitution. *Pambazuka News* (Kenya). Accessed January 7, 2016. http://www.pambazuka.org/en/category/features/66660.

Gnoleba SPO. 2015. *Advances with securing and protecting land rights from a gender perspective.* Conference, Land and Poverty. Washington, DC: World Bank.

Gray L and Kevane M. 1999. Diminished access, diverted exclusion: Women and land tenure in sub-Saharan Africa. *African Studies Review* 42(2): 15–39.

Kaarhus R and Dondeyne S. 2015. Formalising land rights based on customary tenure: Community delimitation and women's access to land in central Mozambique. *Journal of Modern African Studies* 53(2): 193–216.

Kimesera Sikar N. 2014. Women's security of tenure in the context of customary land rights: The case of Maasai women in Tanzania. Conference on Land and Poverty. Washington, DC: World Bank. Accessed January 7, 2016. https://www.conftool.com/landandpoverty2014/index.php?page=browseSessions&form_room=2&metadata=show&presentations=show.

Landesa. 2012. *Women's secure rights to land: Benefits, barriers, and best practices.* Gender Brief Paper. Seattle, WA: Landesa.

Lankhorst M. 2012. *Women's land rights in customary dispute resolution in Rwanda: Lessons from a pilot intervention by RCN Justice and Démocratie.* Brief. Focus on Land in Africa. Accessed January 7, 2016. www.focusonland.com.

Meinzen-Dick RS and Pradhan U. 2002. Legal pluralism and dynamic property rights. *CAPRi Working Paper* 22. Washington, DC: IFPRI. Accessed January 7, 2016. http://ageconsearch.umn.edu/bitstream/55442/2/capriwp22.pdf.

Ndangiza M, Masengo F, Murekatete C and Knox A. 2013. *Rwanda: Assessment of the legal framework governing gender and property rights in Rwanda.* USAID LAND Project. Kigali, Rwanda; Washington, DC: USAID. Accessed January 7, 2016. http://usaidlandtenure.net/sites/default/files/USAID_Land_Tenure_Rwanda_LAND_Report_Gendered_Nature_Land_Rights.pdf.

Ntsimi EMK. 2015. *Gender and land rights in Cameroon: Examining the REFACOF's efforts for the recognition of women's land right.* Conference, Land and Poverty, Washington, DC: World Bank.

Pritchett R. 2015. *Securing women's land rights in customary land tenure through the social tenure domain model (STDM): The case of Mungule Chiefdom, Chibombo District, Zambia.* Conference, Land and Poverty Washington, DC: World Bank.

Scalise E and Kamusiime H. 2015. *Women first approach for empowerment and land justice.* Conference, Land and Poverty. Washington, DC: World Bank.

Sittie JR. 2015. *Advances with securing and protecting land rights from a gender perspective.* Conference, Land and Poverty. Washington, DC: World Bank.

RRI (Rights and Resources Initiative). 2012. *Women and forest in Cameroon: Taking stock of gender in natural resource management in Cameroon.* Brief. Washington, DC: Rights and Resources Initiative. Accessed January 7, 2016. http://www.rightsandresources.org/publication/african-womens-rights-to-forests/.

UN Women. 2014. *World survey on the role of women in development 2014.* Accessed January 17, 2016. http://www.unwomen.org/~/media/headquarters/attachments/sections/library/publications/2014/unwomen_surveyreport_advance_16oct.pdf.

USAID. 2011. Cameroon Country Profile: Property Rights and Resource Governance. Accessed January 7, 2016. http://usaidlandtenure.net/sites/default/files/country-profiles/full-reports/USAID_Land_Tenure_Cameroon_Profile.pdf.

Weah, JTB. 2012 *Women and forest in Liberia: Gender policy and women's participation in the forest sector of Liberia*. Brief. Washington, DC: Rights and Resources Initiative. Accessed January 7, 2016. http://www.rightsandresources.org/publication/african-womens-rights-to-forests/.

Whitehead A and Tsikata D. 2003. Policy discourses on women's land rights in sub-Saharan Africa: The implications of the return to the customary. *Journal of Agrarian Change* 3(1–2): 67–112.

Widman M. 2015. Joint land certificates in Madagascar: The gendered outcome of a "gender-neutral" policy. In Archambault CS and Zoomers A, eds. *Global trends in land tenure reform: Gender impacts*. New York: Routledge. 110–26.

Wierenga M. 2013. Thomson Reuters hosts newsmaker with Liberian President Ellen Johnson Sirleaf. Washington, DC: Reuters. Accessed January 7, 2016. https://tax.thomsonreuters.com/media-resources/thomson-reuters-hosts-newsmaker-with-liberian-president-ellen-johnson-sirleaf/.

11

GENDER DYNAMICS IN ODISHA'S FOREST RIGHTS ACT

Priyanka Bhalla

Introduction and background

Since 1985, developing countries in Asia, Africa and Latin America have increasingly aimed to decentralize natural resources governance and transfer management authority to local communities (Larson and Dahal 2012; Sun et al. 2012; White and Martin 2002). One of the most dynamic policy sectors has been forestry, in which the emphasis has been on legitimizing access to and use of forest land and resources by forest-dwelling groups (Larson and Dahal 2012).

Within forest tenure reform, studies on gender dynamics are overlooked and under-researched (Colfer and Minarchek 2013; Pottinger and Mwangi 2011). Much is written about the *critical* number of women in decision-making bodies as a defining benchmark of women's participation (Agarwal 2001, 2009, 2010) and representation (Sawer 2012). However, understanding what type of *critical actors* or *acts* (defined below) lead to positive outcomes for women (Waylen 2008; Kittilson 2010) in forest tenure-reform processes needs further exploration. There is sparse literature—mostly qualitative in nature and case study based—on how women have participated in the FRA claims process, whether they have received titles or not and how much land they have received (Ramdas 2009; Bose 2011; Das 2014).

The existing literature on critical actors and acts emphasized legislative policy making in the West (Kittilson 2010; Childs and Krook 2009; Waylen 2008), with little attention to women's substantive representation in developing countries, particularly forest tenure reform and decision making at lower levels of governance. This chapter explores these gaps within the Indian context, studying critical actors (individuals), junctures (turning points) and acts (events) that have resulted in positive *processes and outcomes* (Franceschet and Piscopo 2008) for women in the implementation of the Indian Forest Rights Act (FRA), specifically in the southeastern state of Odisha.[1] The distinction between *process-oriented* versus *outcome-oriented* actions, as used here, follows:

> Process-oriented aspects include actions that profile women's issues, regardless of outcome. This may include placing women's issues on the agenda of the relevant committees, introducing gender responsive practices and policies, or networking and alliance-building. Outcome-oriented representation involves the actual transformation of policies and practices.
>
> *(Mwangi 2013)*

Here I ask "*What factors have led to positive processes and outcomes for women in Odisha's Forest Rights Act implementation?*" assessing "critical actors, junctures and acts" (Childs and Krook 2009). Critical actors are individuals or groups able to plan, execute and follow up on gender-friendly interventions in the FRA process. Critical acts are interventions that can "empower women or bring about women-friendly policy change" (Sawer 2012; Dahlerup 1988, 2006). Critical junctures are events or "windows of opportunity" (Kingdon 1984, 1994, 1995), which create an enabling environment for critical acts to occur. Examples may be changes in government, in legislation, or favorable actions by groups or individuals.

Positive processes include addressing gender-specific topics during committee meetings, women participating in claims-verification processes or recommendations by women and men in the post-rights process. I argue that one of the key factors leading to a positive process in FRA implementation is women's meaningful participation. Therefore, this chapter centers on the level of women's participation in FRA committees. Outcomes refer to larger, policy-driven changes, such as an additional article in the FRA legislation, increasing women's decision-making power and membership in the claims committees. The main FRA decision-making bodies analyzed are the claims committees. The focal women include those most affected by FRA implementation—usually tribal women, but also lower- and upper-caste Hindu women. Sub-questions include:

- Who are the critical actors, what are the critical acts, leading to increased women's participation and decision making in the tenure reform processes?
- To what extent and how are women participants and decision makers in the FRA claims committees, specifically at village (Forest Rights Committees), block (Sub-divisional Committee) and district levels (District Level Committee)?

I argue that women's meaningful participation within Odisha's FRA committees has been limited; however, building on my qualitative fieldwork and theoretical strands on "critical mass" and "acts," I found that there have been important examples of critical actors, able to use critical acts and junctures to further gender-friendly processes and outcomes.

1 See Bhalla (in preparation) for more on FRA implementation in Odisha.

BOX 11.1 CRITICAL MASS VERSUS CRITICAL ACTS

"Critical mass" refers to the percentage of women needed (10–40 percent (Sawer 2012), 33 percent being most common) in any policy arena, for women's issues to be placed high on the agenda or for women-friendly policy change to be implemented (Kanter 1977a; Childs and Krook 2009). "Critical actors and acts" refer to individuals or group-led critical actors performing acts in women's interest, with transformative potential (Dahlerup 1988, 2006).

Below, I outline my methodology before providing a brief background on the FRA and women's participation in its implementation. This is followed by a summary of the findings; then analysis and discussion of findings, ending with a presentation of conclusions and policy recommendations.

Methodology

Data collection took place (March 2013–March 2014) in four districts[2] of Odisha and in New Delhi (India), at varying intervals (Figure 11.1). Odisha has three types of forest land tenure, the second highest tribal population in India (62 groups, of which 13 are particularly vulnerable) and is considered "successful" for FRA implementation. My methods included:

1. field observation of the FRA claims process at the state, district and village level;
2. sixteen Focus Group Discussions (FGDs) with FDA-affected community members (seven mixed, eight women only, one men only; either title holders, applying for titles, or aware of the FRA) and,
3. sixteen interviews with FRA claims committee members (village, block and district levels; seven female, nine male)—both male and female members on FRA claims committees were interviewed on an individual basis.[3]

---

2 Mayurbhanj, Sambalpur, Kandhamahal and Nayagarh districts (of 30) were purposively chosen to maximize within-sample difference, based on the following selection criteria: Ecological and socio-economic conditions; proportion and type of resident scheduled caste and tribal populations; district-level FRA status of implementation, and levels of participation by women in the FRA claims committees. In each district, two to four villages were purposively chosen, based on the following selection criteria: villages which have received or applied for individual or community rights titles, proximity to district and sub-divisional offices, multi-ethnic versus homogenous populations, proximity to protected areas and influence of mining activities.

3 Some interviews were conducted by other NGOs in Odisha.

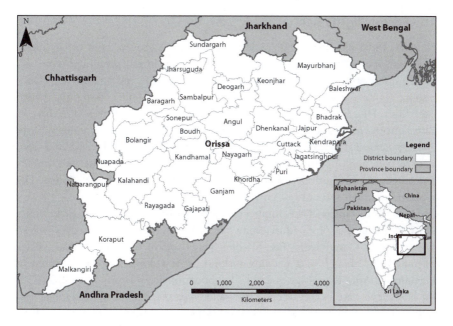

FIGURE 11.1 Map of India, showing Odisha and surrounding states[4]

All interviews were transcribed and coded using NVIVO 10 software. All interviewees were asked about:

- positive and negative perspectives on FRA implementation, including anecdotes of mechanisms that worked well,
- the role and participation of marginalized groups, such as women, the landless and other forest dwellers. There was an emphasis on sharing examples of women's contribution in FRA implementation, and
- views on debates with conservationists, private developers and Maoist conflicts.

In the FGDs, I asked village-level community members about their awareness of the Act, how/if they had applied for claims, what the impact had been and their relationship with the forest department. Questions for the FRC committees delved more specifically into how the committee was constituted, how members were elected, the decision-making pattern for men and women and whether any critical acts are noted (Bhalla in preparation).

---

4 Orissa's name was officially changed to "Odisha" in March 2011. Map source: http://www.mapsoftheworld.com.

The Forest Rights Act

In 2006, India passed the FRA,[5] legitimizing tribal groups'[6] and other forest-dwellers'[7] access to and use of ancestral forest lands. It also promoted their collective conservation and management of forested areas and solidified a claims system, giving the option to apply for both individual and community-owned land titles. Described as addressing "historical injustices" (FRA 2006), this Act came into force in 2008. Its implementation has received mixed reviews (Sarin 2012; Saxena 2011; Ramdas 2009). Interpretation of the Act has differed from state to state, and opposition has been strong among conservationists and forestry bureaucrats. Conservationists fear forest dwellers may further degrade forests. The forest department fears loss of power.

Decision making in FRA implementation consists of four levels, each with a different committee. Individual claims are first collected and verified by village Forest Rights Committees (FRCs), then passed on to the *Gram Sabha* (village assembly) and thence to a decision-making process dominated by government bureaucrats. After the village assembly, a subdivisional-level committee (SDLC) at the block level verifies and accepts claims and passes them to the district-level committee (DLC), which makes the final decisions about claims. FRCs usually have 10–15 committee members, with 4–6 on both the SDLC and DLC.

Gender provisions in the FRA state that FRCs must have at least one-third female membership, the village assembly, at least 50 percent women present, and both the SDLC and DLC must have at least one female member of the three required non-government members. Female committee members are usually not meaningful participants, simply attendees (exceptions discussed below). When claims titles are given to married couples, both husband and wife must be included as primary title holders, and female heads of households must be listed as primary title holder, not her closest male relative (Section 2g, Rule 4(2), Rule 3(1), Rule 5 (c), Rule 7 (c), Section 4(4), Indian FRA).

Despite these provisions, NGOs claim that women's participation in FRA implementation is low at all levels (Saxena 2011; Ramdas 2009). In India, land is mainly owned by men, whether inherited or personally acquired. Patriarchy remains a problem, with male bias in communication, inadequate sharing of

---

5 Officially, "The Scheduled Tribes and Other Traditional Forest Dwellers (Recognition of Forest Rights) Bill 2006."
6 Eighty-four million Indians are considered *Adivasis* or members of "scheduled tribes" (2011 Indian Census). Derived from the British "schedule" of tribes created in 1872, 635 tribal groups are listed in the Indian Constitution, most living in forests and mountains (Gupta 2012; Bose 2011). They fall under four main language/tribal groups: Santhals, Munda, Gond and Bhil. Despite an Indian Constitutional right to customary lands they inhabit, *Adivasi* history has been mired in bitter forest-related battles (Gupta 2012).
7 "Other Forest Dwellers" are non-tribals living in the same areas as tribal community members. They can be scheduled caste members or carry out low-status vocations, such as blacksmiths. Although entitled to apply for FRA claims, their eligibility criteria are different, e.g. they must prove 75 years of residency on the land they are claiming.

information and frequent exclusion of women in policy events; these place women at a disadvantage in decision-making processes (Sarin 1997; Singh 2009; Agarwal 2010).

Globally, women's overall involvement and participation in natural resource management is constrained by "Insecure rights to forest and trees, which constrains their incentives for undertaking sustainable management practices and further limits the range of actions they can take with regard to forest management" (Meinzen-Dick et al. 1997, cited in Sun et al. 2011: 207). In Andhra Pradesh, the titles women received drastically reduced the land they had previously worked, from three to five acres to one-tenth of that size (Ramdas 2009). Similarly, investigations on FRA implementation by a government committee in Odisha and Andhra Pradesh showed that women had little to no awareness of the FRA content or claims process (Saxena 2011). In Uttar Pradesh, women from lower castes, landless women and women from particularly vulnerable tribal groups (PVTGs) are forming their own women's rights committees for forest protection and conservation, disillusioned with their exclusion from early FRA implementation processes (Malik 2012).

A recent analysis of eight district government status reports[8] found little information on female claimants, as records were not disaggregated by sex. Districts with high rejection rates for female claimants (Ganjam and Keonjhar) did not give reasons for the rejections; and in Mayurbhanj, the applications cited as those from female claimants only (i.e. from women-headed households), were in reality titles distributed to married couples in which the woman was listed as a dependent (Das 2014).

In Kandhamahal, a central and heavily forested Odisha district, with a high population of scheduled tribes, women appear more empowered than in other communities, with resulting better chances for meaningful participation. However, as shown below, such conclusions cannot be generalized, even there. Despite these discouraging results, I argue below that opportunities for enhancing women's active participation in the claim process due to critical acts and junctures do exist.

Voices from the field

Community members had mixed feelings about FRA implementation and the factors leading to positive processes and outcomes for women. Four major themes emerged:

1. levels of participation and exclusion of women and men in the FRC, SDLC and DLC,
2. changes in non-timber forest product (NTFP) collection since the Act's implementation,
3. experiences of 'other forest dwellers' with more stringent eligibility criteria, and
4. differential perspectives of men vis-à-vis women on Forest Rights Committees.

8 Courtesy of Sujata Das, former program officer at Vasundhara.

These themes are summarized below and analyzed for critical actors, junctures and acts that can lead to positive processes and outcomes for women in FRA implementation.

Participation and exclusion in the FRA committees

Forest Rights Committees (FRCs)

Women were asked about their level of participation in the claims application process, trainings conducted by NGOs, and selection and membership in the FRCs. Their answers, across the board, were that it was low to non-existent. Initial land verification and the claims application procedure were usually carried out by the FRC president and another FRC member (usually high-caste male community members), supporting staff from an NGO and the relevant government official. In Nayagarh, for example, an older female community leader expressed:

> When the District Welfare Officer came, we were not included, the men of our village did not come get us and we also did not go of our own accord. The awareness-raising meeting was without women. We are the ones protecting the forest, but we still do not feel we have a right/control over it because non-tribals who are committing theft in our forest keep on demanding that we need to show them the land title, give them proof that we have gotten forest rights [under the FRA].

There were exceptions: In Mayurbhanj, an active male district collector tried to ensure more inclusiveness in the process. He organized separate sessions with female community members on land verification, customs of the village and claims content. However, awareness-raising programs and claims procedures were still male and government-driven.

Despite low participation in the awareness-raising programs and claims procedures, women still had more participation in the FRCs than in higher-level committees. This is largely because members have a personal stake in the process, apply for individual titles themselves, or are part of the decision-making process for gaining community rights. FRCs were usually constituted by government-appointed teachers in communities and were almost always chosen by the *Gram Sabha* (village assembly). In some cases, FRC members appointed themselves. No elections were reported for the constitution of the committee (differing from state to state).

The minimum 33 percent quota of female membership imposed on FRCs has been problematic for Bhil tribal women in Rajasthan (Bose 2011). Prior to the FRA village committees, female membership also differed from state to state. In Gujarat, after FRA implementation, each village-level committee is required to have at least one-third female membership, which represents an improvement over previous policy conditions. However, Bhil women interpret the one-third quota as preventing female membership from reaching a majority or parity (Bose 2011). In addition, older, more experienced tribal female members have often been replaced by women more integrated in mainstream Hindu culture, resulting in misrepresentation of cultural and livelihood needs of the tribal population in the process of rule making and enforcement.

FRC members interviewed in Ranpur (in Nayagarh), with a history of strong women's forest protection groups, spoke of their exceptional woman president. She was chosen because of her past leadership posts, committee positions, higher caste, old age and involvement with an NGO, which had been working with their community for many years. This was the only committee with more female (seven) than male (three) members. Neither its men nor women members were elected. Rather, past male leaders nominated themselves, hand-picking female members involved in forest protection in the past, including the female FRC president. The agenda is usually set by the female FRC president, and decisions are taken by consensus after discussion. There is no hierarchy in decision making as there is in other FRCs, where decisions are taken by male members, with women following suit.

As a critical actor, the female president was able to influence and solve a longstanding forest land-boundary issue with another village. Community members of her village collect NTFPs on Village B's forest land. In negotiations for continued access to their land post-FRA, Village B agreed to let only tribal individuals access their land, rejecting use by scheduled castes (SCs). She brought together tribal and SC members of their community with those of Village B. They discussed topics like forest protection, important to both villages. Repeated interaction on shared interests resulted in continued access to the neighboring land, resulting in a positive outcome for all.

Singh, who has written about community forestry in Ranpur and the evolution of women's roles in the forest protection committees, emphasized "democratic spaces across scales" (2009) there, wherein women have different opportunities to exert their influence. She writes:

> . . . change can come from certain locales and spaces: Everyday action and organizing by women to address their common problems may in the process allow them to gain voice and confidence to negotiate space at other scales . . . closer attention to constraints and possibilities at different scales can help practitioners and development workers design strategies that are able to leverage space and foster processes that lead to participation and action at other scales.
>
> *(2009: 66)*

By meeting regularly and identifying troubling common issues, Singh's women's committee succeeded in breaking down cultural and patriarchal barriers, eventually joining the male-dominated executive committees. Singh goes beyond examining the critical mass of women and identifies spaces and scales more likely to create critical actors, acts and junctures.

Interviews with FRCs in heavy mining areas (e.g. Sambalpur and Keonjhar) brought mixed results. Some FRCs in mining areas are better functioning than in areas without such powerful obstacles, because village inhabitants share joint goals. Some villages in Sambalpur have populations displaced by the Hirakud Dam, built in 1966. These populations became landless overnight and have been fighting to regain what they lost. In the 1980s, mining companies started entering these areas to explore coal reserves. Many have illegally acquired land and continue to do so, despite legislation such as the FRA.

In one such village, Puranakhinde, there have been attempts to obtain individual and community rights under the FRA. Their FRC is not just a space to discuss their claims process, but has become a body in which they discuss all important village-related matters. This differs from other study sites, where too many committees trying to address too many issues created inefficiency and a lack of focus.

Officially constituted in 2008 by a community teacher, Puranakhinde's FRC was quickly disbanded: there was strong infiltration by mining interests and not enough tribal and female representation. It was reconstituted after publication of the FRA amendment rules in 2012,[9] a critical juncture for FRA implementation in Odisha. From a total number of seventeen members, six are female, eleven are tribal.

One of the critical actors on this committee is a woman, chosen because she is a vocal community member, a community leader's wife, and has previous experience on other village committees. She has a lower level of forest dependency than poorer women in her village, giving her the luxury of time. She mobilizes other women on the committee, ensuring that all six women are present at each monthly meeting. Despite being a woman of higher caste and wealth in her community, she acts in the interest of both women and men. She shares the community's concerns, as all members experienced displacement and landlessness, and are struggling against takeover by the mining companies.[10]

Members of FRCs are all community members and rarely have official positions with the block-level government. This reduces government bureaucrats' control and provides more room for village-level critical actors to emerge. However, the same cannot be said for the subdivisional- and district-level committees.

Subdivisional- and district-level committees

Unless they were government officials themselves, most female committee members at the SDLC and DLC level are elected representatives. The majority are high-caste, non-tribal, Hindu women, without a personal stake in the process. At the DLC level, they are sometimes hand-picked by the District Collector. Non-governmental male members of the DLC and SDLC were usually informed

9 Prior to the FRA amendment rules, many stated they did not know how to implement the rights written in the legislation. The FRA amendment rules clarified, for example, how committees should be formed and run and how to conduct land verification.

10 Information supplemented by Ms. Puspanjali Satpathy, Vasundhara.

of meeting times; however, they often did not receive agenda items, and were excluded from the overall decision-making process conducted by government members. A male DLC member in Keonjhar said:

> There is no consensus or voting in the decision-making process, they just read what they have decided and ask if everybody is in agreement or not. Then they move on. We have no idea how these decisions were taken, so how we are going to oppose?[11]

Another male DLC member in Mayurbhanj considered the FRA amendment rules a critical juncture. Selection of non-governmental DLC members before the publication of the implementation rules was ad hoc and political. Titles were handed out by government officials in order to meet targets, rather than undergoing any formal process. Local political party power-holders appointed themselves, even without knowledge of the FRA.

Despite a more formal selection process, most members still know little about the Act. The one female member on the Mayurbhanj DLC committee[12] is considered a token, with low awareness of the FRA. There are no fixed meeting times and the committee operates in a similar fashion to the DLC in Keonjhar—run by the government bureaucrats, shutting out other committee members in the actual decision-making process. DLC members interviewed across districts agreed that most work in title verification and awareness raising takes place at the subdivisional level and that committee members should receive funds from the government for monitoring of the Act's implementation and community awareness-raising activities.

At the subdivisional level, female members complained that government officials control the committees and are not transparent about timings and locations of meetings, often calling members at the last minute. Often women had little awareness about the FRA, as did non-governmental male SDLC committee members. A female SDLC member in Keonjhar lamented:

> After I became a committee member, I received an invitation only twice. For the first one, I received the official letter only after the meeting was over and for the second one my brother had to take me on his bike as the sub-collector's office is 30 kilometers away from my residence . . . I have attended only one meeting and was not informed about the agenda . . . If we do not have any idea about the agenda and all the decisions have already been taken, how can we participate in the meeting?

Such experiences are not uncommon. Sarin (1997) and Agarwal (2001, 2009), experts on women's participation within India's Joint Forest Management Program (JFM), explored the constraints to women's participation, the heavy patriarchal

---

11 Interview courtesy of Subrat Kumar Nayak, Vasundhara.
12 As of March 2014.

influence and widespread tokenism, whereby women are put on executive committees as an obligation rather than as substantive members with voice and influence.

NTFP collection post-FRA implementation

All the women interviewed (and many men), regardless of background and age, overwhelmingly focused on NTFP collection, with most women citing it during FGDs as their greatest concern. A village in Mayurbhanj, for example, was relocated from a protected forest nearly two decades ago. No FRC had been constituted there, but the district government had given them community forest rights to fulfill targets. A female community member elaborated:

> We do not know what the actual area is of the title, but we will go everywhere to collect. Most of our interaction is with the forest guard and forester; we do not have contact with higher officials. Before [the relocation] we did not have problems collecting and selling NTFPs, but now the jungle is far away, and difficult to get to. In order to continue being allowed to collect NTFPs, we have to share some with the forest guards on the way.

Such findings fit with previous interviews with Bhil tribal women in Rajasthan (2007–09; Bose 2011), which revealed that the FRA has had a negative impact on both decision-making rights and access to forest areas. The new FRA village-level committees are seen locally as disadvantageous because they restrict collective access rights to NTFPs, whereas the old JFM committees did not. Specifically, Bose's respondents mentioned that the FRA committee:

> . . . restricted the collection of Jatropha seeds by not recognizing it as a NTFP because of its high commercial value as a source of biodiesel fuel at local markets. Before the village FRA committee was established, Jatropha seed was regarded as a NTFP and all respondents had access and rights to collect it in large quantities.
>
> *(2011: 226)*

In southern Kandhamal, women from PVTGs found restrictions on NTFPs, making them more dependent on the commercial market. Most women felt that receiving individual and community rights had made little difference to NTFP restrictions: The forest department still has a stronghold. One exception was in northern Kandhamal, where an NGO had worked with the forest, revenue and tribal departments at the district level for over two years to ensure that both individual and community rights were received and that community members considered themselves owners and managers of their land. There were two critical actors: one male NGO program manager from Kandhamal, with detailed knowledge of both village concerns and the district government; and a senior, female government official who worked closely with tribal communities and has the power to implement decisions quickly. The processes and outcomes these critical actors implement are positive for the whole community; the village has become a model for the rest of the state.

In other communities, both men and women still feared the Forest Department (though their relationship with it had improved after receiving titles) and were unable to take complete control over the management of forest land due to the history of state control. A woman in southern Kandhamal commented that more restrictions were placed by the Forest Department after receiving titles:

> First our situation used to be good, there were good trees around us, bamboo used to grow well and there were many NTFPs that we could collect. But now the forest department says that you can only collect NTFPs from the land that falls under the FRA title you received, not from any other forest land.

The most cited women's issue in the FRA claims process is the plight of single women and widows whose claims are often not properly filed or recorded. As seen in past forest tenure reform processes in South Asia (Sarin 2012), widowed or single women may receive claim certificates, with male relatives in practice retaining decision-making control.

Experiences of "other forest dwellers"

Both men and women in the FGDs, but particularly "other forest dwellers" (OTFDs), complained about different eligibility criteria for scheduled tribe members and OTFDs. This brought about community fragmentation and dissonance within FRCs. Most OTFD community members were still waiting to receive their individual rights, or had been told by government officials to drop their applications altogether.

Male and female FRC members interviewed in Keonjhar spoke the most about community fragmentation between tribals and OTFDs in mixed village populations. This fragmentation was apparent in the committee functioning and activities. Two scheduled caste FRC members had attended the committee's first meeting, but felt excluded from the process. They were told they were not eligible to apply for individual title rights. One scheduled caste FRC member said:

> We never have meetings together, except once in the beginning. But they [tribal members of the FRC] never wanted us to get involved in the process. And we also avoided it, as a [government] officer suggested we cannot apply for forest land now.

Another scheduled caste member expressed:

> The FRC president, with some other selected members, only takes decisions, he never consults or involves us . . . they won't consider us members

and they never ask for our opinion. We have no office, no financial or technical help—how then will the committee work effectively?

A female FRC member, also from a PVTG, was frustrated with the functioning of the committee and commented: "I am not satisfied, as we have no meetings. People have possession of five to ten acres of land; however, after the claims process, they received only one to two-acre patches. . . . "

The FRC president remarked that the OTFD (SC) members are uncooperative, being ineligible to apply for claims. He also felt that the committee should be strengthened, but needed monetary and technical support.

Views of male FRC members

Men most often brought up the issue of gaining individual rights to land they had been cultivating. They see this as a symbol of legitimate, state-recognized property rights, giving them access to government-sponsored development programs previously unavailable. In addition, they believe individual rights can prevent interference from non-tribals who have illegally taken land from them in the past. Men, especially FRC members, were concerned about the future—how would they ensure this individual and community land would be well developed? Some complained about receiving too little land or rights to plots that did not match the areas requested in their claims application.

In Kandhamal, two male members (one tribal, the other an older, upper-caste man) expressed the most contentious issues: the exclusion of OTFDs in claims submissions, filing the Right to Information forms to find out about rejected titles and preventing further forest land depletion by urban outsiders with commercial interests. Neither can be considered critical actors for women or gender-related policies. The claims submission process and finally receiving the claims, whether individual or community, seemed to be the primary goal and aim of these FRC members.

The next section discusses the implications of these results; *who* the critical actors are and *what types* of critical junctures and acts help create more positive processes and outcomes for women.

Critical actors, acts and junctures

Within the FRCs, critical women actors include those with a history of community leadership (such as Nayagarh's FRC president) or whose spouses have such a history. Additionally, being higher caste and wealthy help women become critical actors as they have more free time available. These qualities do not ensure they will act for women, but give them more chances to meaningfully participate in committee forums than other members, male or female.

As Bose (2011) shows from her Rajasthan research, the majority of tribal women in the FRCs tend to be higher-caste women mainstreamed into Hindu society and culture. This pattern can be seen at the DLC and SDLC levels in Odisha, but not necessarily in the FRCs, as the tribal population is much more diverse.

At the SDLC and DLC level, being a governmental or non-governmental committee member also matters. As a non-governmental committee member, the chance of exclusion from decision making is much higher, regardless of gender. The biggest constraining factor in the SDLC and DLC is state control, which allows little space for non-governmental actors to influence the process. In addition, political party control, caste and ethnic background of the committee members all play a role.

Within the history of the FRA the main *critical juncture*, not just for women's participation in FRCs, but also for overall implementation of the FRA was the 2012 amendment rules. These clarified many processes, including the constitution of FRA claims committees and land verification processes by different government departments. They also provided clarification of individual versus community rights. This critical juncture spurred the reconstitution of many FRA claims committees, especially at the FRC level.

These amendment rules increased the types of *critical actors* as individuals at higher levels of bureaucratic leadership began taking initiatives to ensure implementation of the amendment rules. Individuals at the District Collector level in both Kandhamal and Mayurbhanj were proactive post-2012, ensuring that joint land verifications not only included all three government departments (tribal welfare, forest and revenue), but that non-governmental actors were involved and that women were interviewed separately when such verifications took place. Including women of varying backgrounds at different levels of the process remains a challenge, though a start has been made.

Critical acts in the context of FRA implementation and within the functioning of the claims committees differ from village to village. Women in leadership positions at the FRC level elicited more confidence in their communities if they initiated acts that affected the whole community, rather than just one segment. In Ranpur, a series of negotiations took place so that multiple communities could continue collecting NTFPs from traditional areas. In Sambalpur, it was a longstanding effort to eliminate landlessness. At the SDLC and DLC level, dominated by government-appointed bureaucrats, potential critical acts, such as gender-disaggregated data collection, occur in isolated pockets of the state, and only if a government order has been issued by higher authorities (e.g. Ministry of Tribal Affairs).

Introduction of women-friendly issues within the FRA claims committees are *process oriented*. There is no strategy to work towards a specific outcome; rather it is an issue-based process, where any village problem, whether gender related or not, is addressed within the committee. Larger *outcome-oriented* developments are not evident in the current implementation of the FRA in Odisha, as any policy-related change would be long term and difficult to implement. However, smaller outcomes are evident (as in Mayurbhanj and Kandhamal): There is some gender-disaggregated data on FRA land titles distributed to women, and efforts from district leadership and civil society to include women in the claims application process. As FRA implementation only began in 2008, analysis of the long-term impacts of these processes will have to wait.

Women's substantive representation in FRA formulation, implementation and evaluation, from the community to the national level, must be addressed. There are entry points for interventions, which could create an environment for enabling both male and female critical actors to contribute more substantially to women-friendly policy change. Some of these interventions are explored below.

Conclusion and recommendations

Meaningful participation of women in the FRA claims committee process at all levels is limited due to a state-dominated forest tenure reform process, patriarchal strongholds at the community level, and continued patterns of tokenism. Nevertheless, there are examples of better-functioning FRA claims committees, especially at the village level, thanks to critical actors who put women-friendly issues on the agenda and mobilize female committee members.

Critical junctures such as the FRA amendment rules ensure that at least one-third female membership exists in most FRCs. Progressive leadership at the district magistrate level has the potential to create critical acts, leading to more positive FRA implementation outcomes for both men and women. Meaningful participation for women has been infrequent and difficult to implement at the subdivisional and district level, especially for non-governmental committee members who are not part of the traditional bureaucratic structure. Longer-term, systemic changes such as those listed in Box 11.2 will have to occur for greater women's participation, and therefore more positive processes and outcomes at the subdivisional and district level.

The following policy recommendations are suggestions in progress:

BOX 11.2 POLICY RECOMMENDATIONS

At the national level:
 Issue updated amendment rules, which:

- Increase women's minimum membership in the FRC, SDLC and DLC from one third to half.
- Mandate that the head of every alternate FRC be a woman.
- Ensure that all non-governmental committee members in SDLC and DLC are not just PRI representatives, but civil society members with proven track records on the FRA.
- Require mandatory serving times for committee members which are not based on national election cycles.
- Integrate a module on gender mainstreaming within FRA training for committee members.
- Disaggregate data by gender, caste, ethnicity and OTFD-related categories.

At the state and district level:

- Exchange best practices and lessons learned.
- Continue such exchanges on land verification for individual and community rights and recording of FRA titles.
- Initiate exchanges on gender mainstreaming and integration of marginalized groups (e.g. women, OTFDs) in the FRA implementation process.
- Distribute Odisha's progressive FRA government circulars to other states.
- Hold consultations at district level where members of the FRC, SDLC and DLC discuss critical acts and explore how these can be similarly implemented in other parts of the district.

Within civil society, there needs to be a greater distinction between NGOs and sustained movements, long in place. It is sometimes difficult to identify NGO actors with genuine knowledge about forest rights/history and the current FRA claims process. Such actors often fail to include women when the verification or awareness-raising process takes place. The few consultations on women's role and participation in the FRA process have usually been at national or state level, with action committees, of indeterminate value, as an outcome. Consultations are needed instead with community women to clarify customary land boundaries, from which female claimants may have been left out, and discuss what future strategies and ideas they envision on forest management. Involving them in rights recognition processes can serve as a confidence builder among women at the micro-level (Colfer and Minarchek 2014).

Coalition building—through which critical actors and acts can emerge—should occur at the subdivisional level; the district level is too large, bringing difficulties with transportation and meeting times. For the actual FRA claims process, the subdivisional government should appoint perhaps two male and female community leaders/FRC members in every village, tasked with holding separate consultations with women and men and ensuring that female claimants are included.

Finally, it is important to place more emphasis on process versus outcome developments, addressing women's meaningful participation and their critical acts in forest tenure reform processes. Further research on key critical actors and critical junctures is needed to understand when and how enabling factors are created. In addition, cross-state and cross-country comparisons monitored for at least ten years, and more quantitative analysis would result in a higher level of substantive analysis.

Acknowledgments

I would like to thank Esther Mwangi and the Center for International Forestry Research (CIFOR) for funding part of my field research in Odisha. I also thank Vasundhara, Jitendra Sahoo, Tushar Dash, Puspanjali Sathpathy and Subrat Kumar Nayak for giving me invaluable guidance and help whilst in Odisha.

Bibliography

Acharya KP and Gentle P. 2006. Improving the effectiveness of collective action: Sharing experiences from community forestry in Nepal. CAPRi Working Paper No. 54. Washington, DC: International Food Policy Research Institute.

Agarwal B. 1997. Environmental action, gender equity and women's participation. *Development and Change* 28(1): 1–44.

——. 2001. Participatory exclusions, community forestry and gender: An analysis and conceptual framework. *World Development* 29(10): 1623–48.

——. 2007. Gender inequality, cooperation, and environmental sustainability. In Baland, JM, Bardhan, PK, and Bowles, S, eds. *Inequality, cooperation, and environmental sustainability.* New York: Russell Sage Foundation; Princeton, NJ: Princeton University Press.

——. 2009. Gender and forest conservation: The impact of women's participation in community forest governance. *Ecological Economics* 68(11): 2785–99.

——. 2010. *Gender and green governance: The political economy of women's presence within and beyond community forestry.* Oxford: Oxford University Press.

Aggarwal A. 2011. Implementation of Forest Rights Act, changing forest landscape and "politics of REDD+" in India. *Resources, Energy and Development* 8(2): 131–48.

Agrawal A. 2001. Common property institutions and sustainable governance of resources. *World Development* 29(10): 1649–72.

—— and Chhatre A. 2006. Explaining success on the commons: Community forest governance in the Indian Himalaya. *World Development* 3(1): 149–66.

—— and Ostrom E. 2001. Collective action, property rights and decentralization in resource use in India and Nepal. *Politics and Society* 29(4): 485–514.

——, Yadama G, Andrade R, and Bhattacharya A. 2004. Decentralisation, community, and environmental conservation: Joint forest management and effects of gender equity in participation. CAPRI Working Paper No. 63. Washington, DC: International Food Policy Research Institute.

Alston L, Libecap G, and Schneider R. 1996. The determinants and impact of property rights: Land titles on the Brazilian frontier. *Journal of Law, Economics and Organization* 12: 25–61.

Bhalla, Priyanka. In preparation. *The Indian Forest Rights Act: A path to greater inclusion in Indian Forests or a detour towards further marginalization?* PhD Dissertation, Lee Kuan Yew School of Public Policy, National University of Singapore, Singapore.

Bose P. 2011. Forest tenure reform: Exclusion of tribal women's rights in semi-arid Rajasthan, India. *International Forestry Review* 13(2): 220–32.

Boserup E. 1981. *Population and technological change.* Chicago, IL: University of Chicago Press.

Bratton K. 2005. Critical mass theory revisited: The behaviour and success of token women in State legislatures. *Politics and Gender* 1(1): 97–125.

Bromley D. 2008. Formalising property relations in the developing world: The wrong prescription for the wrong malady. *Land Use Policy* 26: 20–27.

Bruce J. 1988. A perspective on indigenous land tenure systems and land concentration. In Downs RE and Reyna S, eds. *Land and society in contemporary Africa.* Hanover, NH: University Press of New England.

Carruthers B and Ariovich L. 2004. The sociology of property rights. *Annual Review of Sociology* 30: 23–46.

Childs S and Krook ML. 2009. Analysing women's substantive representation: From critical mass to critical actors. *Government and Opposition* 44(2): 125–45.

Colfer CJP and Daro Minarchek R. 2013. Introducing the "Gender Box": a framework for analyzing gender roles in forest management. *International Forestry Review* 15(4): 411–26.

Cornwall A. 2001. Making a difference? Gender and participatory development. IDS Discussion Paper No. 378. Sussex: Institute of Development Studies.

Dahal G, Larson A, and Pacheco P. 2010. Outcomes of reform for livelihoods, forest condition and equity. In Larson A, Barry D, Dahal G, and Colfer CJP, eds. *Forests for people: Community rights and forest tenure reform*. London: Earthscan. 183–209.

Dahlerup D. 1988. From a small to a large minority: Women in Scandinavian politics. *Scandinavian Political Studies* 11(4): 275–97.

Das S. 2014. An analysis of RTI information relating to the women and FRA. *Vasundhara Internal NGO Report*. Bhubaneswar, Odisha.

Davidson-Hunt K. 1996. Gender and forest commons of the western Indian Himalayas: A case study of differences. Paper presented at "Voices from the Commons," the Sixth Annual Conference of the International Association for the Study of Common Property, June 5–8, Berkeley, CA.

De Soto H. 2000. *The mystery of capital*. New York: Basic Books.

Ensminger J and Rutton A. 1991. The political economy of changing property rights: Dismantling a pastoral commons. *American Ethnologist* 18: 683–99.

FRA (Forest Rights Act). 2006. Officially known as The Scheduled Tribes and Other Traditional Forest Dwellers (Recognition of Forest Rights) Act, 2006.

Franceschet S and Piscopo M. 2008. Gender quota and women's substantive representation: Lessons from Argentina. *Politics and Gender* 4(3): 393–425.

Gadgil M and Guha R. 1991. *This fissured land—An ecological history of India*. New Delhi: University of Oxford Press.

Government of Orissa Circular. 2010. Accessed January 7, 2016. http://www.orissa.gov.in/stsc/FOREST_RIGHT_ACT/FAQs_(Circular_Community_Forest_Rights).pdf.

Granovetter M. 1974. Threshold models and collective behaviour. *American Journal of Sociology* 83(6): 1420–43.

Hilyard N, Hedge P, Wolvekamp P, and Reddy S. 2001. Pluralism, participation and power: Joint forest management in India. In Cooke B and Kothari U, eds. *Participation: The new tyranny?* London: Zed Books. 56–71.

Jewitt S. 2000. Unequal knowledge in Jharkhand, India: De-romanticizing women's agro-ecological expertise. *Development and Change* 31(5): 961–85.

Kanter RM. 1977a. *Men and women of the corporation*. New York: Basic Books.

——. 1977b. Some effects of proportions on group life. *American Journal of Sociology* 82(5): 965–90.

Kingdon JW and Thurber, JA. 1984. *Agendas, Alternatives, and Public Policies*. Boston: Little Brown.

——. 1994. Agendas, ideas, and policy change. *New Perspectives on American Politics*: 215–29.

——. 1995. The policy window, and joining the streams. *Agendas, Alternatives, and Public Policies*: 172–89.

Kittilson MC. 2010. Comparing gender, institutions and political behavior: Toward an integrated theoretical framework. *Perspectives on Politics* 8: 217–22.

Larson AM and Dahal G. 2012. Forest tenure reform: New resource rights for forest-based communities? *Conservation and Society* 10(2): 77–90.

Libecap G. 2003. Contracting for property rights. In Anderson and McChesney, eds. *Property rights*. Princeton, NJ: Princeton University Press.

Malik R. 2012. Concept paper for discussion on gender, land and agrarian reform in the state of Uttar Pradesh (UP), India. Presented to the UP Agrarian Reform and Labour Rights Committee, 2012.

Meinzen-Dick R and Mwangi E. 2008. Cutting the web of interests: Pitfalls of formalizing property rights. *Land Use Policy* 26: 36–43.

——, Brown L, Feldstein H, and Quisumbing A. 1997. Gender, property rights and natural resources. *World Development* 25(8): 1305–15.

Meola CA. 2012. The transformation and reproduction of gender structure: How participatory conservation impacts social organization in the Mamirauá Sustainable Development Reserve, Amazonas, Brazil, Doctoral Dissertation, Development Sociology. Cornell University, Ithaca, NY.

Mitra A. 2008. The status of women among the scheduled tribes in India. *Journal of Socioeconomics* 37: 1202–17.

Mwangi E. 2007. Subdividing the commons: Distributional conflict in the transition from collective to individual property rights in Kenya's Maasailand. *World Development* 35: 815–34.

——. 2013. CIFOR case for support on critical acts and actors.

——, Meinzen-Dick R, and Sun Y. 2011. Gender and sustainable forest management in East Africa and Latin America. *Ecology and Society* 16(1): 17. Accessed January 7. 2016. http://www.ecologyandsociety.org/vol16/iss1/art17/.

National Committee on FRA Report. 2010. December. New Delhi: Government of India.

North D. 1990. *Institutions, institutional change and economic performance.* New York: Cambridge University Press.

—— and Weingast B. 1989. Constitutions and commitment: The evolution of institutions governing public choice in seventeenth century England. *Journal of Economic History* 49: 803–32.

Olson M. 1971. *The logic of collective action: Public goods and the theory of groups.* Cambridge, MA: Harvard University Press.

Ostrom E. 1990. *Governing the commons: The evolution of institutions for collective action.* Cambridge: Cambridge University Press

——. 1999. Self-governance and forest resources. *Occasional Paper No. 20.* Washington, DC: Center for International Forestry Research.

——. 2005. *Understanding institutional diversity.* Princeton, NJ: Princeton University Press.

——, Gardner R, and Walker J. 1994. *Rules, games and common-pool resources.* Ann Arbor: University of Michigan Press.

Pandey D. 1993. Empowerment of women for environmentally sustainable development through participatory action research. In Ahmed S, ed. *Gendering the rural environment: Concepts and issues for practice.* Workshop proceedings, April 23–24. Workshop report #I. Institute of Rural Management, Anand, Gujarat, India.

Pandolfelli L, Meinzen-Dick R, and Dohrn S. 2008. Introduction: Gender and collective action: Motivations, effectiveness and impact. *Journal of International Development* 20(1): 1–11.

Pottinger AJ and Mwangi E. 2011. Special issue: Forests and gender. *International Forestry Review* 13(2): 1–258.

Quisumbing AR, Payongayong E, Aidoo JB, and Otsuka K. 2001. Women's land rights in the transition to individualized ownership: Implications for tree-resource management in Western Ghana. *Economic Development and Cultural Change* 50: 157–81.

Ramdas S. 2009. Women, forest spaces and the law: Transgressing the boundaries. *Economic and Political Weekly* XLIV(44): 65–73.

Sarin M. 1997. Integrating gender and equity sensitive conflict management in community forestry policies. In idem, ed. *Forests, trees and people programme, conflict management programme.* Rome: FAO.

——. 2012. Overview of India's tenure reform, 1992–2012. Rights and Resources Initiative, Washington, DC.

Sawer M. 2012. What makes the substantive representation of women possible in a Westminster parliament? The story of RU486 in Australia. *International Political Science Review.* Accessed January 7, 2016. doi: 0192512111435369.

Saxena CN. 2011. Women's rights to forest spaces and resources. *United Nations Entity for Gender Equality and the Empowerment of Women.* 33–39.

Schlager E and Ostrom E. 1992. Property rights regimes and natural resources: A conceptual analysis. *Land Economics* 68(3): 249–62.

Shipton P and Goheen M. 1992. Introduction: Understanding African land-holding: Power, wealth and meaning. *Journal of the International African Institute* 62: 307–25.

Singh N. 2009. Democratic spaces across scales. Women's inclusion in community forestry in Orissa. In Cruz-Torres M and McElwee P, eds. *Gender and sustainability: Lessons from Asia and Latin America.* Tucson: University of Arizona Press. 50–70.

Sun Y, Mwangi E, and Meinzen-Dick R. 2012. Forests, gender, property rights and access. *Info Brief 47.* Center for International Forestry Research, Bogor, Indonesia.

Torri AL. 2010. Power, structure, gender relations and community-based conservation: The case study of the Sariska Region, Rajasthan, India. *Journal of International Women's Studies* 11(4): 1-19.

Waylen G. 2008. Enhancing the substantive representation of women: Lessons from transitions to democracy. *Parliamentary Affairs* 61(3): 518–534.

Westermann OJ, Ashby J, and Pretty J. 2005. Gender and social capital: The importance of gender differences for the maturity and effectiveness of natural resource management groups. *World Development* 33(11): 1783–99.

White A and Martin A. 2002. *Who owns the world's forests? Forest tenure and public forests in transition.* Washington, DC: Forest Trends and Center for International Environmental Law.

Wiley LA. 2006. The commons and customary law in modern times: rethinking the orthodoxies. In *Land rights for African development.* CAPRi Policy Briefs. IFPRI, Washington, DC.

Yadama GN, Pragada BR, and Pragada RR. 1997. Forest dependent survival strategies of tribal women: Implications for joint forest management in Andhra Pradesh, India. RAP Publication 1997/24. Regional Office for Asia and the Pacific. Food and Agriculture Organization.

Yoder J. 1991. Rethinking tokenism: Looking beyond numbers. *Gender and Society* 5(2): 179–92.

12

TENURE VS. TERRITORY

Black women's struggles in the Pacific lowlands of Colombia

Kiran Asher

Introduction

Tenure is said to be among the key issues to have a bearing on sustainable forest management and gender equity (see the Introduction to this volume). But in a recent occasional paper on gender and forests in the Amazon, Schmink and Gómez-García lament that there is relatively little literature on women, gender and forest management in Latin America, and not a single citation referred to "tenure and/ or property rights in the region" (2015: 2). This would seem odd given that in the late twentieth century, many Latin American states have legally recognized indigenous and Afro-descendants' collective land and resource rights over 200 million hectares of mostly forested lands (Bryan 2012; Larson et al. 2008). Or as I argue, rather than lacunae in the literature, discussions of tenure rights in forested areas of Latin America, including the Amazon, are occurring as part of discussions about territorial rights and territoriality. These latter draw and build on the large and growing literature on land and agrarian reform in Latin America, and how these intersect with women's rights and gender inequality, especially in the context of neoliberal economic policies.[1]

Bryan reviews the growing literature in critical geography that examines what he calls the "territorial turn" in Latin America, which "references a particular conjunctural moment that can be approached historically, indexing a number of broader political, legal, and economic transformations" (2012: 216). Put differently, it marks a particular moment of tension and possibility when state attempts to clarify tenure

---

1 For example, feminist economist Carmen Diana Deere has written extensively on these issues in English and Spanish in mainstream development journals as well as for policy organizations. There are also many publications from organizations such as *Fundación Arias para la Paz y el Progreso Humano* in Costa Rica, and various NGOs and advocacy groups addressing issues of land and resource access for poor and marginalized groups, including women.

rights for the better functioning of markets intersect with rural demands for agrarian and land reform as part of the broader struggles for social justice. Both of these seemingly contradictory processes are part of the neoliberal governance that has been taking place since the 1980s. Within such a context, indigenous and minority groups in Latin America (and beyond) are framing their claims over ancestral lands as part of a broader demand for racial equality and self-determination. Their claims pivot on different understandings of national identity and claims over territory. Women play important roles in these movements even as they organize around their diverse livelihood and gender concerns.

This chapter outlines the experiences of one network of Afro-Colombian women, who organized to secure their livelihoods, validate their ethnic identity, and struggle for territorial rights in the densely forested Pacific lowlands region of Colombia. It draws on my long-term research on Afro-Colombian social movements and black women's key roles in it (Asher 2009, 2007, 2004).[2] My findings about the complex relations between forest-dependent people and community rights parallel insights from other research (Larson et al. 2010; Mogoi et al. 2012). They include a seemingly simple but crucial point: that forests are but one node in the diverse livelihood strategies of these groups. Furthermore, these communities are marked by heterogeneity of interests and identities (of gender, race or culture, occupation, religious affiliation, political ideology, and more). Thus, women's organizing involves negotiating *multiple* social relations, not just those with men (Manfre and Rubin 2012), and is shaped by and shapes broader political economic and socio-cultural dynamics. In what follows I discuss Afro-Colombian women's organizing and how they link to the broader black struggle of ethnic and territorial rights in the Pacific lowlands of Colombia. I conclude with some remarks on key lessons from that work for future research on sustainable forest management, tenure rights, and gender equity.

Economy, ethnicity and the environment in the Pacific lowlands

The Colombian Pacific Littoral is part of the natural resource rich Chocó bio-geographic region extending from southern Panama to northern Ecuador along the Pacific coast. A global biodiversity "hot spot," the region is home to a variety of ecosystems and myriad plant and animal species, many endemic to the Chocó. In the early 1990s, most of this region was yet to be overrun by drug cultivators and traffickers, guerillas or paramilitary forces. It was better known as a supplier of natural resources: timber, gold, platinum, silver, oil and natural gas. Ninety percent of its population is Afro-Colombian.

2 This includes 16 months of ethnographic research with members of women's networks and groups from 1993 to 1999, and subsequently regular visits from 2005 to 2014. I also draw on published and unpublished manuscripts about and by the women's network and by project coordinators to inform this discussion. For further details on the reflexive, feminist methods that inform my long-term conversations with dozens of Afro-Colombian activists, please see my monograph (2009) and longer journal articles.

In 1993, the Colombian government passed a law (Law 70), based on Article 55 of the 1991 Constitution, which recognizes Afro-Colombians as a separate ethnic minority and accords them various rights, including collective titles to their lands. The 1991 Constitution also introduced widespread neoliberal reforms to generate economic development and extensive environmental conservation measures to preserve Colombian biodiversity.[3] In order to implement these laws, the Colombian state launched numerous "sustainable development" initiatives to promote economic growth, conserve the environment and improve local living standards in the Pacific region.

The situation in the Pacific lowlands was marked by much hope, but also by many paradoxes. State officials and Afro-Colombian activists seemed to agree that the Chocó was rich in both "biodiversity" and "cultural diversity," that these diversities were interrelated and needed to be protected. But there was little consensus on what constituted traditional or culturally appropriate practices and what were the boundaries of black collective lands (Asher 2009). Negotiating these differences included coming to terms with the different ways in which "land rights" were defined by the state versus black communities. For the former, implementing land rights meant granting land tenure as part of the state's development programs for the region. Guaranteeing land rights would also satisfy the terms of World Bank loans to clarify property rights in order to reduce conflicts over resources and provide legal incentives for communities to manage their lands sustainably. In contrast, black communities were struggling for broad ethnic and territory rights over the Chocó region, which they consider their homeland.

It was within this context that there was a burgeoning of grass-roots organizing, including by black women at the local and regional levels. In their turn, state entities and NGOs working in the region sought the participation of Afro-Colombian communities, and especially that of black women in their development and conservation interventions. Key among these interventions were *Plan Pacífico* and *Proyecto BioPacífico* (PBP). The former was a large-scale, multilateral, donor-funded project whose aim was to develop the region's natural resources and stimulate economic growth. The latter (PBP) was a five-year biodiversity conservation program launched in 1992 with funding from the Global Environmental Facility and the United Nations Development Programme. With a mandate to devise mechanisms for the protection and sustainable use of regional biodiversity, PBP became linked to the economic aims of *Plan Pacífico*. As knowledge of *Plan Pacífico* and PBP spread, ethnic groups drew on new global discourses of "rights-based development" and "community-based conservation" and began to pressure the state to make good on its legal promises to recognize their rights. In response to such pressure, the collective titling of ethnic lands, local participation and the preservation of traditional

---

3 The economic and environmental reforms in Colombia were at least partially influenced by the concept of "sustainable development," which links issues of economic development with environmental conservation. They are also representative of reforms occurring across Latin America, and later other parts of the world.

knowledge of natural resource management became subsets of economic develop-ment and environmental conservation mandates.

Attention to local needs and gender concerns was related to the mandates of decentralization and participation outlined in the 1991 Constitution, and to the terms of international funding for *Plan Pacífico* and PBP. Development and con-servation entities tried to fulfill these mandates but saw participation as largely a technical and methodological challenge. Afro-Colombian groups, including wom-en's organizations, became involved with development and conservation projects as part of their multiple and contested struggles for black ethnic and territorial rights in the Pacific Lowlands of Colombia.

Black women's organizing: from development cooperatives to autonomous networks

Like women in other parts of the world, black women in the Pacific organize and work collectively around their quotidian tasks. With the waves of modernization in the region, many of the previous forms of organization and social relations began to change (Escobar 1995; Lozano 1996; Rojas 1996). In the first decade of post-World War II economic development, women, especially those in rural areas of the Third World, were considered part of the economically "nonproductive" domestic and subsistence sectors. Within policy circles, they appeared, if at all, as welfare recip-ients or targets of population control and poverty reduction programs (Braidotti et al. 1994). In the early 1980s, these representations gave way to "Growth with Equity" programs, which focused on "harnessing women's labor" for economic growth and integrating women into mainstream development processes. Such programs were strongly influenced by Ester Boserup's (1970) landmark book, *Women's Role in Economic Development*, which demonstrated that Third World women make a consid-erable contribution to productive sectors, especially in agriculture. These programs were framed by the "Women-in-Development" (WID) approach, which aimed to bring women up to par with men and to ensure that they received equal benefits from development.

These general trends were reflected in development plans for the Pacific. In the 1980s, women appeared as beneficiaries in the population, nutrition and rural health projects of integrated rural development initiatives and welfare pro-grams. Key among such programs were Colombian President Belisario Betancur's *Integrated Development Plan for the Pacific Coast* (PLADEICOP) and *El Plan de Hogares de Bienestar* (Plan for Welfare Homes), which was part of the previous President Virgilio Barcos's Social Economic Plan for the Pacific Region. While women were not the focus of "productive" agriculture intensification projects, large numbers of women later became integrated into the agro-industrial sector as low-paid menial workers on shrimp farms and African oil palm plantations. In 1988, probably influenced by WID approaches, an extended PLADEICOP launched a women's component to facilitate women's contribution to the produc-tive sector (Escobar 1995; Lozano 1996; Rojas 1996).

In the 1990s, PLADEICOP received funds from UNICEF and other international donors to encourage rural groups to establish savings and loan cooperatives. The aim was to form large consolidated groups in order to qualify for loans, get technical help to improve artisanal and agricultural production, and obtain institutional support for marketing their products—in short, to increase rural incomes. Several cooperatives, including some made up of women, were formed with PLADEICOP's support. Among them were *CoopMujeres—Cooperativa de Ahorro y Crédito de Mujeres Productivas de Guapi* (Savings and Loan Cooperative of Women Producers of Guapi) and *Ser Mujer—Cooperativa de Ahorro y Crédito* (To be a Woman—Savings and Loan Cooperative) in Tumaco, Nariño.

Up the coast in the port town of Buenaventura, Valle del Cauca, *Fundemujer—Fundación Para El Desarrollo De La Mujer de Buenaventura* (WomenFund—Foundation for the Development of the Women of Buenaventura) emerged in 1989. Women's health, especially pre- and post-natal care for young mothers, was a central concern, but the cooperative also supported women's productive activities.

Given the demographics of the region, most members of these cooperatives were Afro-Colombian, but membership was not restricted to black women. Cooperative members trained in micro-enterprise management, and established rotating, low-interest credit schemes and formal social solidarity networks, including emergency funds to aid women in times of acute domestic crisis. Besides the usual bureaucrats and development experts, a few *mestiza* and black feminists and social activists worked closely with these women's groups, often serving as consultants to local or federal state programs and regional NGOs. In keeping with the broader development aims in the region, the work of cooperatives was directed toward addressing women's basic needs (Rojas 1996). Given the aims and structures of the cooperatives, however, there was little room for their members to discuss issues such as sexism, domestic violence, or ethnic and racial discrimination—common aspects of experience though they were. Lozano (1996) calls these omissions a form of "neutrality" with respect to gender concerns within the cooperatives.

By the time of my research in 1995, women's cooperatives in the Pacific were successfully institutionalized. *CoopMujeres* claimed 122 members and *Ser Mujer*, 220. With 25 women's groups and a total of 800 members, *Fundemujer* was by far the largest cooperative. Income generation through productive activities was the central aim of most cooperatives but, with changes in the region's political economy and ethnocultural dynamics, new concerns began to appear on their agendas. While not quite as close to home, the build-up to the June 1995 United Nations Women's Conference in Beijing and international discourses about gender, development and the environment also had some bearing on the changes occurring in the cooperatives.

An important vehicle of change was the Program for Black Women, a new initiative funded through the Canadian and Colombian governments and *Fundación para la Educación Superior* (FES—Foundation for Higher Education), a Cali-based development NGO. The coordinators of this program were professionals with extensive experience working with gender and women's issues, and a strong social conscience. In 1994, the program began working with women's groups in the Chocó, funding projects on health, self-esteem, gender and ethno-cultural identity, sustainable natural resource use and management, and income-generating activities. The program's approach was implicitly informed by the Gender and Development (GAD) framework, which was replacing the integrationist WID perspective and aimed to "empower" women to become key decision makers in the household and community. Framed with critical insights from feminists and development professionals in the global North and South, GAD aimed to transform existing gender relations by addressing women's practical (everyday, immediate—*tener*) needs and their strategic (long-term, political—*ser*) gender interests and identities (Braidotti et al. 1995; Moser 1993; Kabeer 1994).

On April 3, 1995, I accompanied the project coordinators on their visit to *CoopMujeres* to brainstorm ideas for projects under the Program for Black Women. After introductions, one of the coordinators gave an overview of FES and its various social and gender projects. Then the director of *CoopMujeres* spoke about the cooperative, its membership, functions and future needs. Many members wanted technical training to learn about obtaining credit, keeping accounts, and increasing the production and sale of their goods. Others chimed in with suggestions for workshops on women's health, family relations, political participation and citizenship. One member asked if the cooperative could sponsor a talk about women and the meaning of Afro-Colombian ethnic identity.

Joining the discussion, the project coordinator observed how gender often seemed to be a secondary element in conversations about development, and floated the idea of self-esteem or women's rights workshops. The director of *CoopMujeres* responded:

> We already *know* our rights. Now we need to learn how to *obtain* our rights, teach other people about our rights. We need to educate our men about women's rights. Last year we celebrated Father's Day and many of our *compañeros* came. This year we are trying to make each member win her partner's support and bring him to the workshop.

Many of the members noted that their *compañeros* (in this context, the term means husbands or domestic partners) accorded them more respect when they brought more income into the household. For cooperative members, more egalitarian relations with their *compañeros* were tied to economic security and income generation.

After an animated exchange about the achievements and aspirations of the cooperative, members reached a consensus regarding its next goals: to concentrate on income-generating activities but also to expand the collective aims of the group to include exploring members' identities and rights as Afro-Colombians and women.

On the following day I met with cooperative members again. Among them was a schoolteacher, who had recently joined *CoopMujeres* because it was "organized, grounded and actively helped single women, single mothers, poor women, and heads of families." But she confessed that, after hearing about Law 70 and black rights, she was curious about "the fuss about black women." When I asked her about her own understanding about being a black woman, she said, "We are black women, joyous but still enslaved, still afraid. We still need to learn to value our dialects, our religion, our dances."

The schoolteacher's *compañeras* (in this context, the term means comrades, sisters-in-struggle, friends), including those from cooperatives in other Pacific coastal states, expressed similar sentiments. They wanted to continue to focus on income-generating activities, but also to expand the collective aims and aspirations of the groups to include attention to "black women's identities" and the struggle for ethnic and territorial rights. With an established organizational and institutional history behind them, Afro-Colombian women's organizing began taking on new forms and significance in the 1990s in the context of the changing political economy and cultural politics in the region.

The recognition of black ethnic and territorial rights under Law 70 created new opportunities and spaces for black women to articulate their concerns. The new *Plan Pacífico* also had resources slated for ethnic, gender and environmental projects, but these were still in draft stage and were yet to take effect. In the meantime, black women activists held region-wide meetings in 1990 and in 1992. At one of the meetings, they established the *Red de Mujeres Negras* (Black Women's Network) with aims to:

> . . . establish communication and solidarity between different women's and mixed organizations, to promote women's organizations through education and empowerment, to strengthen ethnic identity, to study the realities and the needs of women and to make women aware of the management and sustainable use of natural resources and the environment.
>
> *(Balanta et al. 1997: 40)*

Network participants agreed to abide by the principles of "autonomy, affirmation of ethnic-cultural identity, respect, responsibility and conscience" (Rojas 1996). More an idea than an institutional structure, the network took shape as member organizations conducted activities in its name. Among the activities were different kinds of attempts to secure forest-based livelihoods and consolidated collective land rights decreed under Law 70 of 1993. In the section that follows, I outline in some detail the activities of the *Red de Organizaciones Femeninas del Pacífico Caucano (Matamba y Guasá*, Network of Women's Organizations of Cauca), one group of organized black women from the state of Cauca in the Pacific Lowlands of Colombia.

"Seeing through the eyes of black women"—engendering ethnic and territorial struggles

The *Matamba y Guasá* Network of Women's Organizations of Cauca was established in 1996 to "consolidate their [black women's] struggles and to help them communicate with each other" (Red Matamba y Guasá 1997). Matamba y Guasá members engaged in a number of activities: growing food and medicinal plants, promoting informal and formal education, establishing health care and housing projects, helping to implement Law 70, the law that recognizes the ethnic rights of black communities. *Afro-Colombianas* claim that it is through these tasks that they fundamentally support the cultural, development and conservation activities in the Pacific region as black women.

Like rural women in many parts of the world, black women in the Colombian Pacific Lowlands work within and beyond the household. Black women's domestic and community chores include subsistence farming, taking goods for sale to the market, making *guarapo* and *viche* from sugarcane, and many other tasks.[4] Members of a group from Río Guapi, a key river in the Cauca state, note that it is around such work that women organize:

> The organizational force of black women comes from work, from life itself. When women go in search of *el chocolatillo* or to fish, they often leave for up to five days.[5] Others stay in the house. For example, ones who are pregnant or unwell stay back with all the children: If I go, I leave the children with the neighbor and she takes care of them. If I have an older daughter she takes care of all the children, including the neighbors' children. That is the tradition and it becomes a form of work.
>
> *(El Hilero 1998: 16)*

As noted above, women in several Pacific towns had formed small groups around their "productive" activities (such as baking, sewing, selling fish and produce) in the 1980s with the help of Colombian state programs. However, many women, especially from rural areas along the extensive network of Caucan rivers, could not participate in state programs because of the remote locations of their homes. The renewed general impetus of black organizing in the 1990s catalyzed the formation of many black women's self-help groups in rural areas. Referring to how one such group began, members note that it

> . . . emerges from a long history. It has some specific objectives: The recuperation of food and medicinal plants to ensure the subsistence and health needs of their community, and also to conserve biodiversity. The other aim is to strengthen organizational skills and training, especially among rural women.
>
> *(El Hilero 1998: 16)*

During several conversations with me, a key member of one of the groups and the regional coordinator of Matamba y Guasá told me that rural and urban black women wanted to get together to share their experiences and to broaden the organizational base. This desire led them to establish Matamba y Guasá in 1996 to bring together "women defending their ethnic and territorial rights, and working for the welfare of their families and their communities" (Red Matamba y Guasá 1997).

4 Guarapo and Viche are types of liquor from sugar cane, brewed usually by women.
5 *Chocolatillo* is a plant used in basket making.

By 1999, Matamba y Guasá consisted of 74 groups. Each group undertook activities based on its members' needs and experiences. The groups from Río Guapi promoted the use of plants from their *azoteas* and developed menus of traditional dishes; the groups in the Río Timbiqui region concentrated on raising pigs and chickens for food and sale, while the groups in Río Saija focused on extracting traditional products from local food crops (such as molasses from sugarcane). Other groups in the network formulated projects to build houses, promote primary health care, find better transportation to and from regional markets, and obtain basic education for black women and their children.

Matamba y Guasá drew the attention of state agencies and NGOs to the key role Afro-Colombianas could play in sustainable development and conservation enterprises. Subsequently, several Matamba y Guasá undertakings began receiving varying degrees of logistical and financial support from these entities. For example, "productive" activities such as raising chickens, or processing food crops for sale were sponsored by local development efforts. Under their mandates to conserve the region's biodiversity, national and international conservation projects supported Afro-Colombiana attempts to recuperate native food and medicinal plants. Because of their initiatives and efforts, Matamba y Guasá members were also called upon to participate in two key efforts to implement Law 70: a project to demarcate property boundaries and confer collective land rights, and another to outline community development plans in consultation with the local populace. In these ways, black women become key players in many regional organizations and activities such as forming community councils, leading ethnic rights workshops, and calling meetings to recognize and preserve ethnic diversity.

Engaging in these diverse activities meant navigating multiple networks of social and political relations. For instance, women's struggles are not always against men, especially since many of the women are heads of their households. Matamba y Guasá members also noted that they neither think of themselves as subordinate nor consider their work antagonistic to that of their menfolk. Yet, during my interview with coordinating committee members, one of them said, "It took a lot of work, especially for those who have men in their household." Of the 74 groups linked to Matamba y Guasá, 70 are for "women only." Men are welcomed to these groups, noted the *compañeras*, as long as they "behaved themselves." While men are allowed to participate in group activities and meetings, only women make decisions.

Members of the coordinating committee reflect their astute understanding of power relations within their struggles in the following statement about Matamba y Guasá's relations with black organizations in the region:

> We meet them in public spaces but we maintain our characteristics. We interact and reach a consensus but we do not want to get involved in clientelist networks. Rather than obtaining representation and power, we look for spaces of participation for black women.

Afro-Colombianas maintain black solidarity in public. However, they maintain organizational autonomy and remain critical of "obtaining power and representation" through the "clientelist networks" of mainstream party politics in Colombia.

Matamba y Guasá members also understand that their alliances with state and non-governmental entities are similarly fraught with tension. For example, one member who participated in state-sponsored "ethnoscience" workshops spoke positively of the information and knowledge exchanged during these workshops. But she also expressed skepticism about the utility of national "biodiversity data-banks" for local communities. She continued:

> We do not trust too many institutions and agencies. We speak with you because we know you especially through Fundación HablaScribe [a regional NGO]. But we prefer not to get involved in things we do not understand or with people and groups that we do not care about.

Recognizing that there are differences of needs and strategies among the groups linked to the network, each group made its own decisions about seeking funds for their activities. That is, black women's groups also had an astute understanding of how power in representation worked through development and conservation organizations.

There are differences among Afro-Colombian women—they are aware of them, they have chosen to identify commonalities among themselves and to form alliances across them, they reject that black women are primarily "victims" who need aid from the outside, they have a strong understanding of their self-worth, and they are keenly aware of the power of words, discourses and representation. These issues are underscored in the vision and political perspective of the organizations of the Network published in the 1997–98 Annual Bulletin of the Network:

> It is important to clarify that the meeting spaces [of the Network] are generated and constructed by us, with our own initiatives. We have been struggling for recognition of women in our region and to overcome [the obstacles to recognition]. Activities such as ours, imply sacrifices, imply surrender to make our dreams come true and to achieve the proposed objectives. Beginning from these principles today we are ready to identify ourselves as women and come together as a gender, to recognize our similarities and differences . . . We want to be considered protagonists of our lives and of our world.
>
> *(Red Matamba y Guasá 1997–98: 15–16)*

Attention to black women's activism within the political economic and socio-cultural context of the Colombian Pacific serves as reminder of the heterogeneity of black women's identities and interests.

Postcolonial feminists and feminists of color in the west (Mohanty 1991, Asher 2004) have repeatedly pointed out that for these women as for Third World women, struggles for gender rights are intertwined with issues of racism, capitalism (including its latest globalization variant) and nationalism. These insights help flag how black women's concerns and the responses of *Afro-Colombiana* networks emerged within the context of multifaceted, intertwined and mutually constitutive relations of power—of gender (as women), of race or culture (as blacks), of class (as poor people), and of location (as rural, Pacific residents), at a time when blacks were granted special rights and black social movements were in the process of "ethnicization" to translate new laws into concrete results. Not surprising, each struggle affected the others: meeting basic needs remained a central concern of black women's cooperatives because prevalent economic models destroyed or failed to provide adequate livelihoods. Nor had the broader black struggles yet provided concrete economic alternatives; productive activities therefore remained central to women's concerns. At the same time, as local leaders of the Network noted, these political economic concerns were linked to the marginalization and exploitation of Afro-Colombians. That *Afro-Colombianas* were aware of the multiplicity and complexities of these realities and were negotiating them skillfully is evident from the astute remarks of Network members. Stressing that the primary aim of the Black Women's Network was "revindication of ethnicity, gender and appropriation of territory," black women's groups attempted to strengthen the organizational links between and among black communities, and establish political alliances beyond the region (El Hilero 1998; Red Matamba y Guasá 1997, 1997–98). *Afro-Colombianas* were also keenly aware of their audiences, knowing well what could and could not be articulated within the fractured political economy and cultural politics of the Pacific.

Concluding remarks

Since the 1970s, concerns over the global environmental crisis and the future availability of natural resources for continued economic growth have led to alliances among academics, government agencies, national and international nongovernmental organizations (NGOs) and multilateral banks. Asserting the importance of biodiversity for global human welfare, these alliances focus on generating plans to conserve biodiversity efficiently and effectively, and promote sustainable economic development. Such alliances are taking a renewed importance in the context of twenty-first-century concerns about climate change. The focus on the capacities of women and indigenous communities in managing forests, and in adapting to and mitigating the effects of climate change is also taking on a new salience. But as the discussion of Afro-Colombian activism reveals, there are important differences in the ways the relationships between these groups, the state and the rights and responsibility of each play out. While development entities may think of the last in terms of tenure, social movements draw on broader and more political territorial claims.

Since the end of the 1990s, the political and economic realities of the Pacific Lowlands (as of other parts of Colombia) took a turn for the worse. Local people were increasingly caught in the crossfire of violence unleashed by the increased presence of armed forces (guerillas, paramilitary and military) and drug dealers in the region. Since 1999, an estimated 4 million Afro-Colombians have been involuntarily displaced from their homes. In the context of escalating violence (supported in good part by United States funding for the "War on Terror"), standard explanations such as "political corruption" or the need for "better development models" are used to explain why the basic human rights of Colombian citizens are not met. Nor does a focus on single variables such as "tenure," disconnected from broader dynamics, help assess how local, national and global processes of power are intertwined and impact the land and livelihood struggles of forest-dependent communities such as Afro-Colombians. They are also problematic because they rest on apolitical explanations of development and promote benevolence rather than a critical understanding of black women's capabilities. In short, they replicate colonial discourses.

In this chapter, I have argued for the need to situate black women's struggles geopolitically in order to understand the dynamic nature of domination and resistance, and the uneven and multiple power relations within which women act. I have also stressed the heterogeneity of women's movements, and traced how race and ethnicity intersect with gender, class and other factors to shape Afro-Colombian women's needs and activism. Lastly, I invoked postcolonial feminism to suggest to forestry scholars and gender experts that it is imperative to reflect critically on the desires and methods to conserve forests and better the lives of Third World women. That is, I bring into the discussion of forests and gender the reminder that development and conservation projects are projects of environmental and social change, and therefore political projects embedded within complex and uneven networks of power relations. Understanding these power relations and how they unfold in a particular location and sector are key to gender and forestry research and action.

References

Asher K. 2004. Texts in context: Afro-Colombian women's activism in the Pacific lowlands of Colombia. *Feminist Review* 78: 1–18.

——. 2007. *Ser y tener*: Black women's activism, development and ethnicity in the Pacific lowlands of Colombia. *Feminist Studies* 33: 11–37.

——. 2009. *Black and green: Afro-Colombians, development, and nature in the Pacific Lowlands*. Durham, NC: Duke University Press.

Balanta O, Rodríguez B, Sinisterra S, Quiñonez P, Arroyo L and Dinamizador, E. 1997. Red de mujeres negras del Pacífico: Tejiendo procesos organizativos autonomos. *Esteros* 9 (Febrero): 37–42.

Boserup E. 1970. *Women's role in economic development*. New York: St. Martin's Press.

Braidotti R, Charkiewicz E, Häusler S and Wieringa S, eds. 1994. *Women, the environment and sustainable development: Towards a theoretical synthesis*. London: Zed Books.

Bryan J. 2012. Rethinking territory: Social justice and neoliberalism in Latin America's territorial turn. *Geography Compass* 6: 215–26.

El Hilero. 1998. *Mujeres . . . con color y sabor a Chiyangua: Una entrevista*. Cali, Colombia. 15–17.

Escobar A. 1995. *Encountering development: The making and unmaking of the Third World*. Princeton, NJ: Princeton University Press.

Kabeer N. 1994. *Reversed realities: Gender hierarchies in development thought.* London: Verso.

Larson A, Cronkleton P, Barry D and Pacheco P. 2008. *Tenure rights and beyond: Community access to forest resources in Latin America.* Bogor, Indonesia: Center for International Forestry Research.

——, Barry D, Dahal GR and Colfer CJP, eds. 2010. *Forests for people: Community rights and forest tenure reform.* London: Earthscan.

Lozano BR. 1996. Mujer y desarrollo. In Escobar A and Pedrosa A, eds. *Pacífico ¿desarrollo o diversidad?: Estado, capital y movimientos sociales en el Pacífico Colombiano.* Bogotá: CEREC y ECOFONDO. 176–204.

Manfre C and Rubin D. 2012. *Integrating gender into forestry research: A guide for CIFOR scientists and programme administrators.* Bogor, Indonesia: CIFOR.

Mogoi J, Obonyo E, Ongugo P, Oeba V and Mwangi E. 2012. Communities, property rights and forest decentralisation in Kenya: Early lessons from participatory forestry management. *Conservation and Society* 10: 182–94.

Mohanty CT. 1991. Under Western eyes: Feminist scholarship and colonial discourse. In Mohanty CT, Russo A and Torres L, eds. *Third World women and the politics of feminism.* Bloomington: Indiana University Press. 51–80.

Moser CON. 1993. *Gender planning and development.* London: Routledge.

Red Matamba y Guasá. 1997. Segundo Encuentro-Taller Sub-Regional de Organizaciones de Mujeres del Pacífico Caucano "Matamba y Guasá": Fuerza y Convocatoria de la Mujer del Pacífico Caucano. Timbiqui y Guapi, Cauca: Fundación Chiyangua de Guapi, Grupo de Promoción de Santa Rosa de Saija, Asociación Apoyo a la Mujer de Timbiqui, Asociación Manos Negras de Guapi.

——. 1997–98. Visión y perspectiva política y organizativa de la Red de Organizaciones de Mujeres de Cauca: La Red es una familia numerosa. *Boletin Anual de Matamba y Guasá* 1997–98: 13–17.

Rojas JS. 1996. Las mujeres en moviemiento: Cronica de otras miradas. In Escobar A and Pedrosa A, eds. *Pacífico: ¿desarrollo o diversidad?: Estado, capital y movimientos sociales en el Pacífico Colombiano.* Bogotá: CEREC y ECOFONDO. 205–18.

Schmink M and Arteaga Gómez-García M. 2015. *Under the canopy: Gender and forests in Amazonia.* Bogor, Indonesia: CIFOR.

PART IV

Gender and value chains

13

GENDER AND FOREST, TREE AND AGROFORESTRY VALUE CHAINS

Evidence from literature

Verina Ingram, Merel Haverhals, Sjoerd Petersen, Marlène Elias, Bimbika Sijapati Basnett and Sola Phosiso

Introduction

Forests, trees and agroforests (FTA)[1] contribute to people's well-being in myriad ways. The many non-timber forest products (NTFPs)[2] derived from FTA resources are critical to the livelihoods of approximately 1.4 billion impoverished people in the world (FAO/IFAD/ILO 2010). Adding to the multiple food security, nutrition, energy, health and cultural benefits they provide, NTFPs contribute on average 20–25 per cent of annual household income for people living in and near forests in the developing world (World Bank 2004). The consumption and sale of NTFPs can be important particularly for marginalised groups, such as women, among others, whose limited access to land, credit and other assets hamper their ability to pursue alternate livelihood opportunities (Hasalkar and Jadhav 2004).

Although women have long remained in the shadows of agricultural and forestry research for development, the critical link between gender and forest-based livelihoods is gaining recognition. It has been realised that there is significant gender differentiation in the collection and trade of FTA products, which supports notions that there are distinctive 'male' and 'female' roles associated with FTA chains (Ruiz Pérez et al. 2002; Paumgarten and Shackleton 2011; Purnomo et al. 2014; Sunderland et al. 2014). This has led to widespread promotion of different products, particularly by organisations interested in sustainable development, enhancing gender equity and empowering women (Shillington 2002; Neumann and Hirsch 2000 in Shackleton et al. 2011). The gender aspects of forest and tree product chains are distinguished from agriculture-based chains (Doss 2002; Carr

1 FTA refers to the continuum of wild and managed ecosystems and individual trees from which timber and non-timber forest products are sourced (Ingram 2014).
2 While various definitions of NTFPs exist, they generally refer to products or services other than timber produced in forests (Belcher 2003).

2008; WDR 2008; Verhart and Pyburn 2010; Said-Allsopp and Tallontire 2014; Quisumbing et al. 2014) by two particularities. One is the governance arrangements covering access to the resource and to their markets (Meinzen-Dick et al. 1997; Wiersum et al. 2014). Many forest and tree species are wild, governed as public goods or common property as opposed to agricultural crops, which are often cultivated on private property (formally or customarily owned). This difference raises questions about how FTA products are managed, by whom and how access to and benefits from these resources are arranged. Research has highlighted the role of gender in shaping access, management and use of forest resources and their associated benefits (Mai et al. 2011).

How resources and the ecosystems in which they are embedded are governed affects their sustainability. This is the second particularity, as an FTA product's sustainability depends on factors such as

1. the abundance of the species from which the product originates,
2. anthropogenic and natural factors, such as forest degradation and climate change impacts on species populations,
3. inherent species vulnerability which depends on the part(s) used and harvesting intensity, and
4. a species's tolerance to harvesting (Ingram 2014).

Although there is conflicting information (Sun et al. 2011; Mwangi et al. 2011), some studies have shown that increasing women's participation in forest user groups and decision making results in improvements in the management of forest resources, whether at the community, household, or farm level, as well as enhancing livelihoods (Agarwal 2010a). The failure to support both women's and men's involvement in forestry-related processes and enterprises has squandered opportunities to improve the lives of women and their household members, and to promote the sustainable and equitable utilisation of FTA resources (IFAD 2008; Awono et al. 2010).

Unleashing the potential of FTA products for alleviating poverty, increasing gender equality and promoting ecological sustainability require understanding and engaging with global factors – e.g. policies, market trends and climate change – that affect the nature and extent of women's and men's participation in this sector. This is effectively achieved using a value chain approach, which has gained analytical purchase as a perspective from which to study the articulation of political-economic processes linking diverse geographic regions, people and goods. A value chain symbolises the activities involved in bringing a product from the production base to final consumers, including harvesting, transport, processing, transformation, packaging, marketing, distribution and support services, and disposal (Kaplinsky and Morris 2001). A chain can range from a local to global level and may be implemented by various actors – harvesters, processors, traders, retailers and service providers. These linkages show how livelihood systems at the local level are shaped by both local conditions (Pfeiffer and Butz 2005) and wider political, socio-cultural, ecological and economic factors, including underlying gender norms (Brown and Lapuyade 2001). What results are gender-differentiated roles in households and communities, gendered interests in and dependence on forest resources, gender-differentiated access, control and power to make decisions over FTA (and other) resources, and women-specific constraints in terms of benefit capture from chain participation (Riisgaard et al. 2010; Mai et al. 2011). Gender also intersects with other factors of social differentiation such as class, race, ethnicity, religion and age, to influence the relative bargaining positions of different interest groups (Mai et al. 2011).

Interest in FTA chains has increased over the last two decades, but there has been little consolidation of the relevant data on this topic. In particular, studies focusing on the relationship between FTA chains and gender, the factors that influence this relationship, and the nature of interventions seeking to enhance gender equality in FTA chains, are lacking. A more systematic understanding of the information available, the products and regions studied, and the nature and impacts of interventions can result in better targeted research and interventions in FTA chains.

To address this knowledge gap, we present a review of the literature on gender and FTA value chains with a focus on three research questions:

1. Where do gender differences exist within FTA value chains and what do they consist of?
2. What factors influence these gender differences?
3. What kind of FTA value chain interventions have been made and how can future interventions be more gender equitable?

Methodology

A literature review was conducted, consisting of the steps shown in Figure 13.1.

A multidimensional conceptual framework, summarised in Figure 13.2, was used to guide the review. The sustainable livelihoods approach (Krantz 2001) was combined with value chain analysis (Kaplinsky and Morris 2000) and a gendered lens (Colfer 2014). Using the sustainable livelihoods approach implies that asset-based livelihood and sustainability outcomes of interventions are analysed for the actors involved in the chain. The framework highlights governance, acknowledging that multiple governance structures (customary authorities, statutory government, projects, etc.) may set their own 'rules of the game' (Ostrom et al. 1994) and that hybrid arrangements (chain platforms and networks) can emerge in FTA chains (Ingram 2014). Putting value chain analysis at the core implies an analysis of the direct 'actors' involved at different stages of a chain, their linkages and activities, and indirect stakeholders (state, research, non-government, service providers, etc.) involved in the chains. A gendered lens allows the nature of gender differences and outcomes to be identified and analysed.

The search terms, shown in Table 13.1, were identified based on the research questions, which were drawn from the conceptual framework. The terms were then grouped, based on synonymous, alternative spellings and abbreviations of the central concepts.

First, literature was sought from the sources shown in Table 13.2, systematically using combinations of the search terms in groups 1 to 3, and then adding the terms

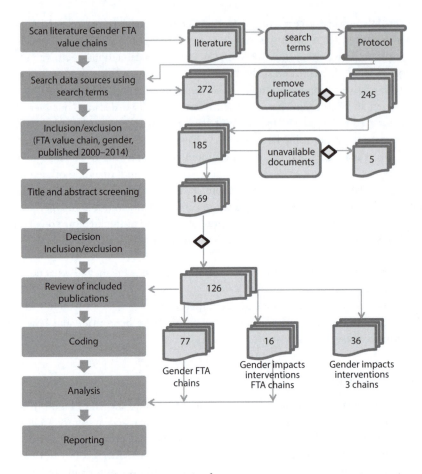

FIGURE 13.1 Steps in the literature review[3]

in group 4. When a large amount of literature was found, the search terms in group 5 were added to refine the search. This resulted in 245 publications.

These publications were assessed for quality and relevance to the research questions, leading to 185 publications being retained.[4] The title and abstract or executive summary of each document was screened to identify publications directly linked to the research questions. If gender and value chains or any of their synonyms or abbreviations were not mentioned or the methods were not explicitly detailed, the publication was excluded. This resulted in 126 publications which were read in full and coded in ATLAS.ti, according to the types of gender differences, where these differences were located along the chain, the type of FTA chain, factors explaining differences, the types of interventions, the geographic location of the origin of the chain, and outcomes or impacts. This evidence was synthesised to respond to each research question.

3 The review protocol is explained in more detail in Haverhals et al. 2015.
4 See Haverhals et al. 2015 for full list.

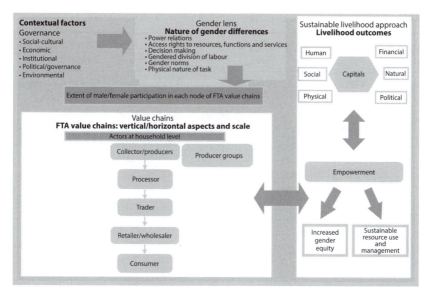

FIGURE 13.2   Conceptual framework underpinning the literature review

Results

Overview of the literature reviewed

The reviewed studies covered a range of chains based on fruits, seeds, nuts, gums, resins, barks, stalks, leaves, branches, fungi and roots – sold for food, feed, medicinal and cosmetic uses, as scents, energy, tools and utensils. The majority (62 per cent) of the documents were peer-reviewed journal articles, 25 per cent were grey literature reports, 10 per cent were books/book chapters, 2 per cent policy briefs and 1 per cent theses. All the publications concerned chains originating in developing countries, with most (58 per cent) covering chains originating in Africa, 26 per cent in Asia and the Pacific, and 16 per cent in Latin America (Figure 13.3). Most studies covering global chains did not address all the stages and locations of the chains. Nearly one-third of the studies described chain interventions.

Gendered differences in FTA value chains

Of sixty publications covering participation in different stages in FTA chains, only 32 per cent state the sex of different participants. The number of women participating in FTA chains was higher than the number of men on a global level, but not on a regional basis (see Figure 13.4).

Most information on participation concerns collectors, particularly in Africa where more women are reported to participate than men. In Latin America, more men reportedly engage in collection. These results mirror Sunderland et al.'s (2014) global study, which found that whilst both men and women participate in collection and processing for trade when a global perspective is taken, men

TABLE 13.1 Groups of search terms used in the literature review

Group 1: date	Group 2: gender	Group 3: value chain	Group 4: FTAs and NTFPs	Group 5: sustainable livelihoods
Literature published between 2000 and 2014	Gender, woman/ women, female, man/ men, male, sex, feminine, masculine, empowerment, power relations	Value chain, VC, Global value chain, GVC, supply chain, commodity, commodity chain, production-to-consumption systems, *filière*	Product(s) – agroforestry, non-timber forest or non-wood forest product, tree, forest	Impact/outcome/ effect on/of: livelihood strategies, social relations, culture, politics, health, poverty, GDP, sustainability, environment, nature, rural development, resources, coping mechanism, adaptation, diversification, transformation, household, income

dominate in Asia and Latin America, and women in Africa. Differences in male and female participation in harvesting are influenced by the physical nature of the task, social restrictions/prescriptions, household responsibilities (such as childcare), and distance to the collecting site. Limitations with respect to tenure (lack of access rights and limited decision-making power over natural resources) were important factors influencing who collects FTA products.

At the processor stage, information on male/female participation was generally lacking. The few cases that did mention participation all indicated female dominance. Participation was mainly influenced by domestic roles, with fewer women participating when the activity was of a very physical nature, involved long distances to the processing site, and/or required access to credit and technology.

TABLE 13.2 Sources of literature

Bibliographic scientific databases	Gateways	Websites of organisations and institutions
Scopus, Web of Science, Gender Studies Database, Social Sciences Citations Index, CAB abstracts, EconLit, SocINDEX, Google Scholar	Eldis, Jolis, 3ie Database of Impact Assessment	UN global impact, CGIAR, CIFOR, Bioversity International, UNDP IFC, ILO, World Bank Research4Development (R4D), KIT, Global Value Chains, SNV FAO, USAID, IFAD

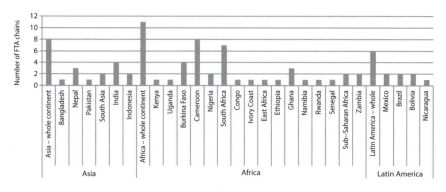

FIGURE 13.3 Location of origin of FTA value chains

At the trader stage, data was also generally lacking on gendered aspects of participation, particularly in Asia and Latin America. In Africa, female dominance was noted. In all cases, women were reported to participate most in small-scale retail trade, with men running larger businesses. The main factors influencing such participation are household responsibilities, distance to trading site, social restrictions/prescriptions, access to capital and literacy level.

Differences in norms surrounding the gendered division of labour were mentioned in 36 per cent of the studies as contributing to women's and men's distinct involvement in chains. This suggests that gendered power relations, particularly within households but also in enterprises, influence women's and men's participation in chain activities and the associated benefits they respectively derive. Inequitable gendered power relations in terms of decision making at household and community scales were mentioned as influencing factors in 14 per cent of the publications.

Gendered contribution to household income from FTAs

Evidence abounds about the contribution of FTA product sales to household income and differences in the revenues and profits gained by men and women. Mentions of gender-differentiated patterns of income generation were found in

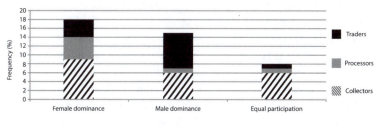

FIGURE 13.4 Gendered participation in stages of FTA value chains

32 per cent of the publications, with men generally earning more than women (see Box 13.1). Across 17 African countries, the sale of FTA products contributed between 22 per cent and 40 per cent of household income (Pouliot and Elias 2013; Pouliot and Treue 2013). The level of income generated by chain actors greatly depended on how, where and when value is added: at the source by managing wild resources, by domesticating FTAs as part of mixed farm–fallow–agroforestry systems, and/or through processing and marketing. No trend was clear, likely due to the disparity in methods and data presented in the publications. Value addition at the source depends on access to forest resources; towards the consumer end of the chain, access to markets is a critical determinant (Wiersum et al. 2014). Besides gender, income is influenced in varying degrees by variables such as age, ethnicity, household composition, marital status, class, and caste (Byron and Arnold 1999; Shackleton and Shackleton 2006 in Sunderland et al. 2014).

BOX 13.1 CHARCOAL VALUE CHAINS IN EASTERN AND SOUTHERN AFRICA: WHERE ARE THE WOMEN?

The charcoal trade is a booming business in the region, benefiting both women and men. Gendered differences are apparent: (1) both men and women participate throughout the charcoal chains but women comprise a small proportion of the actors – estimated at 27 per cent in Uganda (Shively et al. 2010); (2) women dominate both ends of the chain, mainly in seedling and tree management and small-scale retailing, and (3) men dominate the transportation and trade stages, accounting for up to 90 per cent, while women rarely reach 20 per cent in these stages (Delahunty-Pike 2012; Mutimba and Barasa 2005a; Cuambe 2008; Shively et al. 2010).

In Uganda, for instance, women accounted for 6 per cent of transporters in 2008 (Shively et al. 2010). Most of the men involved in the charcoal transport business own vehicles, thus dominating large-scale transportation. Generally, women consider transporting risky, due to inconvenient, risky hours in poor vehicles on poor roads, and illicit behaviour is required to deal with state officials (Sem 2004). Charcoal retailing is dominated by women in most eastern and southern African countries, where their proportion ranges from 57 per cent in Kenya to 90 per cent in Mozambique. Tanzania is an exception with about 16 per cent being female (Mutimba and Barasa 2005; Cuambe 2008; Shively et al. 2010). Typically, benefits are highest in the middle of the charcoal chain, and lowest for producers and retailers at the tail ends. This is shown by the average monthly profits in the chain in Uganda, where women form 6 per cent and 22 per cent of transporters and agents respectively and 69 per cent of retailers (Shively et al. 2010). As women do not engage in the more lucrative stages, they benefit less than their male counterparts.

Often, financial benefits from NTFP sales vary between men and women even when both sexes pursue the same chain activity. For instance, in Cameroon, male eru (*Gnetum* spp.) harvesters earn on average 12 per cent higher profits than women and male bush mango (*Irvingia* spp.) harvesters and exporters, 8 per cent and up to 52 per cent more than women, respectively (Ingram et al. 2014). Moreover, in the female-dominated FTA chains, average annual household incomes from the sale of a given FTA product tended to be lower than in male-dominated chains, with men on average earning 11 per cent higher profits. Higher profit margins were attributed to men's ability to trade in high volumes, and to select the high-value products and stages in which to operate (Ingram, Schure et al. 2014). This finding that women often engage in lower-profit products and men in higher-profit products mirrors studies of agricultural products (Kaaria and Ashby 2001).

Chikoko's 2000 study (in Kiptot and Franzel 2011) of benefits from woodlots in Malawi showed that both women and men harvested multiple products, with no significant difference between male-headed and female-headed households in products harvested. However, male-headed households earned over three times as much income as female-headed households. In the safou (*Dacryodes edulis*) chain, female traders gained lower profit margins per unit than men (Schure et al. 2009). This may be because men sell more per transaction than women and most female traders do not have enough capital to increase their stocks. Asfaw et al. (2013) found that being female positively and significantly affected forest income, whereas other socio-economic characteristics (education, family size and wealth) had no significant effect. Sunderland et al. (2014) found that whether a household was female-headed was consistently statistically significant in influencing household income derived from FTA product sales across regions and that women in female-headed households have larger shares of income from forest products compared to those in male-headed households.

Regional differences were apparent in how much men and women earn from FTA product sales, particularly from processed products. Socio-economic status was also a factor, cutting across gender, influencing people's dependence and engagement in a chain, with poorer households being more dependent on FTA products than richer ones. Benefits were often co-determined by socio-cultural norms and customs, and such socio-economic characteristics as the sex of the household head.

Gendered spending patterns

The literature indicates that increases in women's incomes tend to have greater impacts on food, health and education expenditure and therefore overall household well-being than increases in men's incomes (Guarascio et al. 2013). This review showed that women often engage in harvesting and processing products such as shea (*Vitellaria paradoxa*) in Benin, safou (*Dacryodes edulis)* in Cameroon, bitter cola (*Garcinia kola)* in Nigeria and marula (*Sclerocarya birrea)* in southern Africa in periods when there are few other income-earning alternatives.

This enables households to cover major expenses during seasonal financial shortfalls and generate capital for new activities (Mai et al. 2011; Njie Ndumbe 2013). Overall, women use FTAs to support household needs, while men invest slightly more in farm and other business activities, on major household expenditures and on their personal expenses (FAO 2007). In different regions, women's roles in household decisions on expenditure were positively correlated with improvements in household food and nutrition security (IFAD 2014). How FTA-based income is spent is influenced by other sources of income, household characteristics and individual and household specialisations in specific chain activities. Women's involvement in the trader node, as with female cocoa traders in Bolivia, tends to give them greater control over the income generated (Marshall et al. 2006).

Factors influencing gender differences

The literature indicates that contextual factors influence the gender differences observed in FTA chains. Key economic factors were the effects of new global markets for what were previously subsistence products (Brown and Lapuyade 2001; Gausset et al. 2005) and reforms due to economic crisis and structural adjustment programmes (Brown and Lapuyade 2001). These affect demand and consumption of FTA products, and accordingly their markets, leading to changes in the participation of men and women in the chains.

The most significant socio-cultural factors are closely linked to governance, with cultural norms and customs strongly influencing the participation and activities men and women perform in chains (Rocheleau and Edmunds 1997; FAO 2007; Sikod 2007; Mai et al. 2011). Governance, political and institutional factors are often complex and interlinked. Plural governance, particularly the combination of customary and formal regulatory arrangements, was frequently noted. Societal rules and norms can result in the under-representation of women in the institutions mediating formal governance: government, policy- and law making (Bandiaky-Badji 2011, and this volume; Sikod 2007). However, in household and customary governance arrangements, and in market-based governance arrangements, women have developed strategies which heighten their representation and participation in institutions governing chains (Mai et al. 2011; Guarascio et al. 2013; Sunderland et al. 2014). Gender was just one socio-cultural and demographic variable which socially differentiates men's and women's participation. For example, education affects male/female participation in economic activities (Mai et al. 2011). Gendered divisions of labour and contributions to household income are also influenced by variables such as age, ethnicity, household composition, marital status, class and caste, all of which have varying degrees of influence (Byron and Arnold 1999; Shackleton and Shackleton 2006 in Sunderland et al. 2014).

The most mentioned environmental factors influencing gendered differences in FTA chains were the level of resource degradation, due to anthropogenic and/or natural causes, affecting the quality and quantity of FTA resources available (Belcher et al. 2005; Pfund and Robinson 2005). The literature suggests that women may be more vulnerable than men to the effects of such degradation, because they are more likely to be poor and dependent on natural ecosystems threatened by degradation and climate change and because of the socially and politically driven lack of participation in decision making and access to power. Women's presence in community institutions can improve resource conservation and regeneration, as is the case in India and Nepal (Agarwal 2009). The *buriti* leaf (*Mauritia vinifera*) chain in Brazil (Virapongse et al. 2014) illustrates the consequences of high market demand leading to overharvesting. As the socio-economic situation of harvesters improved, engagement in the chain shifted to less skilled collectors who tended to harvest less sustainably.

Types of interventions in FTA value chains

Figure 13.5 illustrates that from the 32 publications that covered interventions in FTA chains, 66 per cent focused on mushrooms, honey and shea (see also Figure 13.6). Figure 13.7 shows that most interventions were made at the harvester and/ or processor stage.

The vast majority (81 per cent) of interventions in FTA chains reviewed reported a primary focus on gender likely due to the review methodology. Most of the initiatives analysed targeted women exclusively, mainly by stimulating higher female participation in chains and aiming to increase the benefits to women.

Interventions were classified according to the strategies used, shown in Box 13.2. Figure 13.8 shows that most frequently interventions focused on process upgrading achieved through improved technology and marketing advice, infrastructural support, value adding to products and improved marketing. Examples include the mushroom chain in Mexico (Marshall et al. 2006), honey in Rwanda (Matabishi 2012), shea in Mali (Traoré 2002) and NTFP chains in Africa (Shackleton et al. 2011). Upgrading interventions included organising women producers and processors into groups (horizontal upgrading), running businesses, and linking them to traders (vertical upgrading) in the *sabai* grass (*Eulaliopsis binata*) chain in India (Singh 2012) and the shea chain in Senegal (Souare 2002). Product upgrading included support to increase the quality of

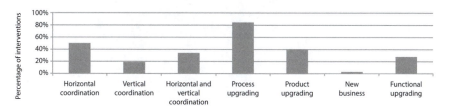

FIGURE 13.5 Interventions per type of FTA chain

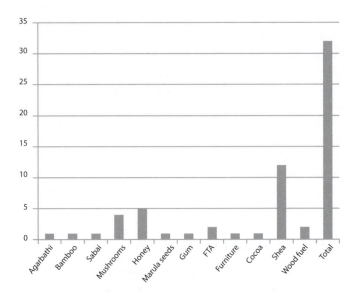

FIGURE 13.6 Target groups of chain interventions

shea butter in Burkina Faso, through improved processing and production techniques (Konate and Ouédraogo 2010), establishing a honey production training
and demonstration centre resulting in quality improvements, new product
markets, and 50 per cent more income for producers in Ethiopia (Sisay 2012).
Over a quarter (28 per cent) of interventions used functional upgrading, such
as skill training in the wood furniture chain in Indonesia (Purnomo et al. 2014)
and the shea chain in Burkina Faso (Ndow 2007).

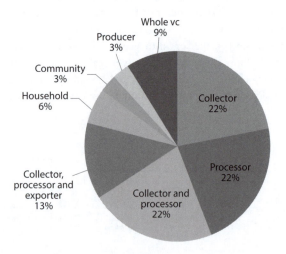

FIGURE 13.7 Primary focus of chain interventions

BOX 13.2 UPGRADING STRATEGIES

Horizontal coordination: between the same type of actors, e.g. harvester groups, trader cooperatives, mixed-gender groups, etc.;

Vertical coordination: between actors in different chain positions, e.g. between individual women or a women's group with their buyers or service providers;

Product upgrading: to more sophisticated products with increased per unit value, e.g. products complying with buyer requirements for higher quality, certification, food safety standards, traceability, packaging, etc.;

Process upgrading: more efficient transformation of inputs into outputs by reorganizing productive activities, e.g. applying new processing technologies, delivering on schedule, reducing waste, etc.;

Functional upgrading: acquiring new functions (or abandoning old ones) that increase the skill content of activities, e.g. grading, primary processing, bulking and storage, transporting; provision of services, inputs or finance, and

New business actors: creating new products and chains from a timber or forest species previously collected for subsistence use.

Sources: Bolwig et al. 2008; GTZ 2007; Mitchell et al. 2009.

Interventions in cocoa and shea chains

Interventions in two chains, cocoa – a cultivated commodity, often in mixed agroforestry and farm systems – and shea[5] – largely wild-harvested in agroforests and woodlands – provide insights into the types of interventions reviewed and their impacts on gender.

West Africa is the source of most cocoa sold on the world market. Cocoa is presented in the literature as a typical 'male crop', due to the physical work involved in cocoa farming, and also because it is a cash crop and most cocoa farms are male owned. Male cocoa farmers generally benefit more than their female counterparts from access to credit, land and markets, and from the social context, where norms generally allow men more control over revenues, decision making and bargaining power. The important roles women play in the cocoa chain in terms of the crop's cultivation, post-harvest processes and quality enhancement have only recently been highlighted (Enete and Amusa 2010; Kumase et al. 2010). The literature also suggests that empowered female cocoa farmers with access to farm inputs can influence decisions about cocoa production activities, and improve bean quality, quantity and incomes from cocoa farming (Vigneri and Holmes 2009). Major chocolate

5 See Elias, this volume, for more on this product.

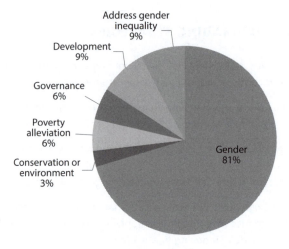

FIGURE 13.8 Type of interventions in chains

manufacturing multinationals and NGOs have increasingly acknowledged the gender gaps that pervade the cocoa chain – particularly in access to training, certification and access to credit and inputs (UTZ 2009; Banerjee et al. 2014). Programmes to address these are generally in the early stages of design and implementation, and so their economic, social and ecological impacts are either not yet apparent (Waarts et al. 2013; Ingram, Waarts et al. 2014) or only just becoming so (IITA 2006).

Numerous interventions to improve female producers' and processors' inclusion and incomes in shea export chains in West and Central Africa have been made in the last two decades. Mainly donor and NGO projects, the interventions were reported to improve women's access rights to shea nuts, improve shea butter quality, increase production volumes, and enhance women's influence in household decision making and bargaining power (Traoré 2002; Sidibé et al. 2012). The literature, however, often lacks independent, detailed impact analyses and generally presents the outcomes for women as a homogenous group, positively influenced by the upgrading strategies applied. Poudyal (2009) and Elias (2010) indicate that some interventions have been captured by elites at the expense of marginalised groups, favouring the empowerment of specific groups of women. In some cases, the success of such interventions has led to men encroaching on what were previously female activities (See IFR Special Issue on Gender and Agroforestry, for additional examples). Shifting power dynamics risk resulting in increased social differentiation, changed household consumption patterns and ultimately the loss of income sources for women (Baden 2013; Wardell and Fold 2013).

Conclusions

Gender differences in FTA value chains

Throughout the literature, relatively little information was found quantifying male/female participation in FTA chains. This information is specifically lacking for the stages downstream of small-scale traders. A bias in the literature was found towards African countries, with 25 of the 40 cases that mentioned specific male/female participation being based in Africa. In general, female dominance was recorded at the collector, processor and small-scale trader stage, but there are strong regional differences. At the collector stage, female dominance only holds in Africa, whereas men dominate in Latin America and Asia. There was a lack of information about the sex of processors involved in FTA chains. At the trader stage, data from Latin America and Asia is lacking. Overall trends emerge that women are mostly active in the upstream and downstream ends of chains, that is, harvesting and small-scale retail trade, and that men tend to run larger businesses.

The nature of gendered differences in participation in FTA chains can largely be ascribed to *social and cultural differences* that influence how chains are governed. These determine gender-differentiated access to, and ownership (tenure) of land, forests, trees, farms, FTA products, labour, technology, credit, information and FTA markets. Because of such differences, women often have fewer or less favourable access rights than men, and if they have such rights, these are often not well defined or enforced. *Differences due to the nature of the product and activity* were also found with regard to the physical demands of chain activities, notably when physically demanding collecting and primary processing activities were required, such as in charcoal production. The time taken to conduct these activities and their location, such as when long forage distances or extended periods away from home were required, were also strong determinants of participation in chains for some products. *Gendered power relations* were noted, mainly at the household level, but also within enterprises, resulting in differentiated benefits to women and men. The ways men and women use revenues from FTA products also differed.

The contextual factors influencing these gender differences

Gendered constraints occur, particularly for women's participation in and benefits from FTA chains, mainly due to socio-cultural, political, economic and environmental factors. The influence of each factor varies depending on the product, geographic region and cultural setting. *Socio-cultural factors* such as cultural norms and customs that tend to disadvantage women strongly influence the work performed, the division of labour in chains and (other) household and economic responsibilities and activities of chain actors. *Governance, political and institutional factors* concerned overlapping customary and formal regulatory arrangements, with societal norms often resulting in the under-representation of women in governance arrangements. Economic factors such as the effects of globalisation of chains and reforms due to economic crises affected global markets for FTA products, and local demand more specifically. The main *environmental factors* mentioned were resource degradation affecting the quality and quantity of FTA resources available. Women were shown in some cases to be more vulnerable than men to the effects of such degradation because of their disadvantaged position and therefore greater dependence on forest ecosystems, and because of their socially and politically driven lack of participation in decision making and power.

Gendered participation in FTA chains is shaped also by other factors of social differentiation such as education, age and ethnicity. *Education* was found to influence participation in and benefits from FTA chains. It is expected that education becomes more influential when moving up the chain. The role of women can be placed in a nested set of relations *cutting across different social constructs* such as ethnicity, of which gender is a prominent but not unique factor.

Making FTA value chain interventions more gender equitable

Most interventions reviewed were implemented at the collector's and/or processor's stage. Although some claimed to specifically target gender relations, in practice they tended to focus on women, rather than on the relations between men and women. Other interventions clearly set out to improve the position of women; these too targeted women separately from their relations to men. Overall, the interventions sought to increase female participation and resulting benefits. The literature indicated that the most successful interventions were those that combined horizontal and vertical upgrading (see Box 13.2). Most interventions (84 per cent) resulted in process upgrading, followed by product upgrading (41 per cent), which often resulted in improvements in product quality, and 28 per cent resulted in functional upgrading.

There are lessons to be drawn from the literature on how to improve gender equality in FTA chains, listed below.

Interventions need to be explicitly gender sensitive:

1. Establishing gender sensitivity in interventions appears to be a critical criterion for success. This implies carefully selecting beneficiaries, intervention partners and developing the gender-mainstreaming capacities of implementing agencies.
2. Technical and social-cultural practices need to be considered when developing interventions, along with the recognition that changing established practices may need a long timescale;
3. Technological changes with market-oriented activities can help transform women's participation in specific chain activities, for example by freeing time;
4. Possible negative social cultural repercussions can result from an increase in women's benefits;
5. Combining vertical and horizontal upgrading achieves better results;
6. Improving women's position in FTA chains and fostering their empowerment requires a multi-pronged approach involving a combination of activities, such as training, technology transfer, increasing their negotiating capacity, developing their business skills and providing them with market information. This can trigger changes by men and women to enable them to self-determine their own upgrading initiatives.

Supporting collective action:

1. Collective action and groups, often critical for achieving the desired gender-equity outcomes, may need support from local leaders and encouragement from projects;
2. Pilots and demonstrations, such as of female leadership in FTA chain activities, can support change.

Consider parallel actions to chain interventions

1. Improve women's literacy levels;
2. Counter discrimination through regulations.

Evidence from the literature reviewed suggests that forest, agroforestry and tree-based chains can be made more gender equitable. Approaches to achieve this include addressing how trade is conducted, power differences in chain activities, as well as the social context which exerts a strong influence (e.g. tenure rights, gender norms, literacy, laws relating to discrimination). Given the focus on processes and outcomes in interventions and discrete time periods, and relative lack of attention to relations between intervention outcomes, impacts and the wider chain context, these lessons should be seen as broad generalisations. Research, practice and policy would benefit from processes that involve local actors in anticipating the results of a planned intervention. This would serve as a benchmark of the extent to which initiatives contribute to more gender equitable chains. Equally, the economic, as well as social and environmental impacts of interventions which target or seek to benefit men and/or women, should be evaluated, preferably ex-ante, during and ex-post, examining both desired and unanticipated impacts.

Acknowledgements

This review was conducted with support from LEI and the Forest and Nature Conservation Policy Group, Wageningen UR and the Consultative Group on International Agricultural Research (CGIAR) Consortium Research Program 6 on Forests, Trees and Agroforestry Livelihoods, Landscapes and Governance. We thank the two anonymous reviewers of this chapter for their constructive advice.

References

Agarwal B. 2009. Gender and forest conservation: The impact of women's participation in community forest governance. *Ecological Economics* 68(11): 2785–99.

——. 2010a. Does women's proportional strength affect their participation? Governing local forests in South Asia. *World Development* 38(1): 98–112.

——. 2010. *Gender and green governance: The political economy of women's presence within and beyond community forestry.* Oxford: Oxford University Press.

Asfaw A, Lemenih M, Kassa H and Ewnetu Z. 2013. Importance, determinants and gender dimensions of forest income in eastern highlands of Ethiopia: The case of communities around Jelo Afromontane forest. *Forest Policy and Economics* 28: 1–7.

Awono A, Ndoye O and Preece L. 2010. Empowering women's capacity for improved livelihoods in non-timber forest product trade in Cameroon. *International Journal of Social Forestry* 3(2): 151–63.

Baden S. 2013. Women's collective action in African agricultural markets: The limits of current development practice for rural women's empowerment. *Gender and Development* 21(2): 295–311.

Bandiaky-Badji S. 2011. Gender equity in Senegal's forest governance history: Why policy and representation matter. *International Forestry Review* 13(2): 177–94.

Banerjee D, Klasen S and Wollni M. 2014. *Market discrimination, market participation and control over revenue: A gendered analysis of Cameroon's cocoa producers.* In 2nd Global Food Symposium. Göttingen, Germany.

Belcher B. 2003. What isn't a NTFP? *International Forestry Review* 5(2): 161–8.

——, Ruíz-Pérez M and Achdiawan R. 2005. Global patterns and trends in the use and management of commercial NTFPs: Implications for livelihoods and conservation. *World Development* 33(9): 1435–52.

Bolwig S, Ponte S, du Toit A, Riisgaard L and Halberg N. 2008. Integrating poverty, gender and environmental concerns into value chain analysis: A conceptual framework and lessons for action research. *DIIS Working Paper No. 2008/16.* Copenhagen: Danish Institute for International Studies.

Brown K and Lapuyade S. 2001. A livelihood from the forest: Gendered visions of social, economic and environmental change in Southern Cameroon. *Journal of International Development* 13(8): 1131–49.

Carr ER. 2008. Men's crops and women's crops: The importance of gender to the understanding of agricultural and development outcomes in Ghana's Central Region. *World Development* 36(5): 900–915.

Colfer CJP. 2014. *An introduction to the gender box*. CIFOR Occasional Paper No. 82. Bogor, Indonesia: Center for International Forestry Research (CIFOR).

Cuambe C. 2008. Woodfuels Integrated Supply Demand Overview Mapping (WISDOM) for Mozambique. In Kwaschik R. *Charcoal and communities in Africa*. Maputo, Mozambique, International Network for Bamboo and Rattan (INBAR). 77–100.

Delahunty-Pike A. 2012. Gender equity, charcoal and the value chain in western Kenya. Working Brief, Nairobi: PISCES.

Doss CR. 2002. Men's crops? Women's crops? The gender patterns of cropping in Ghana. *World Development* 30(11): 1987–2000.

Elias M. 2010. *Transforming nature's subsidy: Global markets, Burkinabè women and African shea butter* [Dissertation] Faculty of Graduate Studies and Research. Montreal, Quebec: McGill University.

Enete AA and Amusa TA. 2010. Contribution of men and women to farming decisions in cocoa based agroforestry households of Ekiti State, Nigeria. *Tropicultura* 28(2): 77–83.

FAO (Food and Agriculture Organization of the United Nations). 2007. *Gender mainstreaming in forestry in Africa*. Rome: FAO.

FAO/IFAD/ILO (Food and Agriculture Organization of the United Nations, International Fund for Agricultural Development and International Labour Organization). 2010. Gender dimensions of agricultural and rural employment: Differentiated pathways out of poverty – Status, trends and gaps. Rome: FAO.

Gausset Q, Yago-Ouattara EL and Belem B. 2005. Gender and trees in Péni, South-Western Burkina Faso: Women' needs, strategies and challenges. *Geografisk Tidskrift-Danish Journal of Geography* 105(1): 67–76.

GTZ (Deutsche Gesellschaft für Internationale Zusammenarbeit). 2007. *The methodology of value chain promotion.* In *Value Links Manual.* GTZ, Division 45 Agriculture, Fisheries and Food and Division 41 Economic Development and Employment.

Guarascio F, Gunewardena N, Holding C, Kaaria S and Stloukal L. 2013. Forest, food security and gender: Linkages, disparities and priorities for action. Accessed 7 January 2016. www.fao.org/docrep/018/mg488e/mg488e.pdf.

Hasalkar S and Jadhav V. 2004. Role of women in the use of non-timber forest produce: A review. *Journal of Social Science* 8(3): 203–6.

Haverhals M, Elias M, Ingram V and Basnett B. 2015. *Gender and FTA value chains: Evidence from the literature.* Bogor, Indonesia: CIFOR.

IFAD (International Fund for Agricultural Development). 2008. *Gender and non-timber forest products: Promoting food security and economic empowerment.* Rome: IFAD.

——. 2014. Gender and rural development brief. In *East and Southern Africa.* Accessed 7 January 2016. www.ifad.org/gender/pub/gender_esa.pdf.

IITA (International Institute of Tropical Agriculture). 2006. Women cocoa farmers in Ghana have something to smile about! *STCP Newsletter* 15: 6.

Ingram V. 2014. Win–wins in forest product value chains? How governance impacts the sustainability of livelihoods based on non-timber forest products from Cameroon. Dissertation, Faculty of Social and Behavioural Sciences, University of Amsterdam, Amsterdam.

——, Schure J, Chupezi Tieguhong J, Ndoye O, Awono A and Midoko Iponga D. 2014. Gender implications of forest product value chains in the Congo basin. *Forests, Trees and Livelihoods* 23(1–2): 67–86.

——, Waarts Y, Ge L, van Vugt S, Wegner L, Puister-Jansen L, Ruf F and Tanoh R. 2014. *Impact of UTZ certification of cocoa in Ivory Coast: Assessment framework and baseline.* Den Haag, The Netherlands: LEI Wageningen UR.

Kaaria SK and Ashby JA. 2001. An approach to technological innovation that benefits rural women: The resource-to-consumption system. In Working document No. 13. *CGIAR systemwide program on participatory research and gender analysis.* Cali, Colombia: Consultative Group on International Agricultural Research, Future Harvest.

Kaplinsky R and Morris M. 2000. *A handbook for value chain research.* Canada: International Development Research Centre (IDRC).

Kiptot E and Franzel F. 2011. Gender and agroforestry in Africa: A review of women's participation. *Agroforestry Systems* 84(1): 35–58.

Konate L and Ouédraogo N. 2010. Inserting the FADEFSO into a value chain: The leverage of control over Burkarina consulting services of SNV, Burkina Faso. *Case Studies.*

Krantz L. 2001. *The sustainable livelihood approach to poverty reduction: An introduction.* Stockholm: SIDA.

Kumase WN, Bisseleua H and Klasen S. 2010. Opportunities and constraints in agriculture: A gendered analysis of cocoa production in Southern Cameroon. Discussion Papers, No. 27. *Courant Research Center: Poverty equity and growth.* Göttingen, Germany: Georg-August-Universität.

Mai YH, Mwangi E and Wan M. 2011. Gender analysis in forestry research: Looking back and thinking ahead. *International Forestry Review* 13(2): 245–58.

Marshall E, Schreckenberg K and Newton AC, eds. 2006. Commercialization of non-timber forest products: Factors influencing success. In *Lessons learned from Mexico and Bolivia and policy implications for decision-makers.* UNEP World Conservation Monitoring Centre.

Matabishi I. 2012. Women and bees? Impossible! In Laven A and Pyburn R., eds. *Challenging chains to change: Gender equity in agricultural value chain development.* Amsterdam: KIT Publishers, Royal Tropical Institute.

Meinzen-Dick R, Brown LR, Feldstein HS and Quisumbing AR. 1997. *Gender, property rights, and natural resources.* Washington, DC: International Food Policy Research Institute.

Mitchell J, Keane J and Coles C. 2009. *Trading up: How a value chain approach can benefit the rural poor.* London: COPLA Global: Overseas Development Institute.

Mutimba S and Barasa M. 2005. National Charcoal Survey: Exploring the potential for a sustainable charcoal industry in Kenya. ESDA (Energy Sustainable Development Africa).

Mwangi E, Meinzen-Dick R and Sun Y. 2011. Gender and sustainable forest management in East Africa and Latin America. *Ecology and Society* 16(1): 17.

Ndow SH. 2007. A story from West Africa: 'Women's Gold' – shea butter from Burkina Faso. In Case study in women's empowerment. Network of African Women Economists (NAWE).

Njie Ndumbe L. 2013. *Unshackling women traders: Cross-border trade of Eru from Cameroon to Nigeria.* Africa Trade Policy Notes.

Ostrom E, Gardner R and Walker J. 1994. *Rules, games, and common-pool resources.* East Lansing: University of Michigan Press.

Paumgarten F and Shackleton C. 2011. The role of non-timber forest products in household coping strategies in South Africa: The influence of household wealth and gender. *Population & Environment* 33(1): 108–31.

Pfeiffer JM and Butz RJ. 2005. Assessing cultural and ecological variation in ethnobiological research: The importance of gender. *Journal of Ethnobiology* 25(2): 240–78.

Pfund JL and Robinson P, eds. 2005. *Non-timber forest products: Between poverty alleviation and market forces.* Bern: Intercooperation.

Poudyal M. 2009. *Tree tenure in agroforestry parklands: Implications for the management, utilisation and ecology of shea and locust bean trees in northern Ghana.* Department of Environment and Politics, York: University of York.

Pouliot M and Elias M. 2013. To process or not to process? Factors enabling and constraining shea butter production and income in Burkina Faso. *Geoforum* 50: 211–20.

—— and Treue T. 2013. Rural people's reliance on forests and the non-forest environment in West Africa: Evidence from Ghana and Burkina Faso. *World Development* 43: 180–93.

Purnomo H, Achdiawan R, Melati M, Irawati RH, Sulthon, Shantiko B and Wardell A. 2014. Value-chain dynamics: strengthening the institution of small-scale furniture producers to improve their value addition. *Forests, Trees and Livelihoods* 23(1–2): 87–101.

Quisumbing AR, Meinzen-Dick R, Raney TL, Croppenstedt A, Behrman JA and Peterman A. 2014. Closing the knowledge gap on gender in agriculture. In Quisumbing A, Meinzen-Dick R, Raney T, Croppenstedt A, Behrman JA and Peterman A, eds. *Gender in agriculture.* New York: Springer.

Riisgaard L, Escobar Fibla AM and Ponte S. 2010. *Gender and value chain development.* Copenhagen: The Danish Institute for International Studies.

Rocheleau D and Edmunds D. 1997. Women, men and trees: Gender, power and property in forest and agrarian landscapes. *World Development* 25(8): 1351–71.

Ruiz Pérez M, Ndoye O, Eyebe A and Ngono DL. 2002. A gender analysis of forest product markets in Cameroon. *Africa Today* 49(3): 96–126.

Said-Allsopp M and Tallontire A. 2014. Pathways to empowerment? Dynamics of women's participation in global value chains. *Journal of Cleaner Production* 107: 114–121.

Schure J, Ingram V, Awono A and Binzangi K. 2009. From tree to tea to CO2 in the Democratic Republic of Congo: A framework for analyzing the market chain of fuelwood around Kinshasa and Kisangani. Paper read at 13th World Forestry Congress, 18–23 October 2009, at Buenos Aires, Argentina.

Sem N. 2004. Supply/demand chain analysis of charcoal/firewood in Dar es Salaam and Coast Region and differentiation of target groups, Tanzania Traditional Energy Development and Environment Organization (TaTEDO), Dar es Salaam.

Shackleton S, Paumgarten F, Kassa H, Husselman M and Zida M. 2011. Opportunities for enhancing poor women's socioeconomic empowerment in the value chains of three African non-timber forest products (NTFPs). *International Forestry Review* 13(2): 136–51.

Shillington LJ. 2002. Non-timber forest products, gender, and households in Nicaragua: A commodity chain analysis. Master's of Science in Wood Science and Forest Products, Faculty of Wood Science and Forest Products, Faculty of the Virginia Polytechnic Institute and State University, Blacksburg, Virginia.

Shively G, Jagger P, Serunkuuma D, Arinaitwe A and Chibwana C. 2010. Profits and margins along Uganda's charcoal value chain. *International Forestry Review* 12(3): 270–3.

Sidibé A, Vellema S, Dembelé F, Traoré M and Kuyper TW. 2012. Innovation processes navigated by women groups in the Malian shea sector: How targeting of international niche markets results in fragmentation and obstructs co-ordination. *NJAS – Wageningen Journal of Life Sciences* 60–63: 29–36.

Sikod F. 2007. Gender division of labour and women's decision-making power in rural households in Cameroon. *African Development* 32(3): 58–71.

Singh AK. 2012. Money doesn't grow on trees: It grows on the ground! In Laven A and Pyburn R, eds. *Challenging chains to change: Gender equity in agricultural value chain development.* Amsterdam: KIT Publishers, Royal Tropical Institute.

Sisay G. 2012. 'Women don't climb trees': Beekeeping in Ethiopia. In Laven A and Pyburn R, eds. *Challenging chains to change: Gender equity in agricultural value chain development.* Amsterdam: KIT Publishers, Royal Tropical Institute.

Souare A. 2002. Overview of the intervention of PROMER in the shea sector in the Département de Kédougou, Senegal. In International Workshop on Processing and Marketing of Shea Products in Africa. Dakar, Senegal CFC, CSE, FAO.

Sun Y, Mwangi E and Meinzen-Dick R. 2011. Is gender an important factor influencing user groups' property rights and forestry governance? Empirical analysis from East Africa and Latin America. *International Forestry Review* 13(2): 205–19.

Sunderland T, Achdiawan R, Angelsen A, Babigumira R, Ickowitz A, Paumgarten F, Reyes-García V and Shively G. 2014. Challenging perceptions about men, women, and forest product use: A global comparative study. *World Development* 64: S56–S66.

Traoré KM. 2002. Strengthening the technical and management capacities of women in the shea sector of Zantiébougou, Mali. In International Workshop on Processing and Marketing of Shea Products in Africa. Dakar, Senegal: CFC, CSE, FAO.

UTZ. 2009. *The role of certification and producer support in promoting gender equality in cocoa production.* UTZ CERTIFIED, Solidaridad-Certification Support Network.

Verhart N and Pyburn R. 2010. The rough road to gender equitable growth: The case of Café de Mujer Guatemala. *Development* 53(3): 356–61.

Vigneri M and Holmes R. 2009. *When being more productive still doesn't pay: Gender inequality and socio-economic constraints in Ghana's cocoa sector.* Rome: FAO, IFAD, ILO.

Virapongse A, Schmink M and Larkin S. 2014. Value chain dynamics of an emerging palm fiber handicraft market in Maranhão, Brazil. *Forests, Trees and Livelihoods* 23(1–2): 36–53.

Waarts Y, Ge L, Ton G and van der Mheen J. 2013. A touch of cocoa, Baseline study of six UTZ-Solidaridad cocoa projects in Ghana. *[LEI] Institute for Agricultural Economics report 2013-048.* Netherlands: Wageningen University & Research Centre.

Wardell A and Fold N. 2013. Globalisation in a nutshell: Historical perspectives on the changing governance of the shea commodity chain in Ghana. *International Journal of the Commons* 7(2): 1–23.

WB (World Bank). 2004. *Sustaining forests: A development strategy*. Washington, DC: World Bank.

WDR (World Development Report). 2008. Agriculture for development: The gender dimensions. *Agriculture for development policy brief*. Washington, DC: World Bank.

Wiersum KF, Ingram VJ and Ros-Tonen MAF. 2014. Governing access to resources and markets in non-timber forest product chains. *Forests, Trees and Livelihoods* 23(1–2): 6–18.

14

UNTAMED AND RARE

Access and power in DRC's emerging luxury bushmeat trade

Gina LaCerva

Introduction

Wild meat in the Democratic Republic of Congo (DRC) is currently undergoing a remarkable transformation—from subsistence protein among poorer rural populations to luxury food among privileged urbanites.[1] With demand growing in DRC's rapidly expanding cities, game consumption has reached unsustainable levels, threatening rural food security and ultimately leading to "the empty forest syndrome" (Redford 1992; Yumoto et al. 1995; Nasi et al. 2011). Although technically legal, there are many aspects of illegality to the trade.[2] Despite nearly fifty years of research and intervention measures, the pace of hunting and international trafficking are both on the rise, and many consider it an intractable problem, rife with military conflict and entangled with the small arms trade (Chaber et al. 2010; Taylor et al. 2015).

Research into the wild meat trade, like other forestry research, has lagged in recognizing the significance of gender in shaping value chains (Fortmann et al. 2008; Rocheleau 2008; Ingram et al., this volume). Women are heavily involved, and there are strong gender divisions of labor, capture of profits, and consumption patterns. DRC's history of sexist colonial practices, conservative Catholic

---

1 In most of the literature, any non–domesticated wild game harvested for food, medicinal and/or cultural use from forested areas is called bushmeat (Ingram 2011). To avoid perpetuating the neocolonial and othering aspects of the term "bushmeat," I will herein use "wild meat" or "game."

2 Illegal aspects include: the killing of animals protected under CITES (Convention on International Trade in Endangered Species of Wild Fauna and Flora); hunting of allowable animals in forests that are off-limits—e.g. inside national parks; hunting out of season; using forged papers; the smuggling of *cartouche 00* shotgun shells; and the illegal ownership of guns. One of the most difficult problems is that each province determines its own hunting regulations. Such incongruent legal frameworks, mixed with longstanding customary laws of land access, have created a complicated tangle of regulations (Ingram 2011).

patriarchy, and years of civil wars have created economic, political and social marginality among Congolese women. As a result, women have long been employed in informal, micro-entrepreneur sectors reliant on interpersonal networks and food production, rather than formal educational skills. As conservationists aim to limit the trade in wild meat, there is a danger of criminalizing the economic sectors in which women continue to have a foothold (Yates 1982; Mianda 2002).

As wild animals—once the feast of the poor—become normalized luxury items primarily available to the rich, complicated questions arise regarding rights of access to traditional food and the appropriate role of conservation organizations. I look beyond the drivers of hunting at the rural landscape to examine demand dynamics, the creation of cultural and economic significance, and embedded gendered relationships as the meat is produced, transported and consumed. By examining gender in tandem with the geographic differences in value, I provide insight into the role of women and men in navigating and constructing these relations of power and access—both contemporarily and historically. I argue that given their deep experience with and knowledge of the wild meat trade, the women involved in provisioning game could play a major role in wildlife conservation.

Methodology

This research relies on an ethnographic study of a wild game value chain, an approach that seeks to understand "insider" symbols and beliefs as distinct entities from "outsider" informatics to uncover how local gender dynamics interface with international institutions and global capital structures (Fortmann 2008; Ingram 2011). It is territorial, network and process oriented. In this way, the physical landscape can be read as a social reality (Dove 1992; Fairhead and Leach 1996; see Figure 14.1).

Primary methods consisted of participant observation, semi-directed interviews, questionnaires, informal interviews with Wildlife Conservation Society (WCS) and World Wildlife Fund (WWF) staff, and market surveys conducted during ten weeks of fieldwork in DRC and Paris in summer 2014 (Table 14.1). Concurrently, WCS undertook an associated market survey in Kinshasa to evaluate overall protein availability/price. Historical documents were analyzed for references to the game trade, gender roles and land tenure issues.

There is no "one" value chain, but the chosen sequence for ethnography was not arbitrary:

1. Salonga National Park is the largest in DRC with the most wildlife and Monkoto is a major hub for conservation and military projects in the region.
2. Mbandaka is a primary trading town where a number of tributaries from Salonga National Park meet the Congo River.
3. Kinshasa is the largest city in DRC and the likely destination of the highest percentage of game from Salonga.

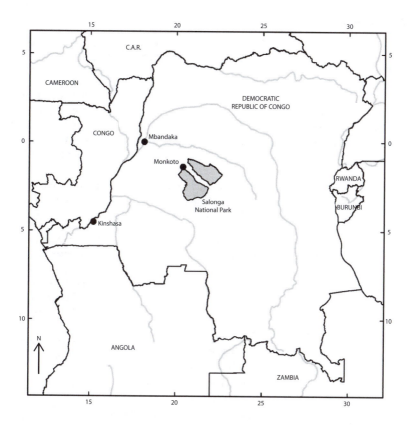

FIGURE 14.1 DRC, showing locations of major nodes in the wild meat value chain

TABLE 14.1 Location and type of formal interviews or observations

Location	Location type	Actor (#)	Gender	Primary information
Monkoto (Salonga National Park)	National Park	Hunters (10) NTFP producers (2) Paramilitary officers— Operation Bonobo (3)	M F M	Driving motivators for poaching; price for meat; terms of transactions (trade, cash, bullets); law enforcement
Mbandaka	Mid-sized city, game transport hub			Market observation; informal interviews with vendors of both genders

(continued)

TABLE 14.1 (continued)

Location	Location type	Actor (#)	Gender	Primary information
Kinshasa	Main site of sale and consumption, DRC's largest city			Demand drivers, scope (price and quantity) of market, changes to trade; Informal consumer interviews in the markets
	Markets	Vendors (20)	19F, 1M	
	Restaurants	Proprietors, chefs, or employees (12)	11M, 1F	Examine the "luxury" market
	Other	Traders (3)	2F, 1M	Define logistical qualities of transport and timing; law enforcement and regulations; consumer preference and demand
		Former park guards (2)	2M	
		Professor at Université de Kinshasa	M	
		Top official at Ministry of Environment	M	
		Air France employees (2)	M	
		Consumers (10)	8 F, 2M	
Paris	Primary international export site	French Customs and Police (*Gendarmerie*) (2)	2M	Understand international trade; consumer preference in Europe; informal market surveys in Chateau Rouge District
		Consumers (2)	1M, 1F	

4.   There have been recent airport raids and concerns in Paris about the import of game, as well as a fairly large expat community from DRC.

Background

Salonga National Park is located approximately 1,160 kilometers to the northeast of Kinshasa.[3] Primarily lowland humid forest, its multiple waterways impart

---

3   At 36,000 square kilometers, it is the largest tropical forest park in Africa and the second largest in the world. It is divided into North and South blocks separated by a corridor of secondary forest and agriculture.

a mosaic quality of swamps and drylands, creating one of the most biologically diverse places on earth (Alexander et al. 1996). Estimates conflict, but for the entire Congo Basin between one and six million tons of wild meat are harvested annually (Brown and Williams 2003; Nasi et al. 2011).

For forest-dwelling Bantu groups, hunting (using snares or bows and arrows) was once seen as an honorable activity that tied the body to the forest in ways both physical and metaphysical. Diets were protein rich and carbohydrate poor. Wild meat was the glue that bound society together, as the larger animals such as forest buffalo were shared communally (Rau 1991). Women had customary rights to certain portions of the meat, with the chief often dividing the catch (Paulme 1963). Cultural norms often restricted consumption of certain species and reserved others for men of high stature.[4]

To understand how hunting became poaching, we must start with how land became owned. The roots of institutional land regulation in Salonga date to the colonial period.[5] Compulsory settlement of nomadic populations into villages along roads facilitated colonial control over a previously scattered rural population. The Congolese were pushed into labor markets through further coercive measures (Hochschild 2009). The Belgians took to sport hunting on horseback while severely curtailing subsistence hunting by local populations (Hopson 2011). With access to forests restricted, and the colonial government monitoring all domestic animals, meat became scarce. Those in authority monopolized it, strengthening it as a symbol of power and potency.

Leopoldville (present-day Kinshasa) became the capital and administrative center of Belgian Congo in 1923. Congolese settlement was restricted—even young male workers were seen as temporary migrants—and a deeply segregated city emerged (Piermay 1997). Because women comprised the majority of the village labor force, particularly in all stages of food production, they were actively dissuaded from moving to urban areas, which were stigmatized as depraved (Paulme 1963). This view was promoted by the Catholic Church and reinforced by a colonial tax of 50 francs annually on *femme libre* (free women) within the city (Gondola 2014). By the 1930s, however, the colonial administration had become concerned by the uneven sex-ratios and "camp-like" atmosphere in Leopoldville. Under the official desire to eradicate prostitution and stabilize worker lives, women were tacitly encouraged to move to the city to establish families or work as *ménagères* ("housekeepers," sometimes with sexual elements; Stoler 2002).

4 For example, the Iyaelima do not eat bonobos as they are seen as having direct links to ancestral and spirit worlds (Hopson 2011).
5 Customary land rights shifted over space and time to account for nomadic lifestyles in pursuit of game and much like shifting cultivation, were a form of game management that ensured certain species were not overhunted (Abernethy 2010; Ingram 2011). When the Belgians arrived, variable land rights were reassigned and legislated to suit colonial needs (Larson and Ribot 2007). In 1956, the Belgians established the Salonga area as the Tshuapa National Park to preserve a large forest tract from being harvested for timber. In 1970, Mobutu Sese Seko took power and expanded Salonga's park to its present-day boundaries (Hopson 2011).

By this period, many women wanted to move to the city, but unequal educational access limited their employment opportunities (Yates 1982; Mianda 2002). Congolese men were trained as doctors, lawyers and administrative workers, while women, even of the upper classes, were encouraged to gain proficiency in domestic activities such as sewing, knitting and housekeeping. This new middle class of évolués meant Congo had a wage labor force twice as large as any other African colony, but women were largely excluded from the rapidly expanding modern economy (Stoler 2002).

Many women found economic independence by becoming micro-entrepreneurs. Prostitution was one route, but numerous others undertook food production or entered into the small-scale trades that men were unwilling to do. The majority of urban men were involved in wage labor with little time for travel, so it was primarily women who undertook occasional trips back to their villages, combining social visits with trade in forest products (Gondola 2014). Official travel permits were required, and frequently permission from their husbands as well, but these constraints compelled women to create transactional relationships with military and colonial officers. Bribes or a share of the profits were given in exchange for unrestricted movement—a practice that still occurs today (DeMerode and Munslow 2006). This mobility allowed women access to game markets and trade networks, and a sort of matrilineal system of knowledge developed that could be passed down to daughters.

Within a few months of colonial independence in 1960, the country fell into a brutal five-year civil war. Kinshasa experienced enormous growth. Mobutu came into power in 1965 and implemented the official state ideology of *Authenticité,* calling for the évolués to dress, speak and eat in an "authentic" Congolese manner. Salaried urbanites were willing to pay good money for traditional meats. Powerful military men were often at the center of this trade, relying on poorly paid foot soldiers and park guards to extort game during periods of high insecurity, and employing their own wives as traffickers (DeMerode and Munslow 2006; Nellemann et al. 2010).

In the early 1990s, war in neighboring Rwanda caused a massive influx of people and small-arms into DRC. Civil war broke out in 1997, and Salonga and many other national parks were overrun by mutinous soldiers, renegade locals and armed refugees (Autesserre 2008).[6] Rebel military leaders stripped park guards of their weapons and stockpiled game and ivory to trade for munitions through strong illicit networks (Vogel 2000; Price 2003). The incursion of AK-47s into the forests changed the pace of hunting, increasing potential returns up to 25 fold (Wilkie and Carpenter 1999); piles of elephant meat could be found for sale out in the open amongst the flies (Stiles 2012). Beginning in 2011, under the moniker "Operation Bonobo," the Congolese paramilitary began patrolling Salonga Park, seizing illegal

---

6 The war lasted until 2002. Almost 5.4 million people died (Coghlan et al. 2007; Hopson 2011). Crisis, starvation and desperation were so rampant that previously held taboos against eating bonobos disappeared (Draulans and van Krunkelsven 2002).

weapons, and arresting suspected poachers.[7] This appears to have led to a decrease in large-scale poaching but may have merely shifted it to less policed areas.[8]

While distribution networks vary, and a river eventually dumps into the sea, so too most game ends up in Kinshasa, as increasing amounts are sold in rural areas rather than consumed directly (Tieguhong and Zwolinski 2009). The majority of meat is smoked for preservation because it may take many months to transport it from forest to city.[9] Although DRC is still primarily rural, in a dramatically short time Kinshasa has become a mega-city of 11 million with a growth rate of 4.2 percent annually (UN 2014).

As the economic and political conditions in DRC deteriorated, large numbers of Congolese of both sexes migrated to Europe (Ngoie and Lelu 2010). Trade networks have developed to transport food products from Africa to the growing European diaspora. Semi-legal container shipments arrive at European ports regularly with vegetables, fish and fruit. Each week, flights from Kinshasa bring immense quantities of smoked game to Europe, but this trade is little studied (Chaber 2009; Chaber et al. 2010).

Results: the geography of value

Despite the complex economics of the wild meat value chain, one trend appears universal: the further game travels from its point of extraction in the forest, the more expensive it becomes (Brashares et al. 2011). In Kinshasa, game costs four to five times more than in villages, and up to three times the price of domestic meat (Wilkie and Carpenter 1999). This is due to many factors but primarily two compounding conditions: (1) worsening road and river transport since the early 1990s have increased transport times,[10] and (2) the forests closest to urban areas have been denuded of major wildlife or deforested for charcoal production and urban sprawl, so game comes from progressively further afield (Abernethy et al. 2010).[11]

Luxury is thus largely a function of the distance—geographic, ecological and social—between consumption and source (Table 14.2). Put another way, this distance can be understood as participants' varied access to and control over the resource at each point along the value chain (Ribot and Peluso 2003; Rocheleau

---

7 ICCN guards (*Institut Congolais pour la Conservation de la Nature*) confiscated over 2,000 pounds of game, more than 120 high-powered firearms, including assault rifles, and 30 suspected poachers were arrested, seven of whom were sentenced to jail (de Wasseige et al. 2013). Elephant populations have increased threefold (CARPE 2013).
8 Wilkie and Carpenter (1999) estimate over US$46 million would be required annually to sufficiently regulate trade in the Congo Basin.
9 Game is a particularly high-value product per kilogram, as it does not degrade much over time and is relatively compact and lightweight (Abernethy et al. 2010).
10 Furthermore, today slender dugout canoes (*pirogues*) remain the only viable option for many river crossings, limiting traffic to bikes, motorcycles and pedestrians (Steel et al. 2008).
11 A fair share of this price is also due to extortion along the chain (DeMerode and Munslow 2006).

TABLE 14.2 The geography of value

	Monkey ~4kg (prices in US$)	Duiker ~4kg	Dist. from source (km)*	Transport method	Actors involved (gender)
Monkoto (Salonga National Park)	$10	$21.50	0–80	On foot or bicycle from forest	Village and market hunters (M)
Mbandaka	$11	$30	320	Bicycle or pirogue	Market hunters (M); traders (F/M)
Inflammable Port (outer Kinshasa)	$18	$65	>1,160	Large barge on Congo River	Traders; wholesale vendors (F)
Grand Marché (central Kinshasa)	$27	$90	>1,16 3 (~3 km from Port)	Vehicle or on foot	Primary and secondary vendors (F)
Restaurant (Kinshasa)	$21.50 (per portion)	$35 (per portion)	>1,163	By airplane (cargo)	*Fournisseurs* (M)
Paris	$110**	No information	>6,062	By airplane (passenger luggage)	Individuals and organized trafficking groups (F/M)

*Distances are approximate.
** Chaber et al. 2010.

2008). While processes of supply and demand matter, it is *the relative differences* between rural and urban regions—between producers and consumers—that actually drives this value creation, both economically and culturally. The higher the gradient of inequality between locations and populations, the greater potential for a luxury market to develop.

Production

The issue of hunting and village livelihoods is strongly gendered (Shackleton et al. 2011). Historically and contemporarily, hunters are nearly unanimously men, and the violence associated with DRC's forests has further restricted access by women.[12] The corridor forest in Salonga is divided by gender—men and women are responsible for

12 Women have repeatedly borne the brunt of the country's upheavals (Hochschild 2009). During the civil wars, rape was a particularly potent and calculated terror tactic, and sexual violence remains high today (Peterman et al. 2011).

different components of harvest. Non-game NTFPs are cultivated and collected by women and have less stigma of illegality (DeMerode et al. 2003).

There are two types of hunters—village and market—although the differences between the two can be murky (Table 14.3). Village hunters are locally based and tend to hunt opportunistically to feed their families and as a component of livelihoods. The villages surrounding Salonga Park hold traditional hunting rights to certain areas of the corridor forest but as these no longer harbor large species (due to over-hunting and habitat destruction), poaching inside park boundaries is widespread but limited to peripheral areas. On the other hand, highly organized groups of market poachers, again nearly exclusively male, set up camps deep in the Salonga forest.

TABLE 14.3 Village vs. market hunters

	Village (subsistence) hunters	*Market (commercial) hunters*
Hunting implement	Bicycle-cable snares, nylons and "traditional" *calibre dooze* shotguns (when they can afford bullets)	AK–47s
Terms of use	Few owned their own guns; guns bartered in exchange for share of the catch	Hunters own guns or hunt on commission. Porters given wild meat as payment (e.g. one monkey for every five they carry out).
Time in forest	Up to a week	Many weeks
Quantities	Return after catching only as many animals as they themselves can carry on foot.	Stockpile many pounds of meat. Stacked in large baskets made from slender bent saplings; hiked out to more accessible parts of forest (can take a week);
Terms of trade	Sold to family or other village members; increasingly stockpiled until a trader (male or female) comes to town, then sold for cash or bartered for soap, cigarettes, salt, or bullets.	Loaded onto modified bicycles and pushed over rutted, single-track dirt lanes and across log bridges (sometimes carrying nearly 50kg of meat) up to 800km over the course of two–three weeks; if a suitable tributary river is accessible, game is loaded onto a *pirogue*.

New programs of militarization are as related to social control as to conservation, entangling both market and subsistence hunters despite having different impacts and value chains (Rowcliffe et al. 2004; Ingram 2011). Those tasked with regulating and enforcing the trade—the poorly and erratically paid paramilitary and ICCN park guards—are frequently implicated in illegal transactions, either directly or by taking bribes (Wilkie 2001). Many others risk their lives to protect the animals (Hart 2011).

Markets

Markets are female-dominated, and women occupy various roles. Administrators are tasked with the overall running of the market, while traders transport game over semi-regular cycles. Others combine this with wholesale operations and have a network of female vendors who sell on commission or loan for them. The majority of game vendors are women, often older, a trend that is consistent with other studies of NTFP markets in Africa (Addo et al. 1994; Ruiz Perez et al. 2002; Elias 2015). A number of young women have recently entered the trade, as they consider it a lucrative employment option. When men are involved, it is often in the context of a family enterprise.

The workforce is also differentiated by specific market, and there is a hierarchy in price and species sold across locations. Markets on the outskirts of Kinshasa are cheaper, and primarily sell inferior meat such as monkey and rodents, or more desirable species at "wholesale" prices.[13] For example, some boats coming from Mbandaka stop outside the city at Maluku Port where vendors sell monkey to local clientele, as well as to other traders who take the product by road (about an hour) to outer markets within Kinshasa (e.g. Cinquantenaire). Kinshasa's main port market (Marché Inflammable) is primarily a wholesale market. The central markets (e.g. Marché Centrale and Marché Liberté) are more expensive, have higher quality, greater variety and more desirable animals. The vendors here are typically older women who have been involved in the trade for many decades. These markets attract a richer clientele of primarily women buyers.

The only freshly butchered game meat for sale was at Marché Liberté. The vendors were male, and I was told the meat arrives by plane from Orientale Province in the northeast of the country. The difference in gender here may have to do with the varying kinds of social capital between genders (Pandolfelli et al. 2007). Selling unsmoked meat requires access to more modern forms of transport and bridging social capital, whereas women tend to excel as wholesalers due to their bonding social capital and longstanding experience within the trade.[14]

---

13 More desirable species include antelope, buffalo and reptiles (usually sold alive). Elephant, chimpanzee and bonobo garner the highest prices, in part due to their illegality (Integrally Protected species), and risk of transport, and because they symbolize prestige.
14 Bridging social capital refers to networks between socially heterogeneous groups. Bonding social capital refers to social networks between homogenous groups.

Women are highly skilled at navigating the complex and often illegal components of the value chain. One study found women working as wholesalers in partnership with military officials (DeMerode and Munslow 2006). Although women may be more susceptible to bribe extortion, they are less likely to be arrested or otherwise punished. At Kinshasa's main port, Marché Inflammable, two sisters owned a market bar that served as headquarters for illegal smuggling of game (and ammunition from the Republic of Congo to hunters in DRC). They explained that they would get illicit meats (chimp or elephant) and ivory hidden inside legal kinds of smoked game or bags of maize. They worked with, and bribed, regulatory officials to ensure delivery of the illegal goods.

There is little information on the ratio of men to women working as international traffickers. However, there is some evidence it is dominated by women and appears to be highly organized (Chaber 2009). Most game seems to arrive in passenger luggage on flights from Africa, often wrapped in nothing more than brown paper.[15] New DNA techniques are garnering interest as a rapid method for detecting species from smoked pieces of meat (Gaubert et al. 2014). Many of the stores where game is allegedly sold are run and staffed by women.[16] The price in Europe can be twenty times the price in the forest, due in large part to the risks and illegality of trafficking (Chaber et al. 2010).

Consumption

In Kinshasa, game is a special rich treat—like caviar or a fancy steak—to be saved and savored.[17] Consumption signifies wealth and class. Just as individual markets and species have become differentiated by price, so too have the ways it is served: as fancy dishes in high-end restaurants, buffet style at restaurants boasting "traditional" food, in family-oriented places alongside fried fish, and to less well-off clientele who eat watered-down monkey stew. Restaurants tend to be male-dominated spaces, as men are more likely to have the money and social freedom to dine out. Restaurant owners, who capture the highest share of profits in the value chain, are primarily men, although some of the older or simpler establishments are owned by women. High-end restaurants are more likely to have male staff and chefs. Because of poor transportation networks and inherent difficulties getting a steady supply of game, restaurants typically have a *fournisseur* (male) located in Mbandaka to regularly and reliably source meat via biweekly passenger and freight flights, thereby bypassing the arduous road and river trip to Kinshasa.

15 This trade is illegal under a variety of European Union food importation laws. Although French Customs and Police (*Gendarmerie*) are primarily concerned with the trafficking of CITES species, some use fears of Ebola and general health and safety to gain political support for harsher controls and more personnel involved in preventing the trade (Smith et al. 2012).

16 With an increased number of raids over the past few years, shop owners in Paris are increasingly concerned about getting caught.

17 Consumption is primarily associated with people whose ancestors were forest-dwelling, such as the Batwa and Bantu. For other groups, fish is the primary cultural food and such people often adamantly deny eating game. Expats often eat game as a novelty food.

Game was seen to confer special health benefits, although this sentiment took on different connotations depending upon gender. When asked why they preferred wild to domestic meat, a number of people (both genders) relayed that because it came from the forest—was not a product of humans—it was unadulterated by artificial preservatives or chemicals and therefore purer.[18] A Congolese wildlife official (male) told me, "We eat it because it makes us strong and intelligent." This was said in contrast to domestic (white) meat, which was not seen as conferring the same benefits of vitality as game (red) meat. In fact, for men, eating domestic meat wasn't considered truly eating "meat" as it was not of high enough quality, taste, or richness. Women were more likely to associate game with more general health-giving benefits. A pregnant woman told me she ate it for the health of her baby. In a video filmed at Orly Airport in Paris, a visibly upset woman was telling the customs official that she was sick and her suitcase of game was her medicine.

Another common theme, shared by both genders, was that game should occasionally be fed to urban children to teach them what life used to be like—specifically as nostalgia for the village life experienced by their elders. One woman in Kinshasa preferred to smoke her own so that the children would also learn how it was prepared. When families visit relatives in the country, they bring them the products of the city, returning home with gifts of smoky game. In Paris, the cost, illegality and irregular supply render game a true luxury restricted to certain classes of people and reserved for celebrations such as weddings.

Fears of overconsumption and lack of money were the primary reasons for self-limiting of consumption. "Eat too much and you might get gout," explained the elites I interviewed, including one woman who told me her doctor had restricted her diet for this reason. A typical response in the urban lower classes was: "If I had money I would eat it." Or "When I have money, I buy it."

Analysis: access and power in a global economy

Each part of this complicated value chain is affected by local, national and international forces, which reach back into the past and extend into an unknowable future. While this may appear to complicate the problem, it also opens up new sites for intervention, solutions and research. One of the least explored avenues is that of women's roles, knowledge and capabilities within the wild meat value chain.

At the source, increased enforcement of hunting regulations, a heavy paramilitary presence, and further engagement and coordination between Regional Governors appears to be helping control hunting, but a lack of funds and continued corruption are primary limiting factors. As game currently provides a large percentage of household livelihoods, improving rural incomes may help curb hunting, as testified by village hunters (Abernethy et al. 2010).[19] There is a group of women in Monkoto working to create more opportunities for NTFP trade, which may

18 Note the resonance with Western conceptions of "organic" and "free range."
19 The forest, and game in particular, is seen as a bank that insulates against economic shocks, whereas domestic animals are more like insurance or savings accounts (Wilkie and Carpenter 1999).

provide an alternative source of income. Yet compared to game, other NTFPs have relatively low value. There is interest at the Congolese Wildlife Department in creating a "legal" game market through domestication of certain wild species. However, in a country with recurrent governance challenges, and where the opportunity cost of hunting remains so low, this approach could potentially further drive the illegal trade by increasing overall demand (Newing 2001; Pailler 2005; de Wasseige et al. 2013).

Furthermore, the exposure to material goods facilitated by NGOs working in these regions often increases demand for cash and consumer goods, and alters cultural norms based on communal sharing of resources. This can have the unintended effect of further promoting hunting, particularly if men have power over where these funds are directed, or if they begin to rely on material status symbols in their desire to perform their role as masculine provider (Robinson and Bennett 2002). "Our money becomes the message," one conservationist told me. While many village hunters told me the cash is used for school and hospital fees, if women do not have access to this income, funds may not benefit the household as a whole (Colfer 2013). Moreover, the jobs brought by NGOs and the military in Salonga are primarily for men, further creating power disparities between genders and affecting women's ability to help regulate hunting.[20]

On the trade side of the value chain, it is important to re-categorize female vendors and traders as successful micro-entrepreneurs, rather than illegal actors in an informal economy (Velde et al. 2006). Considering the complexity of the game value chain, vendors are a particularly important source of knowledge about the dynamics of market conditions, and have been undervalued as potential collaborators or avenues of conservation intervention.[21] The increasing numbers of young women entering the trade reveals broad and unique information about larger economic transformations that would benefit from further analysis. Many of the women I interviewed felt that the trade was more competitive and generally not as profitable as it once was. Most believed this is due not to a reduction in wildlife numbers but rather to: increasing breakdowns in transport, more intermediaries, difficulty in sourcing quality product, and because of DRC's general economic hardship, the general public could no longer afford it. Higher prices did not necessarily represent larger profit margins, and could decrease overall quantities sold or prolong the period required to sell stocks. This particularly affected female vendors who worked as both traders and wholesalers and had semi-regular travel schedules upriver. Conservation organizations were considered problematic to their trade, rather than allies. In some cases, strictly enforcing endangered species bans has had the unintended consequence of promoting desirability and further driving the trade (Nasi et al. 2011).[22]

---

20 Former hunters can become community police, patrolling their own land against outside hunters, and office jobs tend to benefit literate men, keeping women out of the information loop regarding formal forest resource management.

21 The scope and scale of the wild meat network in DRC is difficult to quantify. Market surveys are costly due to irregular transaction periods. Uneven bumps in supply or changes in seasonal abundance can skew results.

22 Today, the risk of transporting elephant meat has meant that the ivory trade has become detached from the meat trade, and most is left to rot in the jungle, increasing its rarity and further driving prices. In urban areas, its consumption is still associated with certain wealthy classes, such as high-ranking military and government officials.

Collective action has been found to be a key to women's empowerment, and fundamental during times of economic, political and social crisis when the state fails to address their needs (Colfer 2013). In part a reaction to the deep gender disparities, DRC has one of the highest number of women's organizations in Africa (Godin and Chideka 2010). Tapping into these networks for conservation purposes has yet to be explored (Coleman and Mwangi 2013). Organizing and educating women involved in the trade might empower them to better police the legality and fight against extortion and corruption within their supply chains. Many of Kinshasa's market administrators are women, and as they are already keeping track of game quantities and sales, they may make extremely useful partners in quantifying the total value, volume and waste within the value chain.

In terms of consumer demand, a variety of female-centered approaches to change cultural norms are being undertaken (Rose 2001; Bowen-Jones et al. 2003). Conservation NGOs are engaging with women vendors to create consumer markets for novel ways of cooking simple proteins like eggs and chicken. This may be particularly effective with younger generations of urban women who have less connection to forest-based diets and might be inspired by Western environmental ethics around food. Social media campaigns and other forms of gendered social marketing could be effective (Bowman 2001). Restaurants, which transform game into a luxury item, seem to be less stigmatized by the conservation community than other actors along the value chain, and could be the site of future efforts. The Catholic Church remains an important institutional structure for many Congolese and could also be a good ally in appealing to the norms and ideals of its constituents. Considering how engrained, ritualized and deeply historical the consumption of game is—and its connotations with health, wealth, purity and nostalgia—there is a long road ahead.

There are far-reaching and understudied consequences as the market becomes more luxury and internationally oriented. As game is often the main investment for NTFP traders, it is an important factor in making it economically feasible to transport other necessary goods from the forests (Abernethy et al. 2010). The international trade also opens up new realms of gendered power relationships vis-à-vis the European state, whereby given their extensive involvement in sale and purchase, Congolese women are potentially unequally disciplined by international enforcement agencies. Moreover, the number of asylum seekers to Europe remains large, with a high proportion being educated women. Existing social networks in the Congolese diaspora have strong influence on behavior and dietary choices of new immigrants, and could help move purchases away from particularly vulnerable species. As Congolese women living in Europe make occasional visits home or wish to support local initiatives from abroad, these transnational relationships can be an important source of influence in both localities (Godin and Chideka 2010).

Conclusion

DRC is sometimes called "the richest, poorest country on earth." Despite its incredible untapped resources, control of the majority of this wealth is concentrated in a few hands while the rest of the country faces low levels of development, significant poverty, and high levels of corruption (Ingram 2011). Increasing Chinese investments are rapidly changing economic and transportation dynamics (Draper et al. 2010). The growing middle class in Kinshasa wants Western comforts, yet wishes to maintain cultural continuity in a rapidly globalizing world. Because of DRC's abundance of ethnic groups, many people maintain strong identity politics through social relations, enterprise associations and food. Although Kinshasa continues to be viewed as providing the best opportunity for social and economic success, it remains an uncertain place to live and typically residents still maintain links to their original villages, often through trade networks. This is particularly true for the wild meat value chain where traders are often related to those in the villages where they purchase game. The influx of cash and resource development into remote areas is having unintended consequences that may further drive demand.

If conservationists want to address the "bushmeat" crisis, women will need to share the center of this change. Future research should include in-depth ethnographic and longitudinal studies of the women involved in the trade and address: (1) How these women see themselves, their power, and their social roles; (2) their openness and desire for alternative livelihoods; (3) best-practices and approaches to impart conservation and regulatory information throughout their social networks. If we approach this problem without alienating those most connected to it, through respect, sensitivity and empowerment, we may have the chance not only to prevent the devastating collapse of DRC's wildlife, but also to improve the social conditions of its long-suffering and incredibly resilient women.

Acknowledgments

This research was conducted under the support of a National Science Foundation Graduate Research Fellowship and the Tropical Resources Institute at Yale University. Additional logistical support came from USAID through the Central African Forest Ecosystem Conservation (CAFEC) program, and the Wildlife Conservation Society. I would like to thank my host at WCS, Michelle Wieland, and my research assistant, Patrick Mukombozi. Jonas Abana Eriksson was an invaluable source of information. I would also like to thank the anonymous reviewers and the editors of this volume for their comments and suggestions. Last but not least, this work would not have been possible without all the women working in the wild meat trade who took the time to speak with me. None of the above bears any responsibility for any remaining errors in the text.

References

Abernethy K, Coad L, Ilambu O, Makiloutila F, Easton J and Akiak J. 2010. Wildlife hunting, consumption trade in the Oshwe sector of the Salonga-Lukenie-Sankuru landscape, DRC. Draft. World Wide Fund for Nature—Central Africa Regional Program Office, DRC.
Addo FA, Asibey EOA, Quist KB and Dyson MB. 1994. The economic contribution of women and protected areas: Ghana and the bushmeat trade. In Munasinghe M and McNeely J, eds. *Linking conservation and sustainable development*. Washington, DC: World Bank. 68–78.
Alexander IJ, Swaine MD and Watling R, eds. 1996. *Essays on the ecology of the Guinea-Congo rain forest*. Edinburgh: The Royal Society of Edinburgh. Vol. 104.

Autesserre S. 2008. The trouble with Congo: How local disputes fuel regional conflict. *Foreign Affairs* 87(3): 95–110.

Bowen-Jones E, Brown D and Robinson EJ. 2003. Economic commodity or environmental crisis? An interdisciplinary approach to analyzing the bushmeat trade in central and west Africa. *Area* 35(4): 390–402.

Bowman K. 2001. Culture, ethics and conservation in addressing the bushmeat crisis in West Africa. In Bakarr M, Fonseca G, Mittermeier R, Rylands A and Painemilla K, eds. *Hunting and bushmeat utilization in the African rain forest: Perspectives toward a blueprint for conservation action.* Washington, DC: Conservation International. 75–84.

Brashares JS, Golden C, Weinbaum K, Barrett C, and Okello G. 2011. Economic and geographic drivers of wildlife consumption in rural Africa. *Proceedings of the National Academy of Sciences* 108(34): 13931–6.

Brown D and Williams A. 2003. The case for bushmeat as a component of development policy: Issues and challenges. *International Forestry Review* 5(2): 148–55.

CARPE. 2013. Anti-poaching operation makes DRC park safe for elephants and people. USAID. Accessed January 7, 2016. http://carpe.umd.edu/Documents/2013/Anti_poaching_factsheet_March_2013.pdf.

Chaber AL. 2009. Investigation of the African bushmeat traffic in France: A threat to both biodiversity and public health. *U.K. Bushmeat Working Group.* Accessed January 7, 2016. static.zsl.org/secure/files/al-chaber-bushmeat-traffic-1458.pdf.

——, Allebone-Webb S, Lignereux Y, Cunningham A and Rowcliffe M. 2010. The scale of illegal meat importation from Africa to Europe via Paris. *Conservation Letters* 3(5): 317–21.

Coghlan B, Ngoy P, Mulumba F, Hardy C, Bemo VN, Stewart T, Lewis J and Brennan R. 2007. Mortality in the Democratic Republic of Congo: An ongoing crisis. Accessed January 7, 2016. http://www.rescue.org/sites/default/files/resource-file/2006-7_congo MortalitySurvey.pdf.

Coleman E and Mwangi E. 2013. Women's participation in forest management: A cross-country analysis. *Global Environmental Change* 23(1): 193–205.

Colfer CJP. 2013. The "Gender Box": A framework for analysing gender roles in forest management. *CIFOR Occasional Paper* 82. CIFOR: Bogor, Indonesia.

DeMerode E and Munslow B. 2006. Species protection, the changing informal economy, and the politics of access to the bushmeat trade in the Democratic Republic of Congo. *Conservation Biology* 20(4): 1262–71.

——, Homewood K and Cowlishaw G. 2003. Wild resources and livelihoods of poor households in Democratic Republic of Congo. *Wildlife Policy Briefing.* Accessed January 7, 2016. http://www.odi.org/sites/odi.org.uk/files/odi-assets/publications-opinion-files/3310.pdf.

de Wasseige C, Flynn J, Louppe D, Hiol Hiol F and Mayaux PH, eds. 2013. *The state of the forest 2013.* Commission des Forêts d'Afrique Centrale (OFAC/COMIFAC) and the Congo Basin Forest Partnership (CBFP).

Dove M. 1992. The dialectical history of "jungle" in Pakistan: An examination of the relationship between nature and culture. *Journal of Anthropological Research* 48(3): 231–53.

Draper P, Disenyana T and Biacuana G. 2010. Chinese investment in African network industries: Case studies from the Democratic Republic of Congo and Kenya. In Cheru F and Obi C, eds. *The rise of China and India in Africa.* London and New York: Zed Books. 112–15.

Draulans D and van Krunkelsven E. 2002. The impact of war on forest areas in the Democratic Republic of Congo. *Oryx* 36(1): 34–40.

Elias M. 2015. Gender, knowledge-sharing and management of shea (*Vitellaria paradoxa*) parklands in central-west Burkina Faso. *Journal of Rural Studies* 38: 27–38.

Fairhead J and Leach M. 1996. *Misreading the African landscape: Society and ecology in a forest-savanna mosaic.* Cambridge: Cambridge University Press.

Fortmann L, ed. 2008. *Participatory research in conservation and rural livelihoods: Doing science together*. Oxford: Blackwell.

——, Ballard H and Sperling L. 2008. Change around the edges: Gender analysis, feminist methods and sciences of terrestrial environments. In Schiebinger L, ed. *Gendered innovations in science and engineering*. Stanford, CA: Stanford University Press. 79–95.

Gaubert P, Njiokou F, Olayemi A, Pagani P, Dufour S, Danquah E, Nutsuakor ME, Ngua G, Missoup AD, Tedesco PA et al. 2014. Bushmeat genetics: Setting up a reference framework for the DNA typing of African forest bushmeat. *Molecular Ecology Resources* 15(3): 633–51.

Godin M and Chideka M. 2010. Congolese women activists in DRC and Belgium. *Forced Migration Review* 36: 33.

Gondola C. 2014. Popular music, urban society, and changing gender relations in Kinshasa, Zaire (1950–1990). In Grosz-Ngate M and Kokole O, eds. *Gendered encounters: Challenging cultural boundaries and social hierarchies in Africa*. New York: Routledge. 65–87.

Hart T. 2011. Celebration and mourning in Kinshasa. Accessed 6 March 2015: http://www.bonoboincongo.com/2011/10/19/celebration-and-mourning-in-kinshasa.

Hochschild A. 2009. Rape of the Congo. *New York Review of Books*. August 13, 2009.

Hopson M. 2011. The wilderness myth: How the failure of the American national park model threatens the survival of the Iyaelima tribe and the bonobo chimpanzee. *Earth Jurisprudence* 1: 61–102.

Ingram V. 2011. Forest-poverty-commodity links in the Congo basin: A value chain perspective. *International Conference: Nature™ Inc? Questioning the Market Panacea in Environmental Policy and Conservation*. The Hague, Netherlands: ISS. Accessed January 7, 2016. http://www.iss.nl/fileadmin/ASSETS/iss/Documents/Conference_presentations/Nature_Inc_Verina_Ingram.pdf.

Larson A and Ribot J. 2007. The poverty of forestry policy: Double standards on an uneven playing field. *Sustainability Science* 2(2): 189–204.

Mianda G. 2002. Colonialism, education, and gender relations in the Belgian Congo: The *évolué* case. In Geiger S, Musisi N, and Allman JM, eds. *Women in African colonial histories*. Bloomington: Indiana University Press. 144–63.

Nasi R, Taber A and Van Vliet N. 2011. Empty forests, empty stomachs? Bushmeat and livelihoods in the Congo and Amazon Basins. *International Forestry Review* 13(3): 355–68.

Nellemann C, Redmond I and Refisch J, eds. 2010. The last stand of the gorilla – environmental crime and conflict in the Congo Basin. A rapid response assessment. United Nations Environment Programme, GRID-Arendal.

Newing H. 2001. Bushmeat hunting and management: Implications of duiker ecology and interspecific competition. *Biodiversity and Conservation* 10: 99–118.

Ngoie G and Lelu D. 2010. *Migration en République Démocratique du Congo PROFIL NATIONAL 2009*. Geneva: Organisation Internationale pour les Migrations (OIM).

Pailler S. 2005. The necessity, complexity and difficulty of resolving the bushmeat crisis in West-Central Africa. *Journal of Development and Social Transformation* 2: 99–107.

Pandolfelli L, Meinzen-Dick R and Dohrn S. 2007. Gender and collective action: A conceptual framework for analysis. *Collective Action Working Paper* 64: 57.

Paulme D. 1963. *Women of tropical Africa*. Abingdon and New York: Routledge.

Peterman A, Palermo T and Bredenkamp C. 2011. Estimates and determinants of sexual violence against women in the Democratic Republic of Congo. *American Journal of Public Health* 101(6): 1060–67.

Piermay J. 1997. Kinshasa: A reprieved mega-city? In Rakodi C, ed. *The urban challenge in Africa: Growth and management of its large cities*. New York: United Nations University Press. 236.

Price S. 2003. *War and tropical forests: Conservation in areas of armed conflict*. Binghamton, NY: Food Products Press.

Rau B. 1991. *From feast to famine: Official cures and grassroots remedies to Africa's food crisis.* London and Atlantic Highlands, NJ: Zed Books Ltd.

Redford KH. 1992. The empty forest. *Bioscience* 22: 412–22.

Ribot JC and Peluso NL. 2003. A theory of access. *Rural Sociology* 68(2): 153–81.

Robinson J and Bennett E. 2002. Will alleviating poverty solve the bushmeat crisis? *Oryx* 36(4): 332.

Rocheleau DE. 2008. Political ecology in the key of policy: From chains of explanation to webs of relation. *Geoforum* 39: 716–27.

Rose AL. 2001. Social change and social values in mitigating bushmeat commerce. In Bakarr M, Fonseca G, Mittermeier R, Rylands A and Painemilla K, eds. *Hunting and bushmeat utilization in the African rain forest: Perspectives toward a blueprint for conservation action.* Washington, DC: Conservation International. 17–20.

Rowcliffe JM, de Merode E and Cowlishaw G. 2004. Do wildlife laws work? Species protection and the application of a prey choice model to poaching decisions. *Proceedings of the Royal Society B: Biological Sciences* 271(1557): 2631–6.

Ruiz Perez M, Ndoye O, Eyebe A and Ngono DL. 2002. A gender analysis of forest product markets in Cameroon. *Africa Today* 49(3): 97–126.

Shackleton S, Paumgarten F, Kassa H, Husselman M and Zida M. 2011. Opportunities for enhancing poor women's socio-economic empowerment in the value chains of three African non-timber forest products (NTFPs). *International Forestry Review* 13(2): 136–51.

Smith KM, Anthony SJ, Switzer WM, Epstein JH, Seimon T et al. 2012. Zoonotic viruses associated with illegally imported wildlife products. *PLoS ONE* 7(1): e29505.

Stiles D. 2012. Elephant bushmeat: New threat to elephants. *SWARA* July–September 2012. 38–41.

Steel L, Colom A, Maisels F and Shapiro A. 2008. The scale and dynamics of wildlife trade originating in the south of the Salonga-Lukenie-Sankuru landscape. *WWF-Democratic Republic of Congo.* (Draft – not for circulation).

Stoler AL. 2002. *Carnal knowledge and imperial power: Race and the intimate in colonial rule.* Berkeley: University of California Press.

Taylor G, Taylor GJPW, Scharlemann M, Rowcliffe N, Kümpel MBJ, Harfoot JE, Fa R, Melisch EJ, Milner-Gulland S, Bhagwat KA, et al. 2015. Synthesizing bushmeat research effort in West and Central Africa: A new regional database. *Biological Conservation* 181: 199–205.

Tieguhong JC and Zwolinski J. 2009. Supplies of bushmeat for livelihoods in logging towns in the Congo Basin. *Journal of Horticulture and Forestry* 1(5): 65–80.

UN (United Nations). 2014. *World urbanization prospects: The 2014 revision, highlights* (ST/ESA/SER.A/352). Department of Economic and Social Affairs, Population Division.

Velde DW, Rushton J, Schreckenberg K, Marshall E, Edouard F, Newton A and Arancibia E. 2006. Entrepreneurship in value chains of non-timber forest products. *Forest Policy and Economics* 8: 725–41.

Vogel G. 2000. Conflict in Congo threatens bonobos and rare gorillas. *Science* 287:2386–7.

Wilkie DS. 2001. Bushmeat hunting in the Congo Basin—A brief overview. In Bakarr M, Fonseca G, Mittermeier R, Rylands A and Painemilla K, eds. *Hunting and bushmeat utilization in the African rain forest. Perspectives toward a blueprint for conservation action.* Washington, DC: Conservation International. 17–20.

—— and Carpenter JF. 1999. Bushmeat hunting in the Congo Basin: An assessment of impacts and options for mitigation. *Biodiversity and Conservation* 8: 927–55.

Yates B. 1982. Church, state and education in Belgian Africa: Implications for contemporary third world women. In Kelly G and Elliott C, eds. *Women's education in the Third World: Comparative perspectives.* Albany, NY: SUNY Press. 127–51.

Yumoto T, Maruhashi T, Yamagiwa J and Mwanza N. 1995. Seed-dispersal by elephants in a tropical rain forest in Kahuzi-Biega National Park, Zaire. *Biotropica* 27: 526–30.

PART V

Longstanding and emerging gendered issues

15

GENDERED KNOWLEDGE SHARING AND MANAGEMENT OF SHEA (*VITELLARIA PARADOXA*) IN CENTRAL-WEST BURKINA FASO[1]

Marlène Elias

Introduction

Shea trees (*Vitellaria paradoxa* C.F. Gaertn.) provide essential benefits to local people and ecosystems in 18 countries across the semi-arid African savanna (Hall et al. 1966). Shea trees are prized for their nutritive fruit, medicinal properties and hardwood, but especially for the butter that women extract from their nuts (Burkill 2000). This shea butter represents the primary source of dietary fat for many agriculturalists living in the species' range and serves important economic and cultural uses (Lykke et al. 2002). Internationally, shea butter is a prized ingredient in cosmetics and confectionaries, which offers significant and growing economic prospects for countries such as Burkina Faso, where shea nuts rank fourth among national export commodities (MEF 2011). Due to its myriad purposes, farmers have selectively protected the species when clearing their fields for agriculture, resulting in agroforestry parklands where shea grows in nearly pure stands (Boffa 1999).

Within West Africa, the importance of shea largely rests in the highly gendered context of its production and trade. Women collect and process shea nuts into butter and sell the nuts and butter locally (Chalfin 2004; Elias and Carney 2007). Nonetheless, shea agroforestry is arguably guided by the specialized and interlocking knowledge repertoires and practices of women and men, who both value and use the species. The gendered dynamics surrounding the selective conservation and management of shea trees can offer insight into intra-household decision-making processes and strategies shaping local agroforestry systems. This insight is important for building upon the range of local knowledge of shea, improving targeting of local

1 This chapter is based on: *Journal of Rural Studies,* Vol 38, Elias, M. Gender, knowledge sharing and management of shea (*Vitellaria paradoxa*) parklands in central-west Burkina Faso, 27–38, Copyright (2015), with permission from Elsevier, as per the requirements of the journal.

actors to develop locally relevant and equitable options for sustainable shea tree management, and influencing decisions that can improve the species' conservation.

Hence, the aim of this chapter is to examine the gendered ethnobotanical knowledge, management, and conservation of *V. paradoxa* in the province of Sissili, Burkina Faso, where shea represents the most prevalent parkland tree species. As Howard (2003: 33) contends, 'some researchers have attempted to develop conceptual frameworks to assess which factors motivate indigenous or peasant farmers to conserve biodiversity, but to date these have neglected to consider gender relations as potentially significant.' What is more, 'examining gender helps us to understand how other forms of social difference influence rural environmental management, not just as "proxy," but because other differences such as age, wealth or origins operate in gender-differentiated ways' (Leach 1994: 22).

This chapter draws attention to how women and men agriculturalists know and use shea trees, and to how knowledge sharing among spouses can inform decision making and sustainable management strategies for the species.

The chapter begins with a review of the literature on gendered use and management of trees, before turning to Amartya Sen's intra-household bargaining model of 'cooperative-conflict' that can foster an improved understanding of the gendered shea agroforestry system. Following a description of the study's context and methodology, I analyse the main findings related to the gendered dimensions of shea tree knowledge, management and conservation. I then examine the implications of these findings and the value of Sen's notion of 'social connectedness' for better conceptualizing natural resource management systems, before providing brief concluding remarks.

Gendered ecological knowledge and arboreal management

In smallholder farming systems, gender is 'a critical variable in shaping resource access and control, interacting with class, caste, race, and ethnicity to shape processes of ecological change' (Rocheleau et al. 1996: 4). Historically rooted and context-specific norms and belief systems prescribe 'appropriate' behaviour for men and women and a gendered division of labour that guides resource use (Leach 1994; Rocheleau et al. 1996). Consequently, gender specialization occurs in the collection and processing of most types of forest products (Sunderland et al. 2014). Customary and formal laws, which are also gendered, structure rights to forests and trees and influence incentives and capacity to manage tree resources (Fortmann and Bruce 1988; Quisumbing et al. 2001). Men and women from the same household or community may have access to trees located in different spaces or even to different parts of the same tree. Hence, they often collect and use different products or gather the same products in different spaces (Rocheleau and Edmunds 1997; Howard 2003). Gendered tenure regimes affect management strategies, as insecure access and control of forest and tree resources limit women's incentives to plant and manage trees to which their long-term rights are tenuous (Fortmann et al. 1997; Howard 2007).

This gendered division of labour, use and access to forest products, and these multiple layers of institutions influence the knowledge and skills women and men acquire about tree resources (Howard 2003; Ayantunde et al. 2008; Dovie et al. 2008). Factors such as kinship, age, specialization and 'motivation, ability and opportunities to learn' more generally intersect with gender to affect the acquisition and transmission of this local ecological knowledge (Boster 1986: 434). Gendered patterns of ecological knowledge and resource use reflect and reinforce a gendered valuation of tree species, traits and products, with preferences for tree species also differing according to other factors of social differentiation, such as marital status or age (Bonnard and Scherr 1994; Cavendish 2000).

Gender relations also affect decision-making processes related to natural resources. Women are generally shown to be more constrained than men in their ability to make decisions related to natural resource management. This may be due to gender norms or technological biases, among others, that hinder their participation in formal decision-making forums such as forest user groups (FUGs) or in household-level decision-making processes (Rocheleau and Edmunds 1997; Agarwal 2001).

Recognizing these gendered specificities in forest and agroforestry settings is critical for understanding women's and men's respective contributions, opportunities and constraints in tree management processes. Yet, the ways in which women and men often collaborate – including through knowledge sharing – to help each other fulfil their gender-specific responsibilities are equally significant and remain poorly explored in most studies, which portray gendered spheres of activity as distinct and separable (Turner et al. 2000). This is an important oversight as collaboration may represent a key feature of sustainable resource management strategies.

Sen's model

Key notions from Amartya Sen's (1990) model of intra-household bargaining can support an analysis of how women and men both specialize and cooperate in shea tree agroforestry. In his revised model of household economics, Sen conceives of the household as the site of coinciding and competing interests where women's and men's spheres of production, consumption and decision making are separate but interrelated.[2] Although women and men manage activities in a specialized manner, an integrated view of how their activities sustain and support each other is needed to understand production processes.

Sen argues that to fulfil their gender-specific responsibilities, household members cooperate when they believe that doing so is more favourable to them than non-cooperation. Various cooperative arrangements may exist and differentially favour some household members over others. According to Sen (1990: 465), 'the nature of the cooperative arrangements implicitly influences the distributional parameters and

---

2 This contrasts with Becker's (1965) household-welfare-function model – referred to as the 'common preference' or 'unitary' model – that sees household members as sharing a common set of preferences and a single decision-making logic.

the household's response to conflicts of interest.' The arrangement that is ultimately pursued is negotiated based upon each member's relative bargaining power, which primarily depends on the person's 'breakdown position' or the alternatives awaiting him or her in the absence of cooperation (see also Folbre 1986; Udry 1996; Carter and Katz 1997). Socio-economic changes, such as new economic opportunities arising from new markets or technologies for shea, can alter the state of this breakdown position and shift the balance of power within the household, thereby affecting collective and individual production and welfare (Agarwal 1997; Doss 2001).

Of particular value here is the emphasis Sen places on the existence of separate domains of specialization that are articulated within the household and inform intra-household decision making, and the 'social connectedness' of husband and wife who:

> . . . live together under the same roof – sharing concerns and experiences and acting jointly. This aspect of 'togetherness' gives the gender conflict some very special characteristics. One of these characteristics is that many aspects of the conflict of interest between men and women have to be viewed against the background of pervasive cooperative behaviour.
>
> *(Sen 1990: 147)*

This complex backdrop of gendered specialization, sharing, cooperation and conflict (or divergent interests) sets the stage for understanding the local shea agroforestry system in Burkina Faso.

Contextualizing and undertaking the study

Study sites

This study was carried out in the town of Léo and the villages of Lan and Prata in Burkina Faso's province of Sissili, which occupies the country's central-west region (Figure 15.1). Lan and Prata sit on the Ghanaian border and are situated nine kilometres south and 49 kilometres east of Léo, respectively. The three sites are in a shea tree-rich zone and due to their proximity to the Ghanaian border, they have long been integrated in local and international shea butter markets. Their female residents are members of one of the largest unions of shea butter producers in the country, the Fédération NUNUNA.[3]

Local annual rainfall ranges between 800 and 1000 millimetres, with peak rainfall occurring between May and October (Fontès and Guinko 1995). The region sits within the southern Sudanian zone and is species-rich compared to the northern parts of the country. Over the past twenty years, an influx of migrants from the

---

3 The Federation NUNUNA was created within the context of international aid projects promoting women's empowerment through international shea butter sales. At the time of the study, it was called the *Union des groupements de productrices de produits karité de la Sissili et du Ziro* and comprised over 3,000 producers organised in 33 village women's groups. The Federation is headquartered in the town of Léo and encompasses members from the surrounding villages of Lan and Prata, among others.

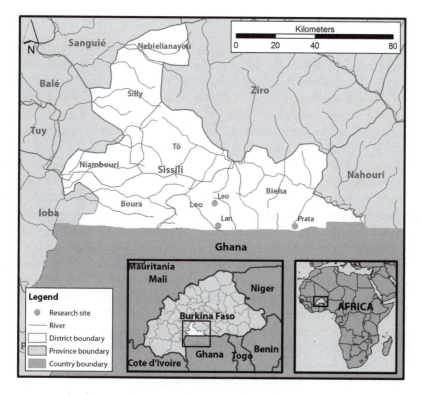

FIGURE 15.1 Study sites in Burkina Faso

country's northern and central areas has joined the province's indigenous Gurunsi inhabitants. These migrants, predominantly of Moose ethnicity, now locally out-number the Gurunsi (Howorth and O'Keefe 1999).

Contemporary Gurunsi and Moose agriculture is based on household production and silviculture, integrating some animal husbandry with subsistence and cash-crop cultivation (Ouédraogo 2003). Extensive agriculture is practiced with limited access to plough-animal traction, and soil fertility is maintained by rotational fal-low, despite a shortening of fallow periods (Gray 2003). Agroforestry and gathering of non-timber forest products (NTFPs), including shea fruit, are integral to local livelihoods. Women are the primary collectors and processors of shea nuts in both Gurunsi and Moose cultures.

Customary land ownership, which remains the norm in Burkina Faso, is lineage-based (Reenberg and Lund 1998; Ouédraogo 2003). Social institutions and taboos regulate access to tree species such as shea. Migrants hold weaker rights to shea trees than indigenous residents, who can claim the products of valuable trees growing on land they have 'loaned' to newcomers. Among both ethnic groups, the planting and felling of shea trees are customarily restricted. Burkina Faso's forestry laws also forbid cutting shea trees without a permit, but as in other parts of West Africa, the state lacks the resources to enforce these laws (e.g. Wezel and Haigis 2000).

Methods

Data for this study were collected between October 2006 and March 2007 with 90 women and 78 men of Gurunsi and Moose descent. Thirty women from each site (Léo, Lan and Prata) were randomly selected from the roster of the Fédération NUNUNA, in which most able-bodied adult, married Gurunsi and Moose women from local villages hold membership. This represents over one-quarter of the women living in Lan and Prata, and follows Brewer et al.'s (2002) observation that twenty to thirty research participants generally suffice to identify a coherent set of items using free listing techniques, as was done in this study, when individuals share a reasonable degree of knowledge about the topic at hand. The women participants' husbands were also interviewed to assess the gender dynamics at play within the shea agroforestry system. Each participant was interviewed individually and privately.

Interviews were conducted using free listing techniques to collect data without initial prompts. This basic ethnographic tool consists of asking respondents to list as many 'Xs' as they can, where 'X' refers to a cultural domain, such as names of trees, uses for products, and so on (Weller and Romney 1988; Brewer et al. 2002). In this way, respondents elicited and discussed any, and as many, points as they wanted for every question posed. Following Gatewood (1984), the number of responses participants gave in this way, or their 'free listing capacity' (Brewer 1995: 108), was then used as a loose indicator of their knowledge of the issues at hand.

Questions primarily centred on participants' knowledge of the factors influencing shea yields, characteristics of quality shea nuts, and uses for the shea tree and its derivatives. As uses for shea butter are extensively documented and tend to outshine the species' other applications, participants were asked to focus on the tree's non-shea butter-related functions. Questions additionally revolved around shea tree management practices and the selective conservation of shea trees. Open-ended, follow-up questions were then asked to 'thicken' the discussion of the role the shea tree plays in the livelihoods and farming systems of participants, their shea tree management practices, skills and underlying knowledge repertoires.

Men's and women's responses were open coded manually according to recurring themes (Stemler 2001). A master list of key thematic categories was created and the frequency of responses that fell within each established category was calculated separately for women and men. Differences and overlaps across gendered responses were qualitatively analysed both in terms of these frequencies and in terms of the range of categories cited by women and men participants. Following Friedman et al. (1986), fidelity levels (FL) were calculated using the formula $FL = (I_p \times 100)/I_U$, where I_p is the number of participants who independently cited a specific variant (e.g. a specific use for shea) for a given domain of analysis (e.g. uses of shea) and I_U is the total number of responses provided for all variants in a given domain of analysis (e.g. the total number of responses cited by participants for all the different uses for shea). The FL indicates the degree of consensus among participants concerning the variants within a given domain of analysis. FLs were calculated separately for women and men. This data was interpreted in tandem with the interview data that explained the reasons behind the observed patterns.

TABLE 15.1 Participant women's and men's endogenous perceptions of the factors influencing shea yields

Factor		Per cent of times cited		Effect on yields
		Women	Men	
Climatic factors	Strong winds	13	8	↓
	High rainfall	5	8	↑
Biological factors	Semi-parasitic plants	38	31	↓
	Predatory insects/fungus	3	2	↓
Anthropogenic factors	Farming practices	10	39	↑
Tree traits	Flowering behaviour (abundant, post-Harmattan)	28	10	↑
	Canopy characteristics (extensive, full leaf flush)	15	20	↑
	Size	10	22	↑/↓
	'Gender': male tree	3	4	↓

Gendered knowledge repertoires

Shea yields and nut traits

Participant women and men showed a detailed understanding of the factors influencing shea yields, including those stemming from the environment in which the tree grows and those associated with specific tree characteristics. They correlated shea yields with the same climatic, biological and anthropogenic factors, albeit in different proportions (see Table 15.1). There was high consensus among women and men that semi-parasitic plants are the most serious problem afflicting shea trees in the study sites (FL = 30.6 and 21.8 for women and men, respectively) and that the strength and timing of winds affects the ability of trees to maintain their flowers and eventually fruit (Table 15.2). Nearly 40 per cent of the men and 10 per cent of the women interviewed also noted that anthropogenic practices affect shea yields, as shea trees in farmed fields are larger, benefit from nutrient inputs from fertilizers, and are protected from pillaging squirrels and bush fires.

Men and women participants further recognized that the top three traits to influence an individual tree's yields were its flowering and leafing behaviour as well as its size (Table 15.2). All female participants and 95 per cent of male participants claimed to know which shea trees provide the best yields in their fields, with some respondents ascribing names to these trees. Seventy per cent of male participants believed that shea trees offering good yields remain the same from year to year, whereas the same proportion of women believed that good producers vary interannually. These perceptions have implications at the time of shea tree selection since, as noted below, productivity is a prime factor influencing the decision to retain or remove shea trees during land clearance.

TABLE 15.2 Variation in knowledge, practices and selection preferences related to shea across gender groups

Category		Variant	Women (n = 90)		Men (n = 78)	
			F	FL	F	FL
KNOWLEDGE	**Factors affecting productivity**	Environmental/ anthropogenic				
		Semi-parasitic plants	35	30.6	24	21.8
		Winds	12	10.2	6	5.5
		Farming practices	9	8.2	31	27.7
		Rainfall	5	4.1	6	5.5
		Predatory insects	2	2.0	2	1.4
		Tree characteristics				
		Flowering behaviour	25	22.5	8	6.8
		Leafing behaviour	14	12.3	15	13.7
		Size	9	8.2	17	15.0
		Gender	2	2.0	3	2.7
		ΣF	113	–	112	–
	Uses for shea tree and derivatives	Medicine	56	41.5	37	31.9
		Shade	25	18.5	12	10.3
		Firewood	22	16.3	8	6.9
		Nutrition (fruit)	18	13.3	31	26.7
		Soap production (Shea nut residues)	5	3.7	3	2.6
		Fertilizer (leaves)	3	2.2	12	10.3
		Fodder (semi-parasitic plants)	3	2.2	3	2.6
		Fire ignition (Shea nut residues)	2	1.5	2	1.7
		Micro-climate improvement	1	0.7	2	1.7
		Construction	0	0.0	3	2.6
		Insect/fungal repellent (bark, burned)	0	0.0	2	1.7
		Honey production (beehives in tree)	0	0.0	1	0.9
		ΣF	135	–	116	–

MANAGEMENT PRACTICES	F	FL	F	FL
Cultivate/weed beneath tree	51	42.9	59	37.6
Optimize tree spacing	28	23.5	20	12.7
Parasite removal	18	15.1	34	21.7
Pruning	13	10.9	25	15.9
Build mound around tree	6	5.0	10	6.4
Dig water retention ditch around tree	1	0.8	2	1.3
Fertilizer application in field	1	0.8	2	1.3
Refrain from bark removal	1	0.8	0	0.0
Control field burning	0	0.0	5	3.2
ΣF	119	–	157	–
SELECTION FACTORS				
Yield	30	30	32	17.0
Spacing	30	22.0	26	13.7
Shading effects	30	22.0	26	13.7
Size/age	22	16.0	16	8.7
Vulnerability to parasites	22	16.0	12	6.6
Fruit and nut characteristics	3	2.0	5	2.9
ΣF	135	–	118	–

Legend: n = number of participants; F = number of times response was cited; ΣF = total number of responses cited by gender group in given category; FL = fidelity level.

Nearly all women participants – who are the ones who process shea nuts into butter – were also attuned to the characteristics of shea nuts yielding superior quantities of quality shea butter; namely small size, closed shell, hard, reddish colour and windfalls (nuts collected from the ground) that are commonly found in uncultivated, bush fields. Despite the fact that men, as per local gender roles, do not produce shea butter, two-thirds of male interviewees recognized that hard, reddish, fallen shea nuts are best for making shea butter, provided that they are properly transformed. Since shea butter is commonly prepared within the homestead, men may be present during certain processing stages, such as during shea nut sorting. Women and men explained that, at this time, men may discuss issues such as shea nut quality with their wives; especially since the price of shea nuts and butter, and men's interest in the trade, have increased.

The majority of women respondents (70 per cent) noted that quality shea nuts used for making butter consistently come from the same trees, whereas a handful of male participants explicitly felt that this is not a man's concern. One middle-aged Gurunsi man explained that 'sometimes I see women walk by certain shea trees without collecting their nuts so I know they must not be of good quality. However, I don't know which trees give good nuts, and this is not my problem.' Male respondents mostly showed interest in fruit taste, with all male participants claiming to know the trees producing sweet-tasting fruit. Participants did not perceive any correlation between shea fruit and nut traits.

Shea uses

Aside from shea butter and fruit, farmers rely upon the species' bark, roots, latex and leaves to meet their daily needs. Male and female participants reported similar uses for the species, which require an intimate knowledge of the properties and processing of its derivatives. Both women and men most frequently cited the species' role in traditional pharmacopeia (FL = 41.5 and 31.9 for women and men, respectively) (see Figure 15.2 and Table 15.2), such as in treating diseases like dysentery and malaria. Women, who are the ones to process the cures prepared with shea tree leaves, bark, latex and roots, were more articulate about the tree's medicinal properties and the preparation and administration of shea-based remedies. Men were also aware of the diseases shea derivatives could cure, but could not describe how these cures are prepared.

Men and women emphasized the species' role in the provision of shade and nutrition. The high FL value (26.7 and 13.3, for men and women, respectively) illustrates the nutritional importance of shea fruit, which ripens during the agricultural period at a time when food reserves are running low. One-third of male respondents cited shea fruit consumption as the primary reason they conserve shea trees in their fields. Women stressed the importance of shea as firewood, which logically follows the local gendered division of labour that ascribes gathering firewood to the female sphere of activities. Shea wood may even be used as firewood for shea butter processing. Neither women nor men emphasized the species' importance in

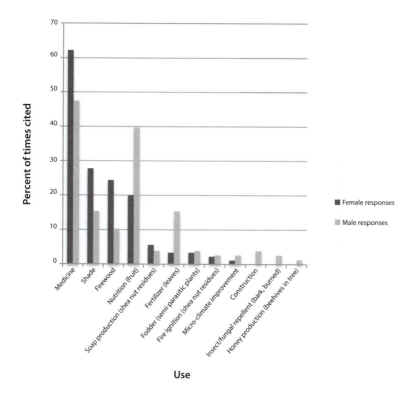

FIGURE 15.2 Uses for the shea tree and its derivatives, according to women and men participants

providing ecological services aside from the soil-fertilizing effects of its leaf litter. Many respondents were adamant that shea is *the most* important local species, and nearly all men and women felt that its conservation is of critical importance.

Shea tree management and conservation

Guided by the ecological knowledge outlined above, farmers have adopted a form of 'protoculture' – or management of valued trees that are not deliberately planted (Boffa 1995) – to improve the shea tree's vigour and productivity. Hence, shea parklands reflect not only the species' natural regeneration patterns, but also the management and conservation practices of female and male agriculturalists who value the species' myriad functions.

Although men are the visible managers of shea trees in West Africa, carrying out field burning and tree culling (Maranz and Wiesman 2003; Chalfin 2004), most women respondents demonstrated a detailed understanding of shea tree manage-ment practices (Figure 15.3 and Table 15.2). They described the same management techniques as their spouses, and there was a high degree of consensus across genders

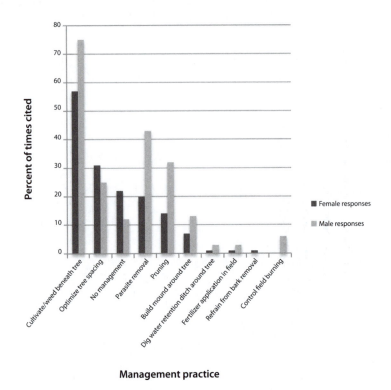

Management practice

FIGURE 15.3 Women and men's shea tree management practices in the study sites

around the top four practices, namely cultivating and weeding beneath shea trees (FL = 42.9 for women, 37.6 for men), optimizing the spacing of shea specimens (FL = 23.5 and 12.7 for women and men), removing semi-parasitic plants (FL = 15.1 to 21.7 for women and men) and pruning (FL = 10.9 for women and 15.9 for men). On average, men named more management practices than their female counterparts (2.1 versus 1.5 responses per participant, respectively), likely because they are generally the ones to remove parasites and carry out field burning or pruning.

Coupled with the facts that it is not a planted species, does not depend on costly fertilizers for growth, and is well adapted to local ecological conditions, the shea tree offers relatively high returns on labour. Accordingly, all female and male interviewees conserve shea trees in their fields, selectively retaining specific individuals with favourable characteristics and eliminating unwanted specimens. Women and men cited similar, interrelated factors – namely yield, spacing and shading effects – as the primary variables influencing the selection process (Figure 15.4 and Table 15.2). The proportion of men and women reporting each of these factors closely matched, as did the FLs calculated for each gender group.

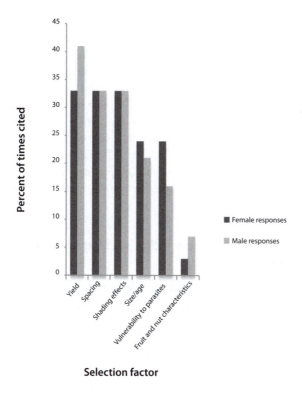

FIGURE 15.4 Factors guiding shea tree selection among women and men participants

In new fields, some male interviewees stated that they assess shea yields for as many as four years before eliminating the weakest producers. Alternatively, participants rely on visible tree characteristics, including shape and size/age, to guide their choices. Obtrusive trees and/or branches are removed, and given the ambiguous relationship between tree size/age and productivity, some participants retain older trees while others conserve younger individuals. A specimen's vigour or resistance to parasitism also affects its chances of being retained.

When asked whether they participate in the shea tree selection process, just over half of female participants stated that their husband single-handedly selects shea specimens for conservation. The others claimed to discuss the selection process with their husband, stating that women recognize the best producers. Hence, although men may ultimately choose which specimens to eliminate, female spouses influence their decision. In a woman's personal field, she alone decides which shea trees to conserve.

In turn, nearly three-quarters of male participants maintained that they discuss the selective conservation of shea trees with their wives before eliminating any specimen, as they claimed that women are more knowledgeable about shea tree yields and shea nut quality. The remaining quarter of respondents explained that women are not always in the fields to identify shea trees at the time of land clearance – a man's task – and that this hinders their participation in the selection process.

In sum, female and male participants cited – in closely matching frequencies – identical factors guiding the selection of shea trees in farmed fields. The majority of participants reported selectively conserving shea trees in a process guided by female and male knowledge of tree yields, as well as by their preferences for specific shea tree productivity, spacing and shading patterns. The implications of these findings for the ways we conceptualize gendered experiences with natural resources are discussed below.

Discussion: intra-household knowledge sharing and shea tree management

Five important points arise from an analysis of the gendered shea agroforestry system. First, both local women and men hold rich, if at times unrecognized, ecological knowledge of shea trees. Overall, this botanical knowledge corresponds with ways of using, processing and managing the species in other regions of its range, and is congruent with scientific findings about the species (e.g. Bayala et al. 2003; Boussim et al. 2004; Okullo et al. 2004; Gwali et al. 2011; Poudyal 2011). Participants also recognized tree traits affecting productivity – such as the tree's leafing behaviour or its 'gender' – and practices, such as the creation of earth mounds or water-retention ditches, that have yet to receive research attention. This information can guide future research for development efforts promoting the sustainable management of shea parklands.

A second significant point is that, although gender differences in ethnobotanical knowledge have been widely reported in Africa (e.g. Wezel and Haigis 2000; Ayantunde et al. 2008), such differences were not generalized in this study. Consistent with the local gendered division of labour, men and women listed the tree's various uses in distinct frequencies. Nonetheless, the range of their responses and the top-ranking responses they provided for each domain of analysis showed strong overlap. This can be explained by the central and constant importance the tree occupies for all village residents, who draw upon it for daily subsistence. As Boster (1986) explains, the greater the shared cultural importance of a species, the more knowledge about the species will be shared intra-culturally. Similar observations have been made in Uganda, where women's and men's classification knowledge of *V. paradoxa* ethno-varieties strongly matched (Gwali et al. 2011) and in south-central Burkina Faso, where no consistent gender differences were found in Gurunsi identification of locally important species (Kristensen and Balsev 2003). Although women's knowledge of shea nut quality exceeded men's, two-thirds of male interviewees were surprisingly knowledgeable about the topic, citing the same traits for quality shea nuts as their female counterparts. Given that men do not process shea nuts into butter, this area of knowledge cannot be attributed to first-hand experience.

The third point, related to the second, is that although rural African men are generally perceived to be associated with tree management and conservation, both women and men may be knowledgeable about and involved in these processes. Again, the large areas of overlap in their responses and the similarly high FLs for a given variant across gender groups suggest that women and men value common factors with respect to shea tree selection and that they collaborate in the species' management and conservation. This is also confirmed by participant reports of intra-conjugal consultations during shea tree selection activities. Hence, men may play a preponderant role in the process, but in the case of shea, which is strongly associated with the female sphere of activities, women contribute their knowledge and preferences to the process in a more or less formal manner.

The fourth point, arising from the last two, is the need for a more fluid conception of how traditional ecological knowledge develops and circulates within the household. Despite the gendered nature of resource use, local knowledge of valued resources is shared through daily interactions among spouses. This may occur to meet practical goals, such as developing resource management strategies, or simply for the sake of conversation between spouses. It is, in fact, difficult to imagine that in subsistence-oriented communities, husband and wife would not discuss matters of importance to their daily subsistence, such as resource use. In the shea case, some of the congruence between male and female responses can likely be attributed to such knowledge sharing – related to shea nut quality, for instance – among spouses.

It is in this respect that drawing upon Amartya Sen's model of intra-household bargaining can sharpen our understanding of the gendered shea tree agroforestry system. As mentioned earlier, the notion of 'social connectedness' advanced by Sen recognizes that women and men have different spheres of activities and decision making, but challenges the assumption that these gendered spheres are separable. Summing up this idea, Razavi and Miller (1995: 15–16) assert that 'although the gender division of labour involves men and women undertaking different activities, it also entails an intricate and changing system of cooperation and exchange.' Bonnard and Scherr (1994: 89) further state:

> . . . the existence of joint decisions, multiple objectives, and mutual depend-ence within the household makes it difficult to construct generalized rigorous and pragmatic models of household or individual behavior. Even if men are the only ones planting trees, women may still provide the critical watering and pruning labor, and their relevant opportunity costs will greatly influ-ence the survival rates of project trees. While men use a specific species for poles, women might use them for fodder, green manure or special herbs and medicine depending on the species selected. There may be differenti-ated preferences for secondary tree products as well. Joint management and decision-making naturally arise out of these scenarios.

Likewise, Lope-Alzina (2007, 34) describes the selection of crop varieties in Mexico as 'a process of negotiation: Decision-making is identified as a prerogative exercised by both men and women and based on the knowledge and skills that each hold, and the trust and recognition of these by their partners.' This aspect of 'social connectedness' is frequently overlooked in studies describing the relationship between African spouses primarily as one of separation or conflict, and spheres of female knowledge and male management of tree resources as chiefly independent.

Results from this study suggest the need to more readily recognize the knowledge transfers occurring within the household that allow for informed decision making in the realms of natural resource management and conservation. This information sharing is a step towards what Assé and Lassoie (2011: 255) have termed 'gender-inclusive decision-making' on parkland issues, wherein 'female adult members are explicit joint partners in choosing natural resource use and conservation actions.' In Sudano-Sahelian Mali, this type of decision-making culture has been associated with innovative and adaptive parkland soil and tree management practices (Assé and Lassoie 2011). Also of importance is paying attention to factors such as age or seniority within the household that influence the knowledge held and ability to make or contribute to resource management decisions.

Finally, the shea case hints at the rapidly evolving nature of gendered knowledge and resource management practices over time. As shea butter gains international economic value, male spouses are showing interest in shea nut processing, and seeking broader information about the resource from their wives. The local and international political economy are in continuous flux, and so too is intra-household knowledge sharing that influences the collaborative and competing resource use and management decisions of husbands and wives.

Stringing these points together illuminates the ways the different layers—knowledge, decision making and practices – that comprise resource management systems are articulated.[4] Gendered knowledge is acquired and shared to inform decisions about resource management. Members choose which information to share and which to withhold based on their collaborative or competing interests.[5] When a set of resource management outcomes is strongly desirable to both husband and wife, spouses have high incentives to collaborate in and thus exchange relevant knowledge. Decisions, which may ultimately be made jointly or not, then partly hinge on this layer of knowledge acquisition and exchange. The layer of practice, which is most visible and often conveys the impression that men are the primary

---

4 Similarly, Berkes (1999) defines traditional ecological knowledge as having dimensions of knowledge, practice and belief. The gendered norms that underlie the acquisition of gendered knowledge and practice were beyond the scope of this chapter, but comprise yet another layer of resource management systems.

5 This knowledge plays a role in strengthening (or weakening, in its absence) the relative bargaining position of household members. Although the relationship between knowledge and power has been extensively demonstrated by Foucault, among others, the role knowledge plays in determining the relative bargaining position of household members is rarely discussed in intra-household bargaining models.

resource managers, cannot be detached from the more covert knowledge and decision-making dimensions upon which it rests. The gender dynamics that shape these layers are in constant evolution as intra- and extra-household factors that bear upon them fluctuate. Each of these gendered layers comprises the cooperative and conflictive elements described by Sen; and all must be considered together to understand natural resource management strategies.

Conclusion

This study has shown that women and men have distinct but strongly overlapping areas of knowledge about the shea tree and its uses, as well as its management and conservation. This can be partly attributed to their personal experiences with the species, but also to knowledge sharing between spouses. In particular, the notion that men decide in matters of tree management and the visibility they receive as primary resource managers – which justifies their primary involvement in parkland management interventions – overlooks the fact that women's contributions to the process may be significant. The case of shea suggests that such intra-household sharing of knowledge about valuable natural resources may hold greater significance for achieving resilient resource management strategies than has been described in previous works on African agroforestry. In particular, recognizing that knowledge, decision making and practices are interlaced and shaped by gendered relations of cooperation and conflict can improve our understanding of resource management processes. This points to the value of fostering inter-gender dialogues within the household and beyond to favour the exchange of knowledge, a joint awareness, and a common vision towards sustainable shea tree management processes that can meet the interests of both women and men. Participatory approaches that work with gender-segregated groups, but bring these together for social learning and knowledge sharing, can contribute to this process. Further research on knowledge sharing within the household – across gender but also age groups – pertaining to natural resource management systems is also in order. Parkland conservation programs must thus engage with men *and* women farmers, given the roles both spouses play in shea tree management and conservation.

Acknowledgements

The author gratefully acknowledges the Social Sciences and Humanities Research Council (SSHRC) of Canada, and the International Development and Research Centre (IDRC) for funding this research. Heartfelt thanks are due to the men and women from Léo, Lan and Prata who generously contributed their time and knowledge to this study; to Reid Cooper, Azizou Yago, Fatimata Traoré, Pamoussa Ouédraogo and Nassiratou Nébié for field assistance, and to Sarah Turner, Stephanie Coen, Carol J. Pierce Colfer, Bimbika Sijapati Basnett and three anonymous reviewers for their valuable comments on the manuscript.

References

Agarwal B. 1997. 'Bargaining' and gender relations: Within and beyond the household. *Feminist Economics* 3(1):1–51.

——. 2001. Participatory exclusions, community forestry and gender: An analysis for South Asia and a conceptual framework. *World Development* 29(10): 1623–48.

Assé R and Lassoie JP. 2011. Household decision-making in agroforestry parklands of Sudano-Sahelian Mali. *Agroforestry Systems* 82: 247–61.

Ayantunde AA, Briejer M, Hiernaux P, Udo HMJ and Tabo R. 2008. Botanical knowledge and its differentiation by age, gender and ethnicity in southwestern Niger. *Human Ecology* 36: 881–9.

Bayala J, Mando A, Ouedraogo SJ and Teklehaimanot Z. 2003. Managing *Parkia biglobosa* and *Vitellaria paradoxa* prunings for crop production and improved soil properties in the Sub-Sudanian Zone of Burkina Faso. *Arid Land Research and Management* 17(3): 283–96.

Becker GS. 1965. A theory of the allocation of time. *Economic Journal* 75(299): 493–517.

Berkes F. 1999. *Sacred ecology: Traditional ecological knowledge and resource management*. Philadelphia, PA and London: Taylor and Francis.

Boffa JM. 1995. *Productivity and management of agroforestry parklands in the Sudan Zone of Burkina Faso, West Africa*. PhD Thesis. Purdue: Purdue University.

——. 1999. Agroforestry parklands in sub-Saharan Africa. *FAO Conservation Guide, No. 34*. Rome: FAO.

Bonnard P and Scherr S. 1994. Within gender differences in tree management: Is gender distinction a reliable concept? *Agroforestry Systems* 25(2): 71–93.

Boster J. 1986. Exchange of varieties and information between Aguaruna manioc cultivators. *American Anthropologist* 88(2): 428–36.

Boussim IJ, Guinko S, Tuquet C and Sallé G. 2004. Mistletoes of the agroforestry parklands of Burkina Faso. *Agroforestry Systems* 60: 39–49.

Brewer DD. 1995. Cognitive indicators of knowledge in semantic domains. *Journal of Quantitative Anthropology* 5: 107–28.

——, Garrett SB and Rinaldi G. 2002. Free-listed items are effective cues for eliciting additional items in semantic domains. *Applied Cognitive Psychology* 16: 343–58.

Burkill HM. 2000 (Edition 2). *The useful plants of West Tropical Africa*. Vols. 4 and 5. Kew: Royal Botanic Gardens.

Carter M and Katz E. 1997. Separate spheres and the conjugal contract: Understanding the impact of gender-biased development. In Haddad L, Hoddinott J and Alderman H, eds. *Intrahousehold resource allocation in developing countries: Methods, models and policies*. Baltimore, MD: The Johns Hopkins University Press. 95–111.

Cavendish W. 2000. Empirical regularities in the poverty–environment relationship of rural households: Evidence from Zimbabwe. *World Development* 28: 1979–2003.

Chalfin B. 2004. *Shea Butter republic*. New York: Routledge.

Doss CR. 2001. Designing agricultural technology for women: Lessons from twenty-five years of experience. *World Development* 29(5): 2075–92.

Dovie BK, Witkowski ETF and Shackleton CM. 2008. Knowledge of plant resource use based on location, gender and generation. *Applied Geography* 28: 311–22.

Elias M. 2015. Gender, knowledge-sharing and management of shea (*Vitellaria paradoxa*) parklands in central-west Burkina Faso. *Journal of Rural Studies* 38: 27–38.

—— and Carney J. 2007. African shea butter: A feminized subsidy from nature. *Africa* 77(1): 37–62.

Folbre N. 1986. Hearts and spades: Paradigms of household economics. *World Development* 14: 244–55.

Fontès J and Guinko S. 1995. *Carte de la végétation et de l'occupation du sol au Burkina Faso (Notice explicative)*. Institut du Développement Rural (IDR), Ouagadougou: Université de Ouagadougou.

Fortmann L and Bruce JW. 1988. *Whose trees? Proprietary dimensions of forestry*. Boulder, CO: Westview Press.

———, Antinori C and Nabane N. 1997. Fruits of their labors: Gender, property rights, and tree planting in two Zimbabwe villages. *Rural Sociology* 62(3): 295–314.

Friedman J, Yaniv Z, Dafni A and Palewitch D. 1986. A preliminary classification of the healing potential of medicinal plants, based on a rational analysis of an ethnopharmacological field survey among Bedouins in the Negev desert, Israel. *Journal of Ethnopharmacology* 16(2–3): 275–87.

Gatewood JB. 1984. Familiarity, vocabulary, size, and recognition ability in four semantic domains. *American Ethnologist* 11: 507–27.

Gray L. 2003. Investing in soil quality: Farmer responses to land scarcity in southwestern Burkina Faso. In Bassett T and Crummey D, eds. *African savannas: Global narratives and local knowledge of environmental change*. Oxford: James Currey. 72–90.

Gwali S, Okullo JBL, Eilu G, Nakabonge G, Nyeko P and Vuzi P. 2011. Folk classification of shea butter tree (*Vitellaria paradoxa* subsp. *nilotica*) ethno-varieties in Uganda. *Ethnobotany Research and Applications* 9: 243–56.

Hall JB, Aebischer DP, Tomlinson HF, Osei-Amaning E and Hindle JR. 1996. Vitellaria paradoxa: *A Monograph*. Publication Number 8. School of Agricultural and Forest Sciences, Bangor: University of Wales.

Howard PL. 2003. Women and the plant world: An exploration. In Howard PL. ed. *Women and plants: Gender relations in biodiversity management and conservation*. London: Zed Books. 1–47.

———. 2007. Are there customary rights to plants? An inquiry among the Baganda (Uganda), with special attention to gender. *World Development* 35(9): 1542–63.

Howorth C and O'Keefe P. 1999. Farmers do it better: Local management of change in southern Burkina Faso. *Land Degradation & Development* 10: 91–109.

Kristensen M and Balslev H. 2003. Perceptions, use and availability of woody plants among the Gourounsi in Burkina Faso. *Biodiversity and Conservation* 12(8): 1715–39.

Leach M. 1994. *Rainforest relations*. London: Edinburgh University Press.

Lope-Alzina D. 2007. Gendered production spaces and crop varietal selection: Case study in Yucatan, Mexico. *Singapore Journal of Tropical Geography* 28: 21–38.

Lykke AM, Mertz O and Ganaba S. 2002. Food consumption in rural Burkina Faso. *Ecology and Food Nutrition* 41(2): 119–53.

Maranz S and Wiesman Z. 2003. Evidence for indigenous selection and distribution of the shea tree, *Vitellaria paradoxa*, and its potential significance to prevailing parkland tree patterns in sub-Saharan Africa north of the Equator. *Journal of Biogeography* 30(10): 1505–16.

MEF (Ministry of Economy and Finance of Burkina Faso). 2011. *Investing in the future: Accelerated growth and sustainable development. Business environment, investment opportunities and public-private partnership*. Ouagadougou: MEF.

Okullo JBL, Obua J and Okello G. 2004. Use of indigenous knowledge in predicting fruit production of shea butter tree in agroforestry parklands of north-eastern Uganda. *Uganda Journal of Agricultural Science* 9: 360–66.

Ouédraogo M. 2003. *New stakeholders and the promotion of agro-silvo-pastoral activities in southern Burkina Faso: False start or inexperience? Drylands issue paper No. E118*. Edinburgh: International Institute for Environment and Development (IIED).

Poudhyal, M. 2011. Chiefs and trees: Tenures and incentives in the management and use of two multipurpose tree species in agroforestry parklands in northern Ghana. *Society & Natural Resources* 24(10): 1063–77.

Quisumbing AR, Payongayong E, Aidoo JB and Otsuka K. 2001. Women's land rights in the transition to individualized ownership: Implications for tree-resource management in western Ghana. *Economic Development and Cultural Change* 50: 157–81.

Razavi S and Miller C. 1995. *From WID to GAD: Conceptual shifts in the women and development discourse*. UNRISD Occasional Paper Series OBP 1. Geneva: UNRISD.

Reenberg A and Lund C. 1998. Land use and land right dynamics: Determinants for resource management options in eastern Burkina Faso. *Human Ecology* 26(4): 599–620.

Rocheleau D and Edmunds D. 1997. Women, men and trees: Gender, power and property in forest and agrarian landscapes. *World Development* 25(8): 1351–71.

——, Thomas-Slayter B and Wangari E. 1996. Gender and environment: A feminist political ecology perspective. In Rocheleau D, Thomas-Slayter B and Wangari E, eds. *Feminist political ecology: Global issues and local experiences*. New York: Routledge. 3–23.

Sen A. 1990. Gender and cooperative conflicts. In Tinker I, ed. *Persistent inequalities: Women and world development*. New York and Oxford: Oxford University Press. 123–49.

Stemler SE. 2001. An overview of content analysis. *Practical Assessment, Research and Evaluation*, 7(17). Accessed 15 April 2010. http://PAREonline.net/getvn.asp?v=7&n=17.

Sunderland T, Achdiawan R, Angelsen A, Babigumira R, Ickowitz A, Paumgarten F, Reyes-Garciâla V and Shively G. 2014. Challenging perceptions about men, women, and forest product use: A global comparative study. *World Development* 64: S56–S66.

Turner NJ, Ignace MB and Ignace R. 2000. Traditional ecological knowledge and wisdom of aboriginal peoples in British Columbia. *Ecological Applications* 10(5): 1275–87.

Udry C. 1996. Gender, agricultural production, and the theory of the household. *Journal of Political Economy* 104(5): 1010–46.

Weller SC and Romney AK. 1988. *Systematic data collection*. Newbury Park, CA: Sage.

Wezel A and Haigis J. 2000. Farmers' perception of vegetation changes in semi-arid Niger. *Land Degradation and Development* 11(6): 523–34.

16

GENDER, MIGRATION AND FOREST GOVERNANCE

Rethinking community forestry policies in Nepal[1]

Bimbika Sijapati Basnett

Introduction

The policy literature surrounding the governance of community forests in the middle hills of Nepal has changed and grown substantially. In the late 1970s and 1980s, the focus was primarily on promoting the effective participation of local communities in the sustainable management of forests. But recently and increasingly, community forestry policies are questioning the undifferentiated view of local communities and the role of gender and social relations in the efficient and equitable governance of community forests. The major actors involved in implementing community forestry – the state, donors and civil society organizations – have been making concerted efforts to mainstream gender and social inclusion in community forestry policies and institutions.

While these are important and commendable developments and reflect a growing momentum for inclusive change in Nepal, I argue that community forestry policy literature still frames 'gender', 'caste' and 'ethnicity' as static social relations, and women, Dalits (people of low castes) and ethnic minorities as uniformly marginalized. Moreover, individuals and communities in rural areas are still assumed to be spatially bounded, tied to their location of residence; and their relationship to forest products remains unproblematized. The latter is of concern in light of the importance of migration, both historically and in the recent past, for rural livelihoods throughout the country.

1 For more information, see Basnett 2011, from which this chapter has been adapted. We are grateful for permission to reproduce this revised version from the *European Bulletin of Himalayan Research*.

Here I draw on primary and secondary analysis of the evolution of community forestry policies, complemented with field research amongst two community forest user groups[2] in the middle hills of Nepal; and question the assumptions about gender relations and rural livelihoods that underpin community forestry policies and practices in Nepal's middle hills. I argue that the view that rural livelihoods are spatially bounded remains divorced from the increasing multilocality of rural livelihoods in Nepal. The case studies serve to illustrate how these dynamics are leading to shifts in men and women's relationships with each other and with the governance of forest resources.

The chapter, divided into three parts, begins by critically assessing the community forestry policy literature from a gender and migration perspective. In particular, it demonstrates the evolution of gender issues in community forestry policy and discourses and the ways in which migration is largely invisible or rendered problematic. The subsequent two sections, case studies, serve to illustrate the complexities of defining gender and social relations in Nepal and the role that migration is playing in re-defining women's and men's entry into and influence over the governance of community forests. The conclusion provides a summary and reiterates the importance of taking migration and gender relations more seriously in community forestry policies.

Gender and migration-based inclusions and exclusions in community forestry policies in Nepal

Evolution of community forestry policies from a gender perspective

Although the National Forestry Plan of 1976 and subsequent legislation marked the beginning of 'people-centred' forest and land use policies in Nepal, gender issues were not explicitly mentioned in community forestry policies until the Master Plan for Forestry in 1987. The major objective of the National Forestry Plan of 1976 was to formally recognize the rights of local communities to manage their own forests, with technical assistance provided by the government. However, as Harper and Tarnowski (2002) point out, in spite of its populist rhetoric, the emphasis of the National Forestry Plan was on the protection, production and 'proper' use of forests. This was in accordance with the government's desire to halt forest degradation and ensure that forests contributed to the development of the national economy. It was assumed that handing over forests to village

2 The field research lasted twelve months, between 2004 and 2006, as part of my PhD dissertation, using a combination of qualitative and quantitative research methods. These included structured household interviews, detailed semi-structured interviews, focus-group discussions, participatory rural appraisal exercises, and participant observation of everyday village life as well as community forestry-related meetings. Concerned about the gender politics of forest governance, I also engaged with feminist concerns over reflexivity and positionality in the research process.

Panchayats[3] would trickle benefits down to those who depended most on forests for their livelihoods.

By the mid-1980s, Pokharel (1997) and Britt (2002) conclude, based on many reports evaluating the performance of the forestry sector, that the condition of the forests handed over to the local Panchayats had not improved, and that the local people who were most dependent on the forests were rarely involved in forest management. Consequently, the Master Plan for Forestry 1988 and its amendment in 1990 stated that forests should go directly to their 'users' and not to the Panchayats; that user groups should be allowed to reap all the benefits of sustainably managing their forests; and that 'women' and 'the poor' should be involved in forest management (HMG/N 1990). However, gender was interpreted as 'women's issues' and women were implicated in forest degradation because they were seen as the main collectors and users of forest produce. It was assumed that by incorporating women in forest management the causes of environmental degradation would be addressed. Although 'women's issues' were mentioned in the Master Plan, they were rarely operationalized in practice (Sijapati 2008).

Since the national conference to celebrate the twenty-fifth year of community forestry in 2003, however, gender issues have become a prominent feature of community forestry policies, and are raised in terms of women's rights. The conference launched the 'second generation issues of community forestry'; recognized 'good governance', 'sustainable management' and 'livelihoods' as three mandates of community forestry, and stated that 'gender and social equity' represented an overarching theme that should be integrated into every facet of community forestry governance. Since the workshop, consolidated efforts have been underway at the national level, through initiatives undertaken by the Ministry of Forestry and Soil Conservation as well as donors, to mainstream gender and social equity in community forestry policies (Sijapati 2008).

The Nepal Government's commitment to promote gender and social inclusion are firmly rooted in the recent Nepal Forest Sector Strategy 2012–22 (FSS). An analysis of the policy document reveals that the Strategy replaces the Master Plan of the Forestry Sector by outlining a new vision, including principles and strategic objectives, as well as targets and priority actions needed to achieve these objectives. 'Gender and social inclusion' are stated as one of the eight fundamental pillars of the strategy. It further states that women, Dalits, 'Janajaits' (ethnic minorities), people with disabilities and other marginalized groups have a right to be included in forestry-related decisions and benefits. It professes to achieve gender inclusion by mandating that 30 per cent of each Community Forest User Group (CFUG)

3 The Panchayat system in Nepal was a pyramidal structure progressing from village-level assemblies to *Rastriya Panchayat* (or national-level Parliament). The system enshrined absolute power of the monarchy and kept the king as head of state with sole authority over all government institutions. In the initial phases of community forestry, national forests were devolved to local Panchayats, as Panchayat forests or Panchayat Protected Forests. In the former case, a village Panchayat could have up to 125 hectares of degraded forests; and in the latter, up to 500 hectares of existing forests (Pokharel 1997).

committee should be women, and that at least one of the major decision-making positions of the committee should be occupied by women. Although scholars observe that merely increasing women in forest governance systems does not guarantee gender egalitarian rules and outcomes (Nightingale 2002; Cornwall 2003; Agarwal 2010), having such a 'critical mass of women' with a conscious recognition and collective articulation of shared interests helps in furthering outcomes favourable to women (Agarwal 2014).

Phillips (1995) reminds us that struggles over representation of women should not replace struggles over economic inequality between women and men. The FSS moves beyond representation by also pointing to the importance of creating cash incomes and job opportunities for poor and marginalized households. The 'Forest Sector Gender and Social Inclusion Strategy' (2008), which the FSS intends to apply across all forest-sector institutions, goes further and points to a wider range of provisions for ensuring that women and socially excluded groups also benefit from community forestry. Some of the provisions include allocating 30 per cent of the community forestry budget for the poor, women and other excluded groups; guaranteeing special rights to forest products for highly forest-dependent communities; discouragement of auctioning practices in the sale of forest products in forest user groups, and providing opportunities to women and other excluded households for wage work in forest management, among other provisions.

The political economy underpinning contemporary community forestry policies

Such concerns over gender and social inclusion in the community forestry policy literature must be situated in the context of growing donor attention to 'gender and social equity' mainstreaming in community forestry on the one hand, and the increasing politicization of, and demands for, gender and socially based inclusion in Nepali politics on the other. Historically, donors have played a key role in the forestry sector in general, and community forestry in particular (Sijapati 2008). Interviews with various government officials, ranging from senior officials framing community forestry policies at the national level to those implementing them at the local level, confirmed the role of donors in pushing for gendered reform.

Leach (2007) has traced the history of gender concerns in natural resource governance policies in developing countries like Nepal. She finds fundamental shifts over the past three decades in the ways in which 'women' are represented and gender-based issues are integrated into donor and NGO policy documents and reports. During the 1980s, the emphasis was on rationalizing women's inclusion in natural resource governance processes. Consequently, simplistic discourses over women's close relationship with nature promoted by Ecofeminism and Women, Environment and Development proponents were readily received and employed to strategically negotiate greater space for women's participation. Recently (as evident in Nepal), influenced by the growing critique by feminist scholars and development practitioners alike, donors and NGOs are recasting older concerns with women and environment in terms of rights and relations in access and control over property. For instance, it is no longer assumed that just because rural women in many developing countries are in charge of firewood collection that women's interests are synonymous with the goals of environmental conservation. Instead, the emphasis is on understanding women's lack of access to alternative sources of energy, as influenced by power relations within and outside the household.

Furthermore, Nepal has been experiencing an explosion in the number of ethnic and caste-based political parties and social movements. Some would argue that such political changes are a product of the ethnic and caste-based political mobilization championed by the Maoist party to garner support for and wage their class-based struggle against the state (DeSales 2000; Hutt 2003; Thapa and Sijapati 2003). Others suggest that the current political volatility has engendered a political vacuum which provides perfect conditions for these organizations and their demands to flourish (Hangen 2010). Nevertheless, the effect has been greater awareness of and demands for social inclusion in the Nepali state and society alike. National women's advocacy groups, with support from donors such as UN Women, have also exploited the opportunity to demand greater women's representation in the polity, and gender-based reform in property rights, citizenship and more. For instance, prominent women's advocacy groups were at the forefront in successfully lobbying for and securing for women 33 per cent of seats in the Constituent Assembly elected in 2008. Gender unequal laws, such as inheritance and citizenship rights that severely restricted women's claims to parental property and relegated them to the status of second-class citizens, have been successfully challenged and reformed. Gender-based reservations have been demanded in every arena of Nepali polity, including in the forest governance arena.

The extent to which these changes will lead to a more inclusive Nepal or one which is further fractured along caste, ethnic and gender divisions is not yet clear. It can be argued that the trajectories thus far, suggest that the discourse on 'caste', 'ethnicity' and 'gender' has created and reinforced identities, and pitted one group against another. Moreover, gender and social equity have been reinterpreted as greater recognition of and access to state resources for 'women', 'low castes' and 'ethnic minorities' (Sijapati 2008). Neither the historical complexities behind caste and ethnic relations nor the context-specific ways in which gender cuts across these relations have been discussed or articulated in these policy documents and policy discourse. The latter means women and men are situated differently in the diverse socio-economic, political and geographic landscapes of the country. As Tamang (2009) points out, donors, major political parties and women's organizations have all contributed to the production of a homogenous Nepali woman, subjugated uniformly throughout the country, irrespective of her position in caste, class and ethnic hierarchies. Ethnographic studies have documented considerable variations in socio-cultural practices (such as marriage practices, division of labour), with contrasting implications for gender relations and women's relative position (Cameron 1998; Benett 2003; Kondos 2004; Gray 2008).

Donor reports on the implications of the civil conflict on gender and social equity aspects of community forestry serve to illustrate such compartmentalized understandings of gender, caste and ethnic relations. A study carried out by the Nepal Swiss Community Forestry Project, for instance, points out that 'despite difficult conflict situations, Community Forestry User Groups, are practicing inclusive democracy, in which there is increased participation and representation of women and socially marginalized groups' (Pokharel et al. 2005: 1). 'Women' and the 'socially marginalized' are seen as two distinct social groups assuming that the constraints that 'women' as a group face are somehow distinct from those of the socially marginalized. The FSS also uses gender as shorthand for women; and social inclusion/ exclusion is separated from gender. Nightingale (2002: 18) rightly argues that both the government and donors have failed to adequately consider 'locally defined differences between people (men and women, different castes and ethnicities) and the ways in which these differences give people uneven access to resources and control over the community forestry management process.'

Absence of migration from community forestry policy discourses

As many prominent scholars of agrarian change in rural areas have shown, rural livelihoods in the global South are increasingly becoming multi-local and no longer confined to farming and land (Ashley and Maxwell 2002; Razavi 2003; Rigg 2005; Thieme 2008; Hecht et al. Forthcoming). This is particularly evident in the case of Nepal where seasonal out-migration, both within the country and to India have historically been a prominent strategy adopted by rural households for a wide range of reasons. These include seeking to escape state policies and agrarian changes, diversifying their incomes, offsetting capital constraints and, increasingly, responding to the growing economic insecurity resulting from the political conflict in the country and fulfilling aspirations of participation in 'modern life' (Regmi 1978; Caplan 1990; Gill 2003; Sharma 2008). Open border policies between India and Nepal have meant that an estimated 1.3 million Nepali migrants are working in India, of whom about 90 per cent are likely to be men (Sharma 2008). Furthermore, globalization and the expansion of markets have given added impetus to the growing mobility of Nepali workers in search of circular migration for international contract work in the Gulf and South East Asia (Seddon et al. 2001).

Considerable uncertainties exist over the precise numbers of Nepali citizens working abroad, although there is no doubt that circular and seasonal migration constitute one of the largest sources of employment for the country. Anywhere between 590,000 and 3 million Nepali workers are currently working in India – exact figures are difficult to determine because of the above-mentioned open-door policy. Furthermore, the period between 1998/99 and 2010 witnessed a 13-fold increase in the number of Nepali citizens migrating for employment purposes to international destinations other than India (from 27,796 in FYI 1998/1999 to 384,667 in FYI 2011/2012), according to Nepal's Department of Foreign Affairs. In FYI 2011/12, the majority went to Qatar (105,681; 27 per cent of the total migrants) followed closely by Malaysia (98,367) and Saudi Arabia (80,455). These figures only capture documented Nepali workers, that is, only those who sought and were granted approval to work abroad by the Department of Foreign Employment. The Department estimates that an additional 40 per cent of the total documented workers are undocumented, but others claim that this figure could be as high as 200 per cent of all documented workers (Sharma et al. 2014).

Migration for employment purposes is a male-dominated phenomenon. Although the total number of women migrating has been on the rise, women migrants constituted a mere 6 per cent of the total migrant worker population in FY 2011/12 (Sharma et al. 2014). The dominant number of men in seasonal and circular migration patterns derives from a host of factors, including gender segregation of markets, lack of opportunities for women outside the domestic sector, and gender norms at the household and community levels that stigmatize women who migrate abroad for work purposes. The Nepal government's ban on women working abroad until recently, due to the rampant sexual and other abuses they faced working in the Gulf (reported by the mainstream media) also played an important role. In a rare insight into the gender dynamics of migration from the middle hills of Nepal to the cities of India, Sharma (2008) demonstrates that migration and its outcomes are often interpreted as a transition from boyhood to manhood for young migrants and their families. By enabling young men to fulfil their sense of material obligation toward their families, migration reproduces local idioms of masculinity and reinforces male-dominated households. As the following cases demonstrate, however, migration interacts with gender, caste and ethnicity to produce a range of outcomes for women and men situated differently within these hierarchies. These gendered dynamics can, in turn, have contrasting implications for the extent and nature of women's and men's voices and their influence in the processes of forest governance.

Despite the importance of migration for contemporary Nepal, the question of how seasonal and transnational migrations are affecting the governance of community forests remains unaddressed in both community forestry policies and the growing policy-oriented scholarship in this field. Community forestry policy literature defines user households as those living in close proximity to forests who are most dependent on forest products for their livelihoods.

The Forest Sector Strategy (2012–22) is the first policy document to make explicit reference to migration in the forestry sector. The section, 'Visualising the Future', states that there has been a demographic shift from the hills towards the Tarai and urban areas (internal migration), coupled with a gradual loss of economically active people from rural areas. But migration is mentioned just this once in the entire document. This statement risks assuming that (a) migration is a relatively new phenomenon, and (b) that it is a *problem* for forests. As Hobley (personal communication, 2015) rightly observes, policy makers and Nepali professionals working for donors and NGOs view migration as an outcome of the contractual failure between the Nepali state and its citizens; in other words, the state's failure to promote economic development and extend social services to its citizens.

As discussed previously, migration is by no means a new phenomenon in Nepal although the rates and nature of migration have changed significantly. Furthermore, once migration is classified as a problem, efforts to enhance the productive capacity of forests are implicitly viewed as responses/solutions by the Strategy document. This is equally flawed because it assumes that providing forest-related income-generating opportunities can serve as a credible alternative to migration. The income-generating components of the strategy are very forest-centric and make little effort to embrace a more integrated approach. More importantly, unless these initiatives are designed to contribute to a structural economic transformation of rural land-scapes where labour moves from low productivity activities to those with higher productivity and returns (Basnett et al. 2014; Basnett and Bhattacharya 2015), it is unlikely that such piecemeal, sector-specific efforts will suffice.

Moreover, by classifying migration as a problem, opportunities that migration can potentially offer to forest restoration and forest governance may also be over-looked. While evidence of the relationship between migration and forests remains patchy and contradictory in the literature (Hecht et al. forthcoming), scholars have found that migration and remittances can provide alternatives to forest-based live-lihoods and unsustainable resource extraction (Curan and Agardy 2002; Hecht and Saatchi 2007; Peluso et al. 2012). However, as Agrawal and Yadma (1997) have reminded us, much rests upon how well local communities and individuals re-design institutions to manage the flow and reap the potential benefits of migration.

Notwithstanding the complex facets of the vast and growing research on gender and migration, scholars have long argued that migration is an inherently gendered process (Chant and Radcliff 1992). Key areas of research inquiry include differences in engagement of men and women in the processes of migration, the role of intra-household relations and labour-market segmentation in the sex and class selectivity of migration flows, the gendered dynamics between migrants and those left behind, and continuities and changes in gender relations as a consequence of migration. Some analysts are optimistic that opportunities to migrate and/or the absence of men from the household as a consequence of migration alters gender ideologies and enables unprecedented 'voice' and 'choice' for women (Chant 1998; Hadi 2001). Others (Elmhirst 2007; Resurreccion and Khanh 2007) point to the com-plex, gendered negotiations that take place between those who migrate and those left behind and the reproduction of gendered identities, roles and obligations that occurs in spite of migration.

Very few studies have explored the linkages among gender, migration and for-estry in Nepal or globally. While scholars have advanced our understanding of the barriers to women's participation in forest governance and the benefits of their inclusion for women's well-being and forest governance, a very static and geo-graphically bounded understanding of individuals, their relations to forests and their livelihood strategies have been thus far portrayed in the literature (Resurrecciion and Elmhirst 2008). Exceptions include Giri and Darnhofer's (2010) analysis of the relationship between men's out-migration and women's participation in com-munity forestry in Nepal. Drawing on qualitative data collected in two community forest user groups, these authors find that women left behind are more likely to be present in and voice their opinions in general assembly. Although the authors do not investigate the impact this has on forest-related decisions or on the flow of benefits that women and men derive from forests, they do postulate that increased participation in general assemblies is likely to have corollary effects on forest decisions. They acknowledge that women's increased participation is mediated by family size and composition, and that women who do not have an adult male in the house are more likely to participate.

While these are important findings, the case studies below serve to illustrate that the relationship between male out-migration and women's empowerment in community forestry is complex and rooted in the changing socio-cultural fabric of Nepali societies and social relations. Pre-existing relations between women and men; the interface between forest policy and local institutions established to govern forests; and the ways in which gender relations are reconfigured and reproduced in light of migration play critical roles in re-defining who has voice, who benefits and who loses.

Seasonal out-migration and the feminization of community forestry among the Tamangs

> We are more dependent on men than men are on us. We depend on them for work and money. But we have learnt that by cooperating amongst ourselves we can help each other out.
>
> *(Middle-aged Tamang woman, Bhatpole Village, Feb. 2005)*

The village of Bhatpole, in Kabrepalanchok District of Central Nepal, is predominantly inhabited by people of 'Tamang' ethnicity and is located adjacent to settlement hamlets inhabited by other ethnic/caste groups such as 'Jasi-Bahuns', 'Magars', 'Chetris', etc. Tamangs are of Tibeto-Burman origin and constitute one of the largest and most socially and economically marginalized ethnic minorities in Nepal.[4] Tamang households of Bhatpole depend upon agricultural and non-agricultural livelihoods within and outside the village. Because of the dearth of good agricultural land, most households rely on seasonal out-migration to Kathmandu and neighbouring towns and cities to supplement shortfalls. Seasonal migration allows them to return to the villages during peak agricultural seasons (e.g. planting, harvesting paddy) to help with family farm production and engage in daily waged agricultural work for the wealthy Jaisi-Bahun landlords in neighbouring villages. Men were generally away for six months or longer per year.

---

4 According to Höfer's (2004) landmark study on state and society relations in Nepal, 'Tamang' as an ethnic group had not existed in official records until 1932. Other anthropological accounts also suggest that the name – 'Tamang' – applies to quite a diverse group located in Central Nepal. There is no uniform Tamang culture, social structure, or overarching political institutions. People speak different Tibeto-Burman languages, some mutually unintelligible. The Nepali state in the process of state-making constructed the label 'Tamang' to incorporate a diverse populace within the state machinery. During my field research, Tamangs of Bhatpole also recalled many differences in rituals, language and customs between themselves and other Tamangs in Central Nepal (Levine 1987; Holmberg 1989).

Although both women and men expressed interest in migrating, seasonal migrants were predominantly male. In Bhatpole, the male-led patterns of seasonal out-migration were not due to gender imbalances at the intra-Tamang level. As I have discussed elsewhere, socio-cultural norms and practices (such as organization of the household, marriage practices, gender division of labour in family farm production and allocation of intra-household resources) did not necessarily place men and women on equal footing. But, they did embody gender egalitarian principles and practices (Sijapati 2008). For instance, women and men interviewees agreed that both genders could exercise equal voice and autonomy in entering and exiting marriages. The ideology of controlling women's sexuality, which they perceived as being common practice among their high-caste neighbours (Chhetri and Brahmin castes) and as a burden on women, was non-existent. In this regard, male-dominated migration trajectories among Tamangs were due to the gendered segmentation of the markets for Tamang labour as well as the inability of Tamang migrant networks to tap into gender-inclusive markets. These percolated down to the household level and defined who was able to migrate and who had to be left behind (Sijapati 2008).

Encouraged by their families, young women and men had previously migrated in equal numbers to work in Kathmandu's carpet industries. The ideology of controlling women's movement was virtually absent among the Tamangs. Although health and carpet industry safety standards for workers were low, workers were subjected to very little gender-based discrimination in terms of duties assigned and wages paid. After the carpet industry collapsed due to declining export volumes and reduced rates of return in the late 1990s (Graner 2001), much of the informal and casual labour demand in towns and cities was specifically for male labour. Tamang networks had little access to other employment opportunities that were able to absorb both male and female workers. Consequently, most women carpet workers had to return to Bhatpole while men continued to find casual work elsewhere.

The allocation of responsibilities for family farm production and domestic work (e.g. collection of firewood and fodder) was defined by 'availability to work' rather than gender per se. But the gender biases embedded in markets for Tamang labour were being transmitted at the intra-household level and becoming evident in gender inequalities in the division of labour (e.g. childcare, domestic work). Women became disproportionately dependent on men for material and extra-local support. Women from households where male members were seasonal migrants witnessed significant increases in their work burden when men were away. However, rather than being passive spectators, women were also capitalizing on the spaces existing within Tamang socio-cultural practices. They were investing in greater cooperation and collaboration with one another to mitigate the gender-based constraints they faced in their everyday lives. This was particularly evident in the way women organized exchange labour of various kinds in family farm production and domestic labour. These were measured, monitored and reciprocated stringently to mitigate the labour vacuum created during men's absences.

Thus, in Bhatpole, community forestry became a women-led initiative, with women at the forefront of promoting and supporting it. Women comprised nine out of eleven community forestry committee members, including the position of the committee chairperson. Community forestry was viewed as a way of addressing the lack of secure and steady access to forest products commonly faced by Tamang women in men's absence. This lack of access was due to shortages in labour to procure the forest products from common or private land and fluctuations in the income required to purchase forest products. Tamang men generally opted to bring their savings home when they returned from seasonal employment rather than send them intermittently. This compelled the women left behind to manage periods of income shortfalls. Collaborating to manage community forestry became a part of and intertwined with ongoing forms of collective effort. Women drew on pre-existing forms of collaboration to discuss and decide on the rules that should govern forests prior to seeking formal handover from the government, and to define men's role in community forestry. Their rules included user eligibility, forest protection, penalties and which types of forest products were to be appropriated – when and by whom they were to be appropriated were decided when women met to make arrangements about exchange labour. For instance, 'users' were individuals rather than households, and both women and men were eligible to be users. The list of users at the end of the community forestry user group's operational plan was primarily one of women. This enabled women to participate in general assemblies, contest in elections to appoint committee members, and have a voice in rules determining access to forest products, etc.

Women feared that involving men, the vast majority of whom migrated seasonally, as equal partners in the community forestry process would significantly increase the costs of participation, require broadening the scope of community forestry to meet men's interests and priorities, and jeopardize the basis for collective action for community forestry governance. At the same time, women also sought strategic support from men in order to liaise with government officials on their behalf and to help them comply with governmental rules and regulations pertaining to the establishment and functioning of community forest user groups.

After the end of women's involvement in the carpet industry, women conceptualized their life spaces as being separate from but simultaneously linked to those of men. Their spaces were limited to the local (the village, local market, neighbouring villages), whereas men operated in both local and extra-local spaces. Men who seasonally migrated outside the village were viewed as being better able to understand, and interact and bargain with extra-local actors such as forestry officials. For instance, although all Tamang women in Bhatpole could converse in Nepali (the official language), they were less at ease with highly sanskritized, formal/official Nepali. This is needed to be able to understand community forestry-related legal documents and converse with forest officials. As Agarwal (2010) notes, community forestry policies, although implemented at the local level, are framed at the national level and beyond. Women often lack the experience and contacts required to forge extra-local networks and influence institutions at high levels. Furthermore, as Nightingale (2005) points out, in spite of the participatory nature of community forestry policies, the highly technocratic support provided by the (national) Department of Forestry assumes that local people have little knowledge about how to manage community forests and must be taught modern silviculture. This reinforces differences between users based on education, literacy and gender.

In the context of Bhatpole, the 'technocratization' of community forestry led Tamang women to depend on male counterparts with literacy skills and extra-local experience to act as intermediaries between district forest officials and Tamang women users. Men generally agreed to play a supporting role as long as they also benefited (along with women) from secure access to forest products and would not have to contribute their time and labour to community forestry governance. In the process of establishing and managing community forests, new gender hierarchies were created as women relied increasingly on men to act as go-betweens with government officials.

Remittances, class and the invisibilization of women among the Dalits of Gharmi

> The familial pressure to migrate, earn sufficient income, and reinvest in the village so as to end the shackles of poverty and caste oppression is much stronger for a man than a woman.
>
> *(Biswa-Karma male, aspiring migrant to Qatar, May 2005)*

> My husband fought and was beaten in a struggle to get our forests from the Poudyals and KCs. But I also fought. We were like backstage and front stage actors in a *natak* [theatre play]. I provided the necessary support, and my husband represented both of us in the struggle.
>
> *(Biswa-Karma female, married with children, March 2005)*

The inhabitants of the village of Gharmi in Kaski District in Western Nepal are high-caste Poudyals and Khatri-Chettris and low-caste Biswa-Karmas (Dalits), with each group occupying its own settlement hamlet. The majority of Biswa-Karma households are dependent on historical patron-client relationships as well as migration outside the village. People from both high castes relied on those from low castes as a cheap source of labour and those from low castes on those from high castes for their livelihoods. Caste-based practices of untouchability characterized everyday social relations between high and low castes. Many low-caste people also supplemented caste-based systems of livelihood with seasonal migration to the fertile agricultural plains of Nepal and India to take advantage of different agricultural seasons and to find non-farm employment.

As Gill (2003) points out, however, many rural households (like those in Gharmi) are dependent on the same type of seasonal out-migration, where demand for labourers does not change, leading to supply outstripping demand in the flat/agricultural countryside of Nepal and India. This means that in Gharmi only a handful of households were able to accumulate an adequate or sustained income through migration; these households were actively sending their young men to the Gulf countries for two to three years at a time. While this involved much higher costs of migration, it meant considerably greater returns in terms of remittances. Thus, migration was differentiating the Dalit community along class lines and cementing these divisions. Members of the remittance class were reinvesting in the village in the form of land and productive resources, and lowering their economic dependence on caste-based patron-client relations. Many were also influenced by the Dalit struggle taking place in Nepal and were instrumental in mobilizing support against caste-based discrimination upon their return to Gharmi.

Even though migration was seen as the only viable option for reducing household vulnerability and increasing the social and economic standing of individuals in the village political economy, migration was not an option for women. Caste-based ideologies such as women's honour and strict enforcement of the gendered division of labour served to control women's mobility outside the household and village. 'Honour' was associated with local idioms of sexuality, defining what constituted 'appropriate' behaviour for women, and was crucial for maintaining women's (and especially junior women's) restricted position in the household and community. The fundamental contradictions in the changing context of caste-based relations in the village were that many of the Dalit socio-cultural practices (especially those related to the treatment of women) mirrored high-caste practices and continued to be strictly enforced. This coexisted alongside the emerging 'remittance class' who were struggling to end caste-based discrimination.

The District Forest Office-Kaski (DFO) handed over the community forests in Gharmi to low-caste inhabitants after three years of fierce dispute between members of high and low castes over usufruct rights. The initial motivation behind the Dalits' request for the handover of community forests was to gain secure access to forest products and to reduce women's work burden. Biswa-Karma households required forest products such as firewood for cooking, fodder for livestock, organic manure for agricultural production and timber for construction purposes. Collection of forest products was associated with locally defined perceptions of femininity and was therefore considered women's responsibility. These demarcations were strictly observed and any transgressions severely reprimanded by the powerful men. For instance, senior, low-caste male interviewees often took offence at my inquiries over the household consumption of firewood during the fieldwork process. As one put it, 'Why are you asking us? You should ask those who are responsible for cooking.' However, a bitter caste-based battle ensued when those of high caste contested the handover on the grounds that their lineage deity (*kul Deota*) was located in the forest and Biswa-Karmas were barred from entering the 'sacred' forests because of locally defined and sanctioned practices of untouchability. Furthermore, for many high-caste people, relinquishing rights to what was perceived as high-caste property would pave the way for greater demands for caste-based equality in other domains and undermine their power and privilege in the political economy of the village.

In response, the senior and powerful men within the Biswa-Karmas employed community forestry as a platform to launch a caste-based struggle. These were also men who were part of the 'remittance class' and least dependent economically on caste-based patron-client relationships, receiving remittances from sons and brothers who had migrated to the Gulf. Discourses over 'equality', 'rights' and 'citizenship' – which were gaining currency in the newly democratic Nepal – were employed to garner support for the movement and win alliances with politicians in the district and with key movers and shakers in the DFO. In addition, these community leaders went to great lengths to portray a 'unified Biswa-Karma' voice against the high castes, and put considerable social pressure on the women and the poorest members of the Biswa-Karma community to ensure that they participated in the struggle too.

By the time community forestry was handed over to the Biswa-Karmas, the struggle over community forests had far-reaching extra-local consequences. Numerous external actors, such as the police, politicians and senior officials in the DFO-Kaski, were involved in mediating the struggle. The story of the 'struggle of the powerless Dalit community for their rights to access forests' had made headlines throughout the district. Consequently, the governance of community forestry was not merely about securing access to forest products, but had transformed into a village-wide public affair that brought with it extra-local recognition and the flow of development aid. The senior and most powerful members had a vested interest in maintaining control over community forestry and showcasing the community forest as a model of sustainable management. The community forestry committee – a major decision-making body – was reserved for senior men involved in the caste-based struggle. Women were not only excluded from the decision-making process but the rules that were developed focused on protecting forests rather than sustainably utilizing them so as to meet women's basic needs.

Conclusion

The findings of this research suggest that the current policy discourse on community forestry governance in Nepal does not adequately reflect the changing landscape of rural livelihoods and gender dynamics. While 'gender and social equity mainstreaming' as a policy figures prominently in national policies on community forestry, the definition of 'gender' is limited to women's participation, and women are assumed to be a homogenous group that is equally marginalized internally. This chapter has highlighted how gender interacts with wider social relations to situate women and men in varying ways among different social groups; the increasing multi-locality of rural livelihoods; and their implications for women's and men's entry into and influence over the governance of community forests.

In the case of Bhatpole, the chapter has demonstrated the ways in which the predominance of male out-migration shaped intra-household gender dynamics and contributed to the feminization of community forestry governance. By contrast, in the case of Gharmi, male migration contributed to the creation of a 'remittance class', which used community forestry as a platform on which it waged a caste-based struggle, but in the process, inequalities along lines of gender, class and seniority among the Dalits were further entrenched.

For far too long, community forestry policies have operated on the assumption of the physical and social boundedness of rural communities. This chapter has attempted to draw attention to the importance of studying migration (internal and external) as one of the factors shaping social change, and of questioning the present approach to the governance of community forests. The chapter has also sought to highlight the importance of cultural factors, the variety of gender relations that exist from place to place, and migration as one of the many influences on gender relations and forest governance.

References

Agarwal B. 2010. *Gender and green governance: The political economy of women's presence within and beyond community forestry.* Oxford: Oxford University Press.

——. 2014. The power of numbers in gender dynamics: Illustrations from community forestry user groups. *Journal of Peasant Studies* 42(1): 1–20.

Agrawal A and Yadma GN. 1997. How do local institutions mediate market and population pressures on the environment? Forest panchayats in Kumaon, India. *Development and Change* 28: 435–65.

Ashley C and Maxwell S. 2002. Rethinking rural development. *Development Policy Review* 19(4): 395–425.

Basnett, BS. 2011. Linkages between gender, migration and forest governance: Re-thinking community forestry policies in Nepal. *European Bulletin of Himalayan Research* 38: 7–32.

Basnett Y and Bhattacharya D. 2015. *Exploring spaces for economic transformation in the sustainable development goals.* London: Overseas Development Institute.

——, Henley G, Howell J, Jones H, Lemma A, and Pandey PR. 2014. *Structural economic transformation in Nepal: A diagnostic study.* Submitted to DFID Nepal. London: Overseas Development Institute.

Benett L. 2003. *Dangerous wives and sacred sisters: Social and symbolic roles of high caste women in Nepal.* Kathmandu: Mandala Book Point.

Britt CD. 2002. *Changing the boundaries of forest politics: Community forestry, social mobilisation and federation-building in Nepal viewed through the lens of environmental sociology and PAR.* Department of Sociology. Ithaca, NY: Cornell University.

Cameron M. 1998. *On the edge of the auspicious: Gender and caste in Nepal.* Urbana and Chicago: University of Illinois Press.

Caplan L. 1990. *Land and social change in East Nepal: A study of Hindu-tribal relations* London: Routledge and Kegan Paul.

Chant SH. 1998. Households, gender and rural-urban migration: Reflections on linkages and considerations for policy. *Environment and Urbanisation* 10(1): 5–22.

—— and Radcliff S. 1992. Migration and development: The importance of gender. In Chant SH, ed. *Gender and migration in developing countries.* London: Belhaven Press.

Cornwall A. 2003. Whose voices? Whose choices? Reflections on gender and participatory development. *World Development* 31(8): 1325–42.

Curan SR and Agardy T. 2002. Common property systems, migration and coastal ecosystems. *Ambio* 31(4): 303–5.

DeSales A. 2000. The Kham Magar country, Nepal: Between ethnic claims and Maoism. *European Bulletin of Himalayan Research* 19: 1–24.

Elmhirst R. 2007. Tigers and gangsters: Masculinities and feminised migration in Indonesia. *Population, Space and Place* 13(3): 225–38.

Gill G. 2003. Seasonal labour migration in rural Nepal: A preliminary overview. *ODI Working Paper* 2018.

Giri K and Darnhofer I. 2010. Nepali women using community forestry as a platform for social change. *Society and Natural Resources* 23(12): 1261–329.

Graner E. 2001. Labor markets and migration in Nepal: The case of workers in Kathmandu Valley carpet manufactories. *Mountain Research and Development* 21(3): 253–9.

Gray J. 2008. *The householder's world: purity, power and dominance in a Nepalese village.* New Delhi: Oxford University Press.

Hadi A. 2001. International migration and the change of women's position among the left-behind in rural Bangladesh. *International Journal of Population Geography* 7(1): 53–61.

Hangen S. 2010. *The rise of ethnic politics in Nepal: Democracy in the margins.* London: Routledge.

Harper I and Tarnowski C. 2002. A heterotopia of resistance: Health, community forestry and the challenges to state centralisation in Nepal. In Gellner D, ed. *Resistance and the state: Nepalese experiences.* Delhi: Social Science Press. 33–83.

Hecht S and Saatchi S. 2007. Globalisation and forest resurgence: Changes in forest cover in El Salvador. *BioScience* 57(8): 663–72.

——, Yang A, Sijapati Basnett B, Padoch C, and Peluso N. forthcoming. *People in motion, forests in transition: Trends in migration, urbanisation, remittances and their effects on tropical forests and forest-dependent communities.* Bogor, Indonesia: Center for International Forestry Research.

HMG/N. 1990. Master plan for the forestry sector: Revised forestry sector policy. Kathmandu: HMGN/ADB/FINNIDA.

Höfer A. 2004. *The caste hierarchy and state in Nepal: A study of the Mulki-Ain of 1854.* Kathmandu: Himal Books.

Holmberg D. 1989. *Order in paradox: Myth, ritual and exchange among Nepal's Tamang.* Ithaca, NY and London: Cornell University Press.

Hutt, M, ed. 2003. *Himalayan 'People's War': Nepal's Maoist rebellion.* New Delhi: Foundation Books.

Kondos V. 2004. *On the ethos of Hindu women: Issues, taboos and forms of expression.* Kathmandu: Mandala Book Point.

Leach M. 2007. Earth mother myths and other ecofeminist fables: How a strategic notion rose and fell. *Development and Change* 38(1): 67–85.

Levine N. 1987. Caste, state, and ethnic boundaries in Nepal. *Journal of Asian Studies* 46(1): 71–88.

Nightingale AJ. 2002. Participating or just sitting in? The dynamics of gender and caste in Nepalese community forestry. *Antipode* 37(3): 581–604.

——. 2005. 'The experts taught us all we know': Professionalization and knowledge in Nepalese community forestry. *Antipode* 37(3): 581–604.

Peluso NL, Suprapato E, and Purwanto AB. 2012. Urbanizing Java's political forest? Agrarian struggles and the nature of re-territorialisation of natures. In Rigg J and Vandergeest E, eds. *Revisiting rural places: Pathways to poverty and prosperity in South East Asia.* Honolulu: University of Hawaii Press.

Phillips A. 1995. *The politics of presence: The political representation of gender, ethnicity and race.* Oxford: Oxford University Press.

Pokharel BK. 1997. Foresters and villagers in contention and compact: The case of community forestry in Nepal. Unpublished PhD dissertation, University of East Anglia.

——, Paudel D, and Gurung BD. 2005. Forests, community-based governance and livelihoods: Insights from Nepal Swiss community forestry project. *Regional workshop on capitalization and sharing of experiences on the interaction between forest policies and land use pattern in Asia.* Godavari, Kathmandu.

Razavi S. 2003. Introduction: Agrarian change, gender and land rights. *Development and Change* 3(1–2): 2–32.

Regmi MC. 1978. *Thatched huts and stucco palaces: Peasants and landlords in 19th-century Nepal.* New Delhi: Vikas.

Resurreccion BP and Elmhirst R. eds. 2008. *Gender and natural resource management: livelihoods, mobility and interventions.* London: Earthscan.

—— and Khanh HTV. 2007. Able to come and go: Reproducing gender in female rural-urban migration in the Red River delta. *Population, Space and Place* 13(3): 211–24.

Rigg J. 2005. Land, farming, livelihoods and poverty: Re-thinking links in the rural south. *World Development* 34(1): 180–202.

Seddon D, Adhikari J, Gurung G. 2001. *The new Lahures: Foreign employment and the domestic economy of Nepal.* Kathmandu: Nepal Institute of Development Studies.

Sharma J. 2008. Practices of male out-migration from the hills of Nepal to India in development discourses: Which pathology? *Gender, Technology and Development* 12(3): 303–23.

——, Pandey S, Pathak D, and Sijapati Basnett B. 2014. *State of migration in Nepal.* Kathmandu: Center for the Study of Labour Mobility.

Sijapati B. 2008. *Gender, institutions and development in natural resource management: A case study of community forestry in Nepal.* Development Studies Institute. London: London School of Economics and Political Science.

Tamang S. 2009. The politics of conflict and difference or the difference of conflict in politics: The women's movement in Nepal. *Feminist Review* 91: 61–80.

Thapa D. and Sijapati B. 2003. *A kingdom under siege: Nepal's Maoist insurgency, 1996–2003.* Kathmandu: The Printhouse.

Thieme S. 2008. Sustaining livelihoods in multilocal settings: Possible theoretical linkages between livelihoods and transnational migration research. *Mobilities* 3(1): 51–71.

17

REVISITING GENDER AND FORESTRY IN LONG SEGAR, EAST KALIMANTAN, INDONESIA

Oil palm and divided aspirations

Rebecca Elmhirst, Mia Siscawati and Carol J. Pierce Colfer

Introduction

Dramatic and devastating changes in East Kalimantan's forest landscape over recent decades reflect the impact of intensified resource extraction through timber concessions, transmigration settlement and the expansion of agri-business in this part of Indonesia. A number of studies have noted how the intersection of landscape change with shifting politics of forest access and the emergence of new forms of employment beyond the forest are bringing about changes in people-forest relationships and a re-spatialization of livelihood strategies that take shape in specific ways in particular places across Indonesia and elsewhere in Asia (Dewi et al. 2005; McCarthy 2010; Li 2011; Peluso 1992; Obidzinski et al. 2012; Bullinger and Haug 2012; Cramb and Curry 2012; Cramb 2013; Kesaulija et al. 2014). To date, much less is understood about the role played by gender: How men and women within forest communities may experience and respond in different ways, and how gender norms are being reworked or perhaps reinforced in the face of such changes (Colfer 2010). With respect to the changes wrought by large-scale investment in oil palm, an emerging literature sets out some of the gendered impacts in Indonesia. Thus far, contributions by Julia and White (2012) and Li (2015) in West Kalimantan, Elmhirst and Darmastuti (2015) in Lampung, and Elmhirst et al. (2015) in East Kalimantan indicate some common gendered experiences. Differences between sites remain apparent, depending on how communities have been incorporated into the oil palm sector and the extent to which diverse livelihoods remain a possibility.

This chapter contributes further to this embryonic literature by adding a historical perspective developed through revisiting the gender dynamics evident within Long Segar, a Dayak community in East Kalimantan studied in depth over the course of the 1980s and up to 2001 (Figure 17.1; Colfer 2008). Long Segar was founded in the early 1960s by small groups of Kenyah Dayak who had migrated from the remote interior to make a living from swidden rice cultivation, forest products and small-scale timber extraction. Detailed anthropological research by Colfer (2008) revealed the creative ways the community adapted to changing government-sponsored and commercial pressures on the forests and resource tenure, particularly those associated with the timber concession. The people maintained the integrity of their cultural systems, the meaning of their lives in relation to forests, and norms of gender egalitarianism, through which hierarchical relationships between men and women are downplayed. We draw on the insights gained during this period of research and couple this with data from a recent "revisit" by a research team led by Elmhirst and Siscawati, which examines the consequences of the recent transition (since 2004) from forest concession to large-scale investment in oil palm there.

Although there were signs in 2001 that oil palm had been rejected by the community, by 2004 transformation of the landscape began, bringing with it a dramatic and divisive impact on lives, livelihoods and aspirations of different groups of men and women. Ambivalence and division in responses to oil palm have been observed in other studies from East Kalimantan, where, for example, Gönner (2011) documented oil palm-related changes and conflicts in a community on the Mahakam River. Iwan et al. (2009) highlight resistance to oil palm among the Kenyah Dayak Uma' Lung on the Malinau River, whilst Belcher et al. (2004) and Dewi et al. (2005) consider oil palm impacts in the context of livelihoods drawn from non-timber forest products and other economic opportunities in the East Kalimantan districts of Paser and Kutai Barat. Potter (2008) notes that for some communities in Kalimantan more broadly, oil palm continues to be heralded as a potential wealth-creator for smallholders (also the view of some within Long Segar).

In this chapter, we examine the contemporary patterns of these impacts and responses to oil palm expansion, particularly in relation to gendered relationships with forests, swiddens and food production, and in the context of broader livelihood opportunities and changing land tenure. Drawing on Colfer's earlier work, we are able to place contemporary changes in historical and cultural context, whilst also using the benefit of hindsight to reexamine conclusions drawn in relation to gender norms, engagement with the forest, and directions of change. We consider research findings from the 1980s (including gendered responses to the timber concession and wider political economic changes) alongside more recent, albeit less extensive data on responses to commercial oil palm. We aim to shed light on how gendered aspirations emerge for young and old from the context of the contrasting socio-natures represented by timber concessions (in the 1980s and 1990s), transmigration settlement and more recent investments in oil palm in East Kalimantan's "political forest."[1]

---

1 The term "political forest" deliberately draws on Peluso and Vandergeest's (2001) extensive discussions of the role forests have played in state political strategies in South East Asia. The term conveys the assumption of state authority over forests, superseding the rights, claims and practices of forest dwellers.

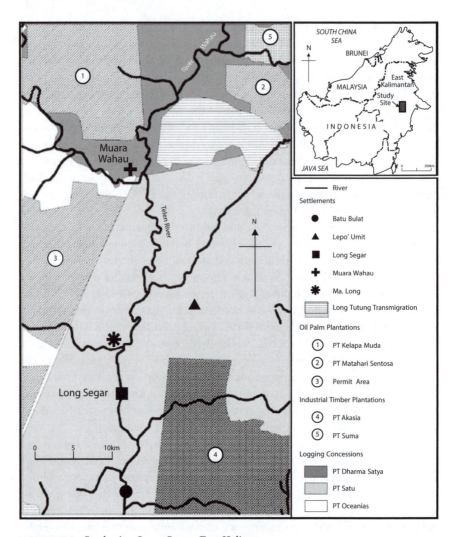

FIGURE 17.1 Study site, Long Segar, East Kalimantan

Revisiting Long Segar: changing contexts

Profound changes in rural Asia have, in recent years, prompted a number of efforts to "restudy" communities that had in the past been studied in depth. Our approach here follows that advocated by Rigg and Vandergeest (2012), whereby a restudy involves a revisit to a previously studied community with the intention of reflecting systematically on change. In our case, we focus on changes associated with the gender dimensions of natural resource access, livelihoods and aspirations. At the same time, we are mindful that in building from Colfer's earlier studies, we are engaged in something more than simply revisiting Long Segar: our work also involves revisiting gender as a concept (Lund et al. 2015) and in dealing with a changing socio-political context where until recently, public discourses around customary entitlements to "state forest" were politically "off limits."

Colfer's research in Long Segar began in 1979 with a 10-month ethnographic study that focused on the lives of the Uma' Jalan Kenyah Dayak, and was undertaken as part of a 1979–80 project entitled "Interactions between People and Forests in East Kalimantan."[2] Located in the lowland area flanking the Telen River, Long Segar was selected to capture the adaptations being made by swidden cultivators in the face of externally induced changes. Such changes included the allocation of forest lands for timber concessions and efforts by government extension workers to discourage shifting cultivation and persuade them to restrict the geographical extent of their farming. Change and adaptation were not new phenomena to this community: they were created by Uma' Jalan who had begun migrating from more remote Long Ampung in the 1960s in response to land scarcity and religious and political differences.[3] Long Segar, which was formally designated as part of a Resettlement Project (*Resetelmen Penduduk*) in 1972, offered the prospect of access to trade goods, to education, medical services and wage labor associated with forest clearing.[4] The Uma' Jalan had formally discarded their animist tradition to become Christian by 1964, with the majority belonging to an evangelical Protestant group, and about a third being Catholic. Many vestiges of their animist system remained, however.

At the time of this early research, the foundation of the Uma' Jalan economic system was upland (swidden) rice cultivation with supplementary gardening, hunting, fishing, forest food and minor forest product collection, as well as intermittent wage/contract labor. Then part of a timber concession, the landscape was dominated by old-growth forest. Trade was via the river, where longboats would moor, buying agricultural surpluses from the community and selling manufactured goods, of which some were sold in various small stores in the community. The community was reachable in two days by long boat and ten hours by speedboat from the provincial capital Samarinda. Colfer's repeat visits over the course of the next two decades chart the transformation of this landscape (Colfer 2008). Transformations included landscape changes associated with the timber concession (granted originally to US logging company Georgia Pacific, later to PT Kiani Lestari) and the establishment of an industrial timber plantation (HTI) to serve the plywood trade, and with this, the establishment of a 10,000-hectare transmigration resettlement site. This brought large numbers of "outsiders" to the area, specifically, primarily Javanese migrants, whose livelihoods, religion and social practices contrasted with those of the Uma' Jalan.

2 Kenyah Dayak make up one of the groups of Kalimantan's Dayak peoples, a group with considerable linguistic and cultural diversity. Uma' Jalan are a sub-group of Kenyah Dayak.
3 "Land scarcity" here simply meant that suitable land for ricefields was now too far from the village to reach on foot within a reasonable amount of time, understood as anything more than two hours; cf. analyses of increasing distances to ricefields (Colfer and Dudley 1993). See Eghenter 1999, for further discussion of the reasons for Kenyah Dayak migrations from the interior to riverine locations like Long Segar.
4 This program sought to bring "development" to Indonesia's indigenous people, specifically in relation to settling swidden cultivators (see Dove 1985).

The arrival of oil palm

In East Kalimantan, oil palm cultivation began on a relatively small scale in 1982 under the Indonesian New Order government's Nucleus Estate Program, managed by the state-run plantation company PTP VI. However, investment has accelerated rapidly in recent years, led by the private sector. East Kalimantan is not among Indonesia's leading oil palm-producing provinces, but the total area under oil palm was recorded in 2013 as having reached 1,115,415 hectares. Of this, the bulk (862,782 ha) comprised large-scale private estates, 22,367 ha of state-run plantations, and 230,266 ha as "smallholder" oil palm (Dinas Perkebunan Provinsi Kalimantan Timur 2014). Oil palm was established across the river from Long Segar by two companies, PT Bukit Subur, and PT Tapian Nadenggan in 2007, replacing the timber concession. Land was acquired by the companies to create nucleus-plasma estates, in which 80 per cent of the oil palm estate remains under the control of the company for the duration of its concession (known as the nucleus), whilst 20 per cent of planted area is redistributed to the community for management as smallholder oil palm (this portion is known as "plasma"). Profits from the plasma are received by those who have been allocated plots once fees and costs of establishing the planted area have been deducted. In Long Segar, plasma redistribution has yet to unfold, but oil palm expansion generally has had multiple effects.

Although swidden rice farming continues in Long Segar, much of this is at some distance from the village itself—community leaders explained how oil palm had taken over much of the area described as "their" KBNK (*Kawasan Budidaya Non Kehutanan*, land designated as non-forest cultivation areas under national land use categories). Livelihoods continue to involve a combination of upland rice, gardening and fishing, but hunting and forest product collection are much less prevalent, reflecting the disappearance of old-growth forest. Instead, the most marked feature of livelihoods is wage labor, now associated with the establishment of large-scale oil palm plantations. River trade remains important, but the road from Samarinda was completed in 2011, reflecting local government efforts to stimulate corporate investments in oil palm. This has allowed vehicle access to Long Segar, except during the rainy season.

In revisiting gender and forestry in Long Segar, our aim has been to investigate the impact of these changes on gender dynamics, and their implications for women's once crucial roles within Uma' Jalan livelihoods. Whilst it is difficult to disentangle the impacts of oil palm from other important economic, socio-cultural and political changes in this community over the last decade, the arrival of this crop has brought specific changes to the gender dynamics of off-farm work (creating wage opportunities for women) and access to productive resources (limiting access to land and forest). We explored how these played out in terms of shifts in gender norms within households and within the community more generally. A key question we addressed was whether the gender egalitarianism described by Colfer in her research was being reworked, as people sought to adapt to new livelihood circumstances, and whether her early prediction of a reduction in women's status has borne out.

Research methodology

The restudy investigation, discussed below, involved fieldwork of much shorter duration than Colfer's original ten months, and therefore we could not establish strong relationships of trust between ourselves and the community, or undertake close observation over the course of a season. Our insights are therefore somewhat restricted, and detailed comparison of ethnographic data between the two studies is not possible. Rather, our approach has been to identify some key themes around gendered resource access, livelihoods and aspirations evident in the earlier work, and explore changes associated with recent large-scale oil palm investments in Long Segar. Data collection for the restudy applied a qualitative methodology designed to highlight the gender-differentiated impacts of oil palm, and how such impacts reflect and potentially reshape gender norms.[5]

Following a short scoping visit, the main data collection was undertaken intensively over a two-week period in October 2014 by a field team comprising male and female interviewers and note-takers, who stayed in the community. This provided an opportunity to supplement formal data collection with informal and unstructured observations and reflections. Data collection included two key informant interviews with local leaders (both male) and eight in-depth individual life stories with men and women recruited to reveal the divergent pathways of those engaging with independent smallholder oil palm cultivation and those affected by large-scale oil palm investments. Six focus group discussions were undertaken, each with around ten participants. The "ladder of life" FGD (Focus Group Discussion) with poor men and poor women focused on gender norms, livelihoods and the effects of oil palm on socio-economic mobility, whilst the "capacities for innovation" FGD sought the views of men and women regarded in the community as successful "entrepreneurs." A final pair of FGDs focused on "aspirations of youth" in the context of oil palm expansion, and each included ten men and ten women aged 16 to 24. Most of those participating in the FGDs described themselves as farmers, although two women "innovators" were involved in non-farm work. Interviews were recorded by hand by the note-taker, and fully transcribed before being coded and analyzed.

Characterizing gender norms and dynamics in Long Segar

Across South East Asia, gender is usually characterized by egalitarianism and complementarity rather than hierarchy of male and female roles, and relative autonomy of women in relation to men (Atkinson and Errington 1990; Ong and Peletz 1995). As with other Dayak groups in Kalimantan, social relationships among the Uma' Jalan are not ordered by a fundamental code of gendered differentiation and there

5 The methodology used builds on a set of data collection instruments developed for the CGIAR global study on "Enabling gender equality in agricultural and environmental innovation" (GENNOVATE) developed by Patti Petesch (2014) with inputs from Paula Kantor, Lone Badstue, Gordon Prain and Jacqueline Ashby. The data instruments were further developed to include questions specific to the gendered impacts of oil palm investments.

is little evidence of norms restricting or enabling activities purely on the basis of gender (Tsing 1990). Indeed, from fieldwork undertaken between 1979 and 1980 in Long Segar, Colfer was able to reveal considerable sex-role flexibility, and little cognitive distance between "male" and "female" as categories (Colfer 2008).

Whilst norms of gender egalitarianism, complementarity and fluidity in men's and women's roles are one thing, gendered practices may reveal some slippage from these dominant representations. In Long Segar, by the 1980s, the potency and over-all prestige enjoyed by men, in political and economic domains, was increasingly found to exceed that of women. In her analysis at that time, Colfer suggested this was partly due to a growing rigidification of once-fluid gender roles in response to wider socio-economic trends. These included the reinforcement of a cash economy, as rice became a tradable commodity and contact increased with less egalitarian ethnic groups, including transmigrants and contract workers from other parts of Indonesia (Colfer 2008). This observation invites an analysis of gender as a process (urged also by Colfer and Minarchek 2013), reiterated and redrawn through its entanglements with different types of social hierarchies, including life course, ethnicity, social class and religion. It is also suggestive of a need to recognize gender not simply as an essentialized and geographically bounded form of knowledge, but also as something produced through widening geographies of production, trade and communication (Ong and Peletz 1995).

With this in mind, a snapshot of gender dynamics in forest-based livelihoods in Long Segar in the early 1980s reveals a number of key patterns. First, in relation to production activities, one of the most important sources of household livelihood was swidden rice cultivation, and on a daily basis this was women's most consistent (and consistently profitable) activity. However, in a departure from traditional condi-tions, the relative accessibility of Long Segar (vis-à-vis their homeland community further inland) meant rice surplus could be sold for cash. This had meant that men had become more involved in rice cultivation due to its value as a commodity and men's role in trade. Dominance of men in the rice trade, despite women's centrality to rice cultivation, resulted from men's generally higher levels of literacy, numeracy and skills in the Indonesian language. This gender distinction was reinforced by Dayak women's avoidance of dealing with men from other ethnic groups, where prevailing stereotypes about their promiscuity had tended to invite sexual harassment and other negative interactions. In this way, the lived experience of gender in relation to swid-den rice farming and marketing was shifting, through the community's engagement with wider markets, reflected in the intersection of gender with ethnic hierarchies.

Secondly, gendered relationships with the forest were also changing. An impor-tant role for Kenyah Dayak women is in the cultivation and gathering of non-rice food stuffs (ferns, bamboo shoots, vegetables, fruits, etc.). As swidden rice farm-ing was becoming more important financially in an increasingly monetized local agrarian economy, Colfer noted that less emphasis was being placed on this as an element in the livelihood system. This led her to question whether this shift could lead to a reduction in women's status in a context where status within the household is related to ability to contribute to material well-being. However, access to forests for non-rice foodstuffs had remained important as an economic safety net in Long Segar.

Thirdly, as has already been mentioned, Long Segar was increasingly becoming a cash economy. Part of the reason Uma' Jalan had moved to this location was because of the availability of wage work, an option available to men and women. However, work in the production of lumber was the preserve of men, who worked for the timber company on a subcontracting basis, an arrangement that allowed them to respond to seasonal agricultural labor requirements on their own land. Off-farm work involving migration was a male activity, congruent with an emphasis on women's productive activity in agriculture.

Fourthly, whilst domestic caring responsibilities were associated with women, there was a blurring in gender roles within this sphere. It was not unusual for men to cook or care for children: decisions around this depended on the need to tend to other responsibilities outside the home, and for women, this included rice production, the mainstay of the community and of ethnic/gender identity.

Finally, in Long Segar, although there were no formal barriers to women's participation in community decision making, in practice decisions affecting the whole community were taken at meetings attended by at least one representative from each household. Any women *could* attend the meeting, but Colfer observed that those who did tended to sit off to one side. This chimes with Tsing's (1990) observations of dispute resolutions amongst the Meratus Dayak, where gender identities were produced and reinforced through men's dramatic customary (*adat*) performances during which women were audiences. In Long Segar, public meetings were called to plan major community activities such as celebrations or responses to visiting officials and to deal with conflicts that were not resolved at the household level. Women's participation was more than token but not equal to men's. A key point of observation that links in part to gender concerns a general lack of public displays of disagreement: openly expressed conflict was rare within the community.

In sum therefore, gender dynamics in Long Segar were, at the time of the earlier study, characterized by complementarity of roles and limited gender-based hierarchy. Whilst there were some tasks associated with women (notably rice farming) and some activities with men (in terms of being the outward-looking "face" of the community, representing their households to the wider community, and engaging in circular migration), these roles were relatively fluid. The emergence of a cash economy, coupled with exposure to other ethnic groups brought about by large-scale developments locally in the 1980s, had contributed to a hardening of gender divisions and for Colfer at the time, the beginnings of a sense of gender hierarchy within prevailing norms.[6] However, in 1999, she had seen encouraging signs that education

6 Colfer noted that Long Segar's women were beginning to experience reduced autonomy and greater dependence on men for reasons that included difficulties maneuvering heavy new technology, a comparative lack of access to moneymaking opportunities, and reduced access to outsiders and their information because of the more exploitative attitudes towards women prevalent among other ethnic groups in the area. Such changes were profound in a social setting where men traditionally left women to manage competently on their own.

might be one avenue through which women's comparatively high status could be maintained. Our curiosity about subsequent directions helped to spur this research.

Oil palm impacts: dimensions of continuity and change

The transformation in the landscape around Long Segar since Colfer's earlier research has been dramatic. Forest land began to be cleared for oil palm in 2007, as plantation concessions were established on state land categorized as non-forest cultivation area (KBNK), areas which many locals regarded as *adat* land. By 2014, two companies were in operation, PT Bukit Subur and PT Tapian Nadenggan, covering 2,000 and 800 hectares respectively in Long Segar. From interviews with community leaders and focus-group participants alike, a prevailing view was that the oil palm companies had a negative impact on resource access and livelihoods, and that the incorporation of the community into a large-scale oil palm system has been "adverse" (McCarthy 2010). Amongst those interviewed for this project, there was little understanding of the process of land acquisition that was undertaken by the companies, nor of the mechanism by which "permission" from the community had seemingly been granted. "Socializations"[7] had taken place to inform the community, but these had tended to involve men only, as representatives of their households. The nucleus-plasma business model of the two oil palm companies described earlier had failed to materialize, with only a few households receiving their plasma allocation, creating considerable community division. Other adversities described in focus-group discussions and interviews related to poor experiences in terms of receiving compensation for trees and standing crops on land acquired for the concessions. As one (male) community leader put it when interviewed: "At first people didn't know anything at all about land. PT Tapian Tadenggan used 700–800 hectares of people's land without giving any compensation. They [community members] didn't know, they just claim for it now."

However, whilst there are many opponents of oil palm, others have welcomed the immediate prosperity the crop has brought for some.[8] Within our study, three farmers (from the innovator focus group and semi-structured interviews with innovators) described how they had begun to reap relatively substantial benefits from having grown oil palm themselves. According to one community leader, further households this year have cleared land to make way for their own crops of oil palm, despite this being seen as "a rich man's crop."[9] As a proportion of the community, however, key informants estimate that less than 10 percent of households are engaged in oil palm in this way, including primarily those with access to large plots (i.e. more than 10 hectares). For the most part, the community's engagement with

7 The Indonesian term, *sosialisasi*, refers to a process whereby people are informed of and persuaded (sometimes forced) to accept policies made higher up in the government.

8 Haug (2014) also notes the varied responses to oil palm within a Benuaq Dayak community in West Kutai, East Kalimantan.

9 Rist et al. (2010) make the point that oil palm can be an attractive proposition for Indonesian smallholder farmers when conditions are favorable.

oil palm is as wage laborers—either clearing land to make way for oil palm, or in maintenance activities (both women and men). The gendered impact of oil palm in this forest community therefore involves (i) pressures on continued access to productive resources, as forest and, in some cases, forest gardens, give way to oil palm; (ii) increasing availability of off-farm wage work, and therefore of the relationships between off-farm work, farm and forest, and (iii) social impacts, particularly the divisiveness that adverse incorporation into oil palm systems has engendered there. Our focus here is on how these changes give rise to continuities and changes in gender norms, how these relate to different generations, and how gender norms play out, as divided aspirations emerge at household and community level, and as oil palm investments alter the social function of the landscape in gendered ways.

Gender, oil palm and swidden rice cultivation

As an expanding body of research suggests, changing access to forest lands has a particularly acute impact on many Dayak communities because of its impact on resource management systems, food security and safety nets provided by forest resources (Colchester et al. 2006; Gönner 2011; Haug 2014; Urano 2014). As the *adat* forest disappeared under mono-cropped oil palm, there have been a number of consequences for gendered livelihoods (as in West Kalimantan—see Julia and White 2012; Li 2015). First, the distance people must travel to reach their swiddens has increased significantly. As one (male) interviewee put it:

> Our field used to be only 500m from the river. Now, whether we like it or not, we have to go to the KBK [*kawasan budidaya kehutanan*—state forest] to farm because "our" KBNK has been converted into oil palm concessions . . . Now it is oil palm.

All members of the community have been affected by the replacement of rice fields by oil palm: "I wouldn't call it improvement, but I call it livelihood shifting, it's not improvement at all. The ones who used to sell paddy for a living, now they work as daily laborers [on the oil palm plantation]" (village official, Long Segar). In focus-group discussions with women, it was evident that the sale of rice in order to access cash, which had been such a key factor in a nascent local cash economy in the 1980s, has been replaced by daily wage work. Women link this specifically to the arrival of oil palm. According to one participant: "The change is that we don't sell paddy anymore like we used to. We sold paddy for our children's transportation. And also to buy sugar." Another continued: "The wage from the oil palm company is used to buy our daily needs."

Rice cultivation remains important, but is grown further from the village on the remaining land. For women, where high social status links to their role in rice cultivation (Colfer 2008), this change has been felt acutely, as distances make it complicated to combine household and agricultural tasks. Despite the Uma' Jalan flexibility in gender role, there remains an expectation amongst men that this domain is ultimately women's responsibility, even as women expressed an expectation that men would participate in domestic work, alongside their other roles. Thus, for both men and women, motorbikes are needed to get to fields, if agriculture and domestic work are to be combined. Wages from plantation work and access to credit (available specifically for purchasing motorbikes) have enabled many households to make such purchases, and for wealthy households, even four-wheeled vehicles are used to access fields.[10] However, their use is gendered. There are no restrictions as such on women riding motorbikes to their swiddens, but difficult terrain makes this a challenge, and particularly older women become dependent on men as drivers. Norms associated with women's competency in securing material well-being independently of men are therefore challenged further by new forms of dependency associated with the realities of getting to distant swiddens.

The removal of forest itself has also affected the community and gender relations—there are now limited options for hunting or gathering of forest herbs/vegetables. Prior to the arrival of oil palm, the forest served the function of improving food security, enabling nutritional and livelihood diversification, particularly for women. Uma' Jalan women said the loss of the forest meant diets had been affected; There was less pork (*babi hutan*, hunted by men) and fish was also a problem. In his research with Benuaq Dayak in nearby Kutai Barat, Gönner (2011) found that the forest provided "waves of opportunities" for communities who would frequently switch from one income source to another, depending on resource availability, market prices, seasonality and so on, and that this was a resilient strategy for coping with external shocks. This is also evident in Colfer's earlier work (2008). In Long Segar, as oil palm has taken over the landscape, women's relationship with the forest has altered, as opportunities such as the production of handicrafts using rattan, bamboo and other forest products have gone. This was described by women in focus-group discussions:

- "There have been a lot of impacts, we can't go fishing anymore. The fish are trapped in little streams . . . ."
- "The forests are taken by the company."
- ". . . and also rattan . . . and also bamboo . . . ."
- "Nobody makes crafts anymore."
- "People used to eat pork but now it's difficult to find pigs."
- "A lot of our land is cultivated [by the plantation company]."

Other kinds of tree crops (e.g. cacao) have fallen in and out of favor, depending on markets and susceptibility to pests. Colfer had noted that rubber was one of the tree crops that Uma' Jalan experimented with in the 1980s and early 1990s, and

---

10 In "ladder of life" FGDs with both men and women, an indicator of households falling below the community's poverty line was NOT having a motorbike (suggested by people themselves).

this tree crop remains important for livelihoods. In GFDs, women ranked rubber as second in importance to paddy. However, amongst wealthier households, rubber is beginning to give way to oil palm as people with access to capital begin to experiment with independent smallholder cultivation.

As one female FGD participant put it, "People used to grow rubber but now more people are trying to grow oil palm. Now people grow both rubber and oil palm." Another continued: "Many people cannot afford to grow oil palm. Rubber doesn't need fertilizer, but oil palm does." Despite these innovations, which affect a relatively small proportion of Long Segar families, the centrality of swidden agriculture as a safety net has remained, and a strong sense of differentiation between richer and poorer families is becoming evident. For most households, "diversification" has taken on a new and rather singular meaning, that is, swidden livelihoods augmented by off-farm wage work at the oil palm plantation.

Changing relationships between forest, agriculture and non-farm work

One of the most significant changes brought by oil palm has been the availability of wage work in Long Segar for both men and women. As the plantations were being established, there was daily wage work for both sexes clearing and planting. Demand for labor meant the capacity to earn cash was vastly increased: around 85 percent of women in Long Segar work for wages at the oil palm plantation. For the most part, this is casual daily work, and women combine this with their other responsibilities by working until 2 p.m., after which they resume their domestic and subsistence activities. During times of labor demand on their swiddens, wage work at the plantation is abandoned, reflecting women's continuing key role in household rice self-sufficiency. As one woman put it:

> I work as a daily-based laborer at the oil palm company. I also cultivate the field. I didn't want to make fields but upon seeing other people make fields it just didn't feel right. Now I plant paddy twice a year.

In focus discussions with men, attitudes towards women's wage work were generally relaxed. Wage work was fine to help the household economy as long as women didn't leave their "obligation," i.e. their rice fields and domestic responsibilities. When asked the same question, women suggested that wage work enabled them to "help their husbands."

The availability of wage work locally, and perhaps also a need to remain available whilst issues over land, compensation and distribution of plasma are clarified, has meant less male circular migrations than was the case when Colfer undertook her research (Colfer 1985). She reported that women were on their own managing agroforestry-based subsistence and family welfare for considerable amounts of time. The requirement for women to manage on their own is less prevalent, and with it, a source of female pride and self-respect has waned somewhat.

Wage labor has enabled households to avoid selling their precious rice crop to raise cash for their children's travel costs to attend school or buy household goods and food items like sugar. The importance of education, especially for girls, is an enduring feature of Uma' Jalan lives in Long Segar, confirming Colfer's (2008) findings from 15 years ago. However, there is some ambivalence in relation to the dependence on oil palm cash wages now being felt, which is emblematic of wider divisions in the community, and deepened by the advent of oil palm.

Expanding a cash economy: divided aspirations in households and the community

Large-scale corporate investment in oil palm in Long Segar has revealed divergent aspirations in households and communities, and opened up a raft of divisions and conflicts more generally. Current conflicts largely relate to the problematic way in which the community was incorporated into oil palm systems. Oil palm companies were granted concessions on the area categorized as "non-forest state land," much of which was understood locally as land belonging to the Uma' Jalan community. The concessions were established as nucleus-plasma arrangements and those receiving plasma land will be entitled to an income from the oil palm, once the costs of establishing it have been paid off. The inequities built in to this system for communities like the Uma' Jalan are well documented in studies elsewhere. Specific problems include a lack of clarity over property rights (between state, community and other actors), difficulties of implementing what is effectively an individualized form of land title in areas where communal forms of entitlement tend to prevail, and a lack of transparency and justice over how plasma is distributed to particular community members (McCarthy 2010; Li 2015).

All of these problems were evident in Long Segar, where a key issue has concerned compensation arrangements and the allocation of plasma to households. This is largely due to a lack of formal recognition of territorialized property rights beyond an "understanding" within the community as to which piece of land is associated with which family. In the absence of formal arrangements, the allocation of plasma has been ad hoc, and has borne little relationship to people's understanding of their entitlements. As one woman put it: "I had a plot of land, the company borrowed it. But the company didn't give me any fee or compensation ... When people were distributing plasma we didn't get any." She went on to explain about her father's role in founding Long Segar and in "opening the area" in the late 1960s. When land was acquired for oil palm, it turned out that her family had been missed off the list for receiving plasma:

> I stayed in Long Segar because my mother is here. I was upset—we were the ones who opened this area but now we didn't get anything. I cried at that time. People take many [i.e. the oil palm company took a lot of land], people get money but I didn't.

Both male and female community members described the plasma as not "fitting" the community, by which they meant that there would not be a sufficient number of plasma allocations for all households to be in receipt of one, because the land area reserved for plasma is limited. Oil palm investments have meant changes in local hierarchies and accompanying hostilities. Another woman described how the names of people from outside the community had been included on the plasma "list." Misunderstandings and a lack of transparency around community entitlements to land have also fueled a land market for those buying into the area as recognition of the value of individual title grows. Concerns about this are shared by both men and women.

The above quotes reveal a strengthening of local consciousness with regard to the importance of their individual entitlements. This has been furthered by those within the community who see oil palm as an opportunity. Although there are significant barriers to successful smallholder investment in oil palm (price of the planting material, fertilizer requirements, irrigation, a marketing network and a relationship with a processing plant), a small number of families in Long Segar are attempting to cultivate oil palm using seeds bought from the nearby transmigration settlement (cf. Potter and Lee 1998, in West Kalimantan). Capacity to engage in smallholder oil palm is facilitated by cash transfers from children now living in the provincial capital, from successful rubber cultivation, or from those who have been working successfully as wage laborers more locally.

One male participant in the research started to plant oil palm (around eight hectares) four years ago on land he had inherited from his parents, and for which he had been able to obtain a letter of land ownership from the village government (not an official land certificate issued by the National Land Agency). This letter of land ownership put his name as "the owner of the land," signaling a new arrangement within local land tenure mechanisms in Long Segar. Capital for getting started came from profits earned from his rubber crop. Smallholder investment in oil palm is primarily being led by men, and it is men who undertake the main labor for this activity. As one female focus group participant put it:

> It was difficult during the early times. Especially when we had to transport the shoots to the land. It was far and we had to ride on a motorcycle. We could only carry six shoots at a time. They were heavy, especially the big ones. It took a long time to plant them. My son took three days off from his work. My husband dug the hole. Oil palm needs a lot of strength.

Those opposed to growing oil palm are skeptical about the likely success of smallholder oil palm (which involves considerable time investment of men and women), but acknowledge it is potentially more sustainable than wage work, for which the time horizon is limited, once trees have grown and the only work is some maintenance of trees and harvesting.

The lure of wage work is seen as problematic by some factions in the community. In focus-group discussions, young women suggested that young men were opting to take up plantation work rather than continue schooling. They suggested there was an economic driver for this, but that it also reflected boys' laziness and lack of engagement with the education system. Young men's responses offered a more generous interpretation: work in the oil palm plantation enabled them to help their parents. Their concern centered on the limited opportunities offered at the plantation. As one young male participant said: "The forest has run out. So I hope the government will be able to embrace our younger generation in order to gain support from the company and also better job opportunities from the company." Girls continue to go to school and aspire to a college education. Employment in oil palm company offices is marked as an aspirational career linked to an emerging "modern" subjectivity, currently alongside but potentially replacing the centrality of rice cultivation for female identity. Interestingly, of the "innovators" interviewed in the study, all reported having daughters working as "staff" (i.e. white-collar jobs) in nearby oil palm companies. This is certainly an aspiration for many, although out of reach for those from lower-income households in this increasingly divided community. The future looks uncertain for those who are looking beyond the wages and newly acquired consumer items. As a member of the community leadership explained, the future will either involve moving away from the forest or it will involve pushing further into as yet uncultivated area categorized as non-forest state land:

> The function of this company which is opening our area is only providing employment to turn us into laborers and create plantations. Being laborers won't make us rich. We don't want to be considered as forest vandals, it's a very negative impression. From what I see now some companies are destroying our environment.

He was cynical about the program of planting a million trees: "I plant a million trees but I cut down hundreds of millions of trees."

Conclusion

The ambivalence in attitudes towards oil palm in Long Segar is symptomatic of the profound impacts corporate investment in the crop is having in this area. Many of those impacts are gender differentiated, and moreover, gender norms have had a hand in shaping how these play out. Many of the gender norms that Colfer identified in the 1980s remain remarkably resilient in the face of large-scale landscape change. The centrality of rice cultivation remains an important pillar, not only of household food security but of feminized identities within the community. Rubber also continues to figure as the community's most important tree crop, and one which for some has yielded the capital needed to diversify further and perhaps even experiment with independent investments in oil palm. Thus although plantation wage work is very important (e.g. in terms of numbers taking it up, of the impact a wage economy has had on consumer behavior), rice and rubber cultivation take precedence, even where peoples' fields are some distance away. The outcome of this, not just in Long Segar but elsewhere in Indonesia where a subsistence principle remains critical, has been efforts on the part of companies to recruit laborers from outside the area. This has brought more "outsiders" from other ethnic groups and religious backgrounds into the area, and with it, new values, and potentially a slight "hardening" of attitudes around women's autonomy, to some extent to prevent harm, in the form of harassment of Uma' Jalan women by outsiders. Evident in some interviews, this is in line with findings from other research in Indonesia where intergenerational relations develop around the protection of young women in the face of perceived threats from those assumed to follow different norms and practices (Elmhirst 2007). Continued investment in girls' education follows from this: enabling a future away from the farm. By considering research findings from the 1980s (when gendered responses to the timber concession and wider political economic changes were becoming evident) alongside more recent, albeit less extensive, data that considers responses to commercial oil palm, we have seen new forms of gendered aspirations emerge from the context of oil palm's specificity. A key finding from this research suggests that substantial differences are opening up between those who have been able to capitalize on the oil palm boom and others, whose lives are increasingly dislocated. The latter find the changes wrought by oil palm to be a significant threat to established livelihoods based on rice, rubber and forest products. In this way, changes associated with oil palm in Long Segar reflect the interplay among gender, class, generation and ethnicity, and in turn, have a hand in reshaping and possibly opening up divisions in a once relatively egalitarian forest-based community.

Acknowledgments

This research was undertaken as part of the USAID-funded CIFOR project on "Economic Choices and Tradeoffs to REDD+ and Low Carbon in Asia," with additional financial support from the Research Council of Norway ("Revisiting gender in development: Complex inequalities in a changing Asia"). Key inputs to the study were provided by CIFOR scientists Bimbika Sijapati Basnett and Krystof Obidzinski. Primary data collection and preliminary analysis was completed by Muhammad Fajri, Musdahliah Handayani, Samuel Hatsong, Hulyana and Deden Setiadi. We are very grateful to the people of Long Segar for their gracious hospitality and their generous engagement with data collection for this study.

References

Atkinson JM and Errington S. 1990. *Power and difference: Gender in island Southeast Asia.* Stanford, CA: Stanford University Press.

Belcher B, Imang N and Achdiawan R. 2004. Rattan, rubber, or oil palm: Cultural and financial considerations for farmers in Kalimantan. *Economic Botany* 58(1): 77–87.

Bullinger C and Haug M. 2012. In and out of the forest: Decentralisation and recentralisation of forest governance in East Kalimantan, Indonesia. *ASEAS – Österreichische Zeitschrift für Südostasienwissenschaften* 5(2): 243–62. Accessed January 7, 2016. doi: http://dx.doi.org/10.4232/10.ASEAS-5.2-4.

Colchester M, Jiwan N, Andiko, Sirait M, Firdaus AY, Surambo A and Pane H. 2006. *Promised land: Palm oil and land acquisition in Indonesia – implications for local communities and indigenous peoples.* Forest Peoples Programme, Sawit Watch, HuMA and ICRAF, Bogor.

Colfer CJP. 1985. On circular migration from the distaff side. In Standing, G. ed. *Labour circulation and the labour process.* London: Croom Helm.

———. 2008. *The longhouse of the Tarsier: Changing landscape, gender, and well being in Borneo.* Phillips, ME: The Borneo Research Council, Inc.

——— ed. 2010. *The equitable forest: Diversity, community, and resource management.* London: Routledge.

——— and Dudley RG. 1993. *Shifting cultivators of Indonesia: Managers or marauders of the forest? Rice production and forest use among the Uma' Jalan of East Kalimantan.* Rome: Food and Agriculture Organization of the United Nations.

——— and Minarchek RD. 2013. Introducing "the Gender Box": A framework for analysing gender roles in forest management. *International Forestry Review* 15: 1–16.

Cramb R. 2013. Palmed off: Incentive problems with joint-venture schemes for oil palm development on customary land. *World Development* 43: 84–99.

——— and Curry GN. 2012. Oil palm and rural livelihoods in the Asia-Pacific region: An overview. *Asia Pacific Viewpoint* 53(3): 223–39.

Dewi S, Belcher B and Puntodewo A. 2005. Village economic opportunity, forest dependence, and rural livelihoods in East Kalimantan, Indonesia. *World Development* 33(9):1419–34.

Dinas Perkebunan Provinsi Kalimantan Timur. 2014. *Komoditi Kelapa Sawit.* Accessed August 25, 15. http://disbun.kaltimprov.go.id/statis-35-komoditi-kelapa-sawit.html.

Dove MR. 1985. The agro-ecological mythology of the Javanese and the political economy of Indonesia. *Indonesia* 39: 1–36.

Eghenter C. 1999. Migrants' practical reasonings: The social, political, and environmental determinants of long-distance migrations among the Kayan and Kenyah of the interior of Borneo. *Sojourn: Journal of Social Issues in Southeast Asia* 14(1): 1–33.

Elmhirst R. 2007. Tigers and gangsters: Masculinities and feminised migration in Indonesia. *Population, Space and Place* 13(3): 225–38.

——— and Darmastuti A. 2015. Material feminism and multi-local political ecologies: Rethinking gender and nature in Lampung, Indonesia. In Lund R, Doneys P and Resurreccion BP. *Gendered entanglements: Revisiting gender in a rapidly changing Asia.* Honolulu: University of Hawaii Press and Copenhagen: NIAS Press.

———, Siscawati M and Basnett BS. 2015. Navigating investment and dispossession: Gendered impacts of the oil palm "land rush" in East Kalimantan, Indonesia. Land Deal Politics Initiative: Perspectives from East and Southeast Asia Paper No. 69. Accessed January 7, 2016. http://www.iss.nl/research/research_programmes/political_economy_of_resources_environment_and_population_per/networks/land_deal_politics_ldpi/conferences/land_grabbing_perspectives_from_east_and_southeast_asia/.

Gönner C. 2011. Surfing on waves of opportunities: Resource use dynamics in a Dayak Benuaq community in East Kalimantan, Indonesia. *Society and Natural Resources* 24(2): 165–73.

Haug M. 2014. Resistance, ritual purification and mediation: Tracing a Dayak community's sixteen-year search for justice in East Kalimantan. *Asia Pacific Journal of Anthropology,* 15(4):357-375. Accessed January 7, 2016. doi: 10.1080/14442213.2014.927522.

Iwan, R. and Limberg G. 2009. Tane' Olen as an alternative for forest management: Further developments in Setulang Village, East Kalimantan. In Moeliono M, Wollenberg E and Limberg G, eds. *The dentralization of forest governance: Politics, economics and the fight for control of forests in Indonesian Borneo.* London: Earthscan/CIFOR. 193–207.

Julia and White B. 2012. Gendered experiences of dispossession: Oil palm expansion in a Dayak Hibun community in West Kalimantan. *Journal of Peasant Studies* 39(3–4): 995–1016.

Kesaulija FF, Sadsoeitoebeon BMG, Peday HFZ, Tokede MJ, Komarudin H, Andriani R and Obidzinski K. 2014. Oil palm estate development and its impact on forests and local communities in West Papua: A case study on the Prafi Plain. Working Paper 156. Bogor, Indonesia: CIFOR.

Li TM. 2011. Centering labor in the land grab debate. *Journal of Peasant Studies* 38(2): 281–98.

———. 2015. Social impacts of oil palm in Indonesia: A gendered perspective from West Kalimantan. *CIFOR Occasional Paper* 124.

Lund R, Doneys P and Resurreccion B, eds. 2015. *Gendered entanglements: Revisiting gender in rapidly changing Asia.* Copenhagen: NIAS Press and Honolulu: University of Hawai'i Press.

McCarthy JF. 2010. Processes of inclusion and adverse incorporation: Oil palm and agrarian change in Sumatra, Indonesia. *Journal of Peasant Studies* 37(4): 821–50.

Obidzinski K, Andriani R, Komarudin H and Andrianto A. 2012. Environmental and social impacts of oil palm plantations and their implications for biofuel production in Indonesia. *Ecology and Society* 17(1): 25.

Ong A and Peletz MG. eds. 1995. *Bewitching women, pious men: Gender and body politics in Southeast Asia.* Berkeley: University of California Press.

Peluso NL. 1992. The political ecology of extraction and extractive reserves in East Kalimantan, Indonesia. *Development and Change* 23(4): 49–74.

Peluso NL and Vandergeest P. 2001. Genealogies of the political forest and customary rights in Indonesia, Malaysia, and Thailand. *The Journal of Asian Studies* 60(3): 761–812.

Petesch P. 2014. Innovation and development through transformation of gender norms in agriculture and natural resource management. Methodology guide for global study. Unpublished report.

Potter L. 2008. The oil palm question in Borneo. In Persoon G and Osseweijer M (eds.). *Reflections on the heart of Borneo.* Leiden: Tropenbos International, Tropenbos Series number 24: 69–90.

——— and Lee J. 1998. Tree planting in Indonesia: Trends, impacts and directions. CIFOR Occasional Paper 18: 1–76.

Rigg J and Vandergeest P, eds. 2012. *Revisiting rural places: Pathways to poverty and prosperity in Southeast Asia.* Honolulu: University of Hawai'i Press.

Rist L, Feintrenie L and Levang P. 2010. The livelihood impacts of oil palm: Smallholders in Indonesia. *Biodiversity and Conservation* 19(4): 1009–24.

Tsing AL. 1990. Gender and performance in Meratus dispute settlement. In Atkinson J and Errington S, eds. *Power and difference: Gender in island Southeast Asia.* Stanford, CA: Stanford University Press. 95–125.

Urano M. 2014. Impacts of newly liberalised policies on customary land rights of forest-dwelling populations: A case study from East Kalimantan, Indonesia. *Asia Pacific Viewpoint* 55(1): 6–23.

18

CONCLUSION

Looking forward in gender and forestry research and praxis

Marlène Elias, Bimbika Sijapati Basnett and Carol J. Pierce Colfer

As this compilation has shown, the breadth of themes and perspectives related to gender and forestry research and action is expansive. The works presented here contribute to filling important gaps in knowledge concerning the barriers that cause gender inequalities in access to tree resources, decision making and sharing of forest benefits. Yet, it provides but a taste of this dynamic and growing field, as thinking around the nexus between gender and forests continues to evolve.

We stand at an exciting juncture; despite the many global challenges we face, there is hope for improving both the world's forests and the well-being of the women and men who use and manage them. The formal schooling of boys and girls creates an opportunity to engage with youth in novel ways and to interest them in the benefits modern science can offer, all the while celebrating and building on the traditional knowledge that has guided forest management processes for millennia and enhancing its inter-generational transmission. The spread of new information technologies and communications media can support these processes, facilitate collective action and activism, and serve as a vehicle to scale out innovations. New public and private trade policies (e.g. corporate social responsibility) are creating opportunities to engage women and men, young and old, in more remunerative, eco-friendly forestry value chains. The proliferation of participatory tools and their application throughout the cycle of research for development enhances the ability of local voices and perspectives to shape the design of locally relevant and equitable forestry programmes, projects, policies and institutions. And initiatives to improve dialogue across interest groups – local women and men, traditional and state authorities, the private and public sectors, and civil society – and across scales can help reconcile conflicts of interest, build synergies and drive progressive socio-ecological transformation.

We hope this book, and its sequel, *The Earthscan Reader on Gender and Forests*, will inform further research and action to explore these development spaces and opportunities to contribute to equitable and sustainable change in forest landscapes.

INDEX